# Springer Texts in
# Electrical Engineering

Consulting Editor: John B. Thomas

# Springer Texts in Electrical Engineering

Robert J. Marks II
Editor

# Advanced Topics in Shannon Sampling and Interpolation Theory

With 91 Figures

Springer-Verlag

New York  Berlin  Heidelberg  London  Paris
Tokyo  Hong Kong  Barcelona  Budapest

Robert J. Marks II
Department of Electrical Engineering, FT-10
University of Washington
Seattle, WA 98195
USA

Library of Congress Cataloging-in-Publication Data
Advanced topics in Shannon sampling and interpolation theory/[edited
  by] Robert J. Marks II.
  p.  cm.—(Springer texts in electrical engineering)
  Includes bilbiographical references and index.
  ISBN-13: 978-1-4613-9759-5
  1. Signal processing—Statistical methods. 2. Sampling
  (Statistics)  I. Marks, Robert J.  II. Series.
TK5102.5.A3325   1992
621.382′2′015194—dc20                                    92-25590

Printed on acid-free paper.

Production managed by Natalie Johnson; manufacturing supervised by Vincent Scelta.
Camera-ready copy prepared from the editor's LaTeX files.
Printed and bound by Hamilton Printing Co., Rensselaer, NY.

9 8 7 6 5 4 3 2 1

ISBN-13: 978-1-4613-9759-5          e-ISBN-13: 978-1-4613-9757-1
DOI: 10.1007/978-1-4613-9757-1

# Acknowledgments

*Advanced Topics in Shannon Sampling and Interpolation Theory* is a companion volume to the text *Introduction to Shannon Sampling and Interpolation Theory* (Springer-Verlag, 1991). The contributors to this volume are to be congratulated on the completeness and clarity of their chapters. This volume is dedicated to them. The bibliography is a collective work of all of the contributors.

A special commendation goes to Dr. Abdul J. Jerri. Dr. Jerri served as a visiting professor in Kuwait. He called me in Seattle from there and reported that his manuscript had been completed. Iraq invaded Kuwait the next day. After a number of days of uncertainty, he called again. He had escaped with only that which he could carry. We are thankful that one of the items he chose to carry was a hard copy of the chapter you see in this book.

This volume would not have been possible without the tireless efforts of Payman Arabshahi. His patience in putting this volume together is beyond my understanding. My family, Jeremiah, Joshua, Marilee and wife Monika, have been monumentally supportive with their encouragement and prayers. I also thank Ruth Wagner for her substantive secretarial assistance.

# Contents

# Contributors

**Martin J. Bastiaans** Technische Universiteit Eindhoven, Faculteit Elektrotechniek, Postbus 513, 5600 MB Eindhoven, Netherlands.

**P. L. Butzer** Lehrstuhl A für Mathematik, RWTH Aachen Templergraben, D–5100 Aachen, Germany.

**Kwan F. Cheung** Department of Electrical and Electronic Engineering, The Hong Kong University of Science and Technology, Clear Way Bay, Kowloon, Hong Kong.

**Franco Gori** Dipartimento di Fisica - Università di Roma "La Sapienza", Piazzale Aldo Moro, 2, I - 00185 Rome, Italy.

**Abdul J. Jerri** Department of Mathematics and Computer Science, Clarkson University, Potsdam, NY 13699, USA.

**Robert J. Marks II** Department of Electrical Engineering, University of Washington FT-10, Seattle, WA 98195, USA.

**Farokh Marvasti** Department of Electronic and Electrical Engineering, King's College London, University of London, Strand, London WC2R 2LS, England.

**Henry Stark** Department of Electrical and Computer Engineering, Illinois Institute of Technology, 3301 South Dearborn, Chicago, IL 60616, USA.

**R. L. Stens** Lehrstuhl A für Mathematik, RWTH Aachen Templergraben, D–5100 Aachen, Germany.

# 1

# Gabor's Signal Expansion and Its Relation to Sampling of the Sliding-Window Spectrum

Martin J. Bastiaans

## 1.1   Introduction

It is sometimes convenient to describe a time signal $\varphi(t)$, say, not in the time domain, but in the frequency domain by means of its *frequency spectrum*, i.e., the *Fourier transform* $\bar{\varphi}(\omega)$ of the function $\varphi(t)$, which is defined by

$$\bar{\varphi}(\omega) = \int \varphi(t)e^{-j\omega t}dt;$$

a bar on top of a symbol will mean throughout that we are dealing with a function in the frequency domain. (Unless otherwise stated, all integrations and summations in this contribution extend from $-\infty$ to $+\infty$.) The frequency spectrum shows us the *global* distribution of the energy of the signal as a function of frequency. However, one is often more interested in the momentary or *local* distribution of the energy as a function of frequency.

The need for a *local frequency spectrum* arises in several disciplines. It arises in music, for instance, where a signal is usually described not by a time function nor by the Fourier transform of that function, but by its *musical score*; indeed, when a composer writes a score, he prescribes the frequencies of the tones that should be present at a certain moment. It arises in optics: geometrical optics is usually treated in terms of rays, and the signal is described by giving the directions (cf. frequencies) of the rays (cf. tones) that should be present at a certain position (cf. time moment). It arises also in mechanics, where the position and the momentum of a particle are given simultaneously.

A local frequency spectrum can be constructed in different ways. One favorite candidate is the *Wigner distribution function* [953. 663. 139. 45. 67, 222, 889. 59]. introduced in 1932 by Wigner in mechanics to describe mechanical phenomena in a so-called *phase space*. The Wigner distribution function is a representative of a rather broad class of bilinear time-frequency functions [226, 227, 222], which are related to each other by linear transformations. Some well-known time-frequency representations — like Woodward's ambiguity function [966, 707, 889], Rihaczek's complex energy density function [768], and Mark's physical spectrum [577] — belong to this class. This broad class of bilinear time-frequency functions is known as the Cohen class [226]; any function of this class is described by the general formula

$$F(t,\omega) \;=\; \frac{1}{2\pi} \iiint \varphi(\tau + \tfrac{1}{2}t')\varphi^*(\tau - \tfrac{1}{2}t')k(t,\omega,t',\omega')$$
$$\times e^{-j(\omega t' - \omega' t + \omega' \tau)} d\tau dt' d\omega'.$$

where an asterisk denotes complex conjugation. The choice of the kernel $k(t,\omega,t',\omega')$ selects one particular function of the Cohen class; the Wigner distribution function, for instance, arises when we choose $k(t,\omega,t',\omega')=1$, whereas $k(t,\omega,t',\omega')=2\pi\delta(t-t')\delta(\omega-\omega')$ yields the ambiguity function. In this contribution we will not consider the Wigner distribution function or any other member out of this class of bilinear time-frequency signal representations.

Another strong candidate for a local frequency description of a signal is the *sliding-window spectrum* — or *windowed Fourier transform* — a generalized version of the *short-term Fourier transform*, which is well-known in speech processing [746, 691]. It is defined as the *cross-ambiguity function* (see [889] and the references cited there) of the signal $\varphi(t)$ and a *window function* $w(t)$ and is constructed in the following way. We multiply the signal by a complex conjugated version of the window function, which is usually more or less concentrated around a certain time moment $t$, say, and determine the Fourier transform of the product, with frequency variable $\omega$, say. Thus we create a function of time $t$ and frequency $\omega$, simultaneously, which might be considered as the local frequency spectrum of the signal.

A local frequency spectrum like the sliding-window spectrum describes the signal in time $t$ and frequency $\omega$, simultaneously. It is thus a function of *two* variables, derived, however, from a function of *one* variable. Therefore, it must satisfy certain restrictions, or, to put it another way: not any function of two variables is a local frequency spectrum. The restrictions that a local frequency spectrum must satisfy correspond to *Heisenberg's uncertainty principle* in mechanics, which states the impossibility of a too accurate determination of both position and momentum of a particle. In this chapter we will show that the sliding-window spectrum is completely determined by its values on the points of a certain time-frequency lattice, which is exactly the lattice suggested by Gabor [313] as early as 1946.

A third candidate for a local frequency spectrum is *Gabor's signal expansion*. In 1946. Gabor suggested the expansion of a signal into a discrete set of properly shifted and modulated Gaussian elementary signals [313]. A quotation from Gabor's original paper might be useful. Gabor writes in the summary:

> Hitherto communication theory was based on two alternative methods of signal analysis. One is the description of the signal as a function of time; the other is Fourier analysis . . . . But our everyday experiences. . . insist on a description in terms of *both* time and frequency . . . . Signals are represented in two dimensions, with time and frequency as co-ordinates. Such two-dimensional representations can be called 'information diagrams', as areas in them are proportional to the number of independent data which they can convey. . . . There are certain 'elementary signals' which occupy the smallest possible area in the information diagram. They are harmonic oscillations modulated by a probability pulse. Each elementary signal can be considered as conveying exactly one datum, or one 'quantum of information'. Any signal can be expanded in terms of these by a process which includes time analysis and Fourier analysis as extreme cases.

Although Gabor restricted himself to an *elementary signal* that had a Gaussian shape, his signal expansion holds for rather arbitrarily shaped elementary signals [50, 55]. We will show that there exists a strong relationship between Gabor's signal expansion and the sampling of the sliding-window spectrum, and that Gabor's signal expansion can be used to reconstruct the signal from its sampled sliding-window spectrum.

In section 2 we will introduce the sliding-window spectrum and show some of its properties. Sampling of the sliding-window spectrum is studied in section 3, and the reconstruction of the signal from the sampling values by means of the Zak transform is shown there. Some examples of window functions are considered in section 4. In section 5 we introduce Gabor's signal expansion and we show an easier way to reconstruct the signal from its sampled sliding-window spectrum. Some examples will be considered in section 6. Propagation of Gabor's expansion coefficients through linear systems and some ideas about the number of degrees of freedom of a signal will be the subject of section 7. In section 8 we will describe an optical means for generating Gabor's expansion coefficients and we will show a link to folded spectrum techniques.

We will restrict ourselves to one-dimensional time signals; the extension to two or more dimensions, however, is rather straightforward. Most of the results can be applied to continuous-time as well as discrete-time signals. We will concentrate on continuous-time signals, but we will state the results for the discrete-time case, if necessary, as well. To distinguish continuous-time from discrete-time signals, we will denote the former with curved brackets and the latter with square brackets: thus $\varphi(t)$ is a continuous-time and $\varphi[n]$ a discrete-time signal. We will use the variables in a consistent manner: in the continuous-time case, the variables $t$, $m$ and $T$ have some-thing to do with time, and $\omega$, $k$ and $\Omega$ with frequency, and the relation $\Omega T = 2\pi$ holds throughout; in the discrete-time case, the variables $n$, $m$ and $N$ have something to do with time, and $\vartheta$, $k$ and $\Theta$ with frequency, and the relation $\Theta N = 2\pi$ holds throughout.

## 1.2   Sliding-Window Spectrum

The sliding-window spectrum $S_w(t, \omega)$ of the signal $\varphi(t)$ is defined as

$$S_w(t, \omega) = \int \varphi(t')w^*(t' - t)e^{-j\omega t'}dt', \qquad (1.2.1)$$

where $w(t)$ is the window function. We note that the sliding-window spec-trum can be considered as the Fourier transform of the product of the signal $\varphi(t)$ and a conjugated and shifted version of the window function $w(t)$. The window function may be chosen rather arbitrarily; mostly it will be a function that is more or less concentrated around the origin. The sliding-window spectrum can then be considered as a windowed or short-term Fourier transform of the signal, which, indeed, can be interpreted as a local frequency spectrum. If the window function is chosen a very narrow function, like a Dirac function, the sliding-window spectrum reduces to a pure time representation of the signal; if, on the other hand, the window function is chosen constant, the sliding-window spectrum reduces to a pure frequency representation. In general, however, the sliding-window spectrum is an *intermediate* signal description between the pure time and the pure frequency representation.

Instead of the definition in the time domain, there exists an equivalent definition in the frequency domain, reading

$$S_w(t, \omega) = \frac{1}{2\pi} \int \bar{\varphi}(\omega')\bar{w}^*(\omega' - \omega)e^{j\omega't}d\omega' \cdot e^{-j\omega t}. \qquad (1.2.2)$$

The factor $e^{-j\omega t}$ causes a slight asymmetry between the definitions (1.2.1) and (1.2.2); if desired, more symmetric definitions result from adding a factor $e^{j\omega t/2}$ to the right-hand sides of these relations.

The sliding-window spectrum of a one-dimensional signal can easily be displayed by optical means. Since this spectrum is the cross-ambiguity func-tion of $\varphi(t)$ and $w(t)$, we can use the optical arrangements that are designed to display such cross-ambiguity functions [785, 594, 595, 41, 761]; we only have to convert the time functions $\varphi(t)$ and $w(t)$ to space functions.

We will give some properties of the sliding-window spectrum, which can be derived directly from the definitions (1.2.1) and (1.2.2). Other proper-ties can be found in the literature on cross-ambiguity functions. (See, for instance, [889] and the references cited there.)

## 1.2.1   INVERSION FORMULAS

Since the definition given in (1.2.1) of the sliding-window spectrum $S_w(t, \omega)$ can be considered as a Fourier transformation of the product $\varphi(t')w^*(t'-t)$ with respect to $t'$, we can easily find a way to reconstruct the signal $\varphi(t)$ from its sliding-window spectrum by simply writing down the corresponding *inverse* Fourier transformation

$$\varphi(t')w^*(t) = \frac{1}{2\pi} \int S_w(t' - t, \omega)e^{j\omega t'} d\omega. \qquad (1.2.3)$$

An inversion formula similar to relation (1.2.3), but based on the definition (1.2.2) in the frequency domain, can be formulated to reconstruct the spectrum $\bar{\varphi}(\omega)$.

There exists another way of reconstructing the signal from its sliding-window spectrum, viz., by means of the inversion formula [139]

$$\varphi(t') \int |w(t)|^2 dt = \frac{1}{2\pi} \iint S_w(t, \omega)w(t' - t)e^{j\omega t} dt d\omega, \qquad (1.2.4)$$

which represents the signal as a *linear combination* of shifted and modulated window functions, with the sliding-window spectrum $S_w(t, \omega)$ as a weighting function. However, this linear combination is *not unique* [139]; indeed, there are many kernels $S(t, \omega)$ that satisfy the relationship

$$\varphi(t') \int |w(t)|^2 dt = \frac{1}{2\pi} \iint S(t, \omega)w(t' - t)e^{j\omega t} dt d\omega. \qquad (1.2.5)$$

One obvious kernel is suggested by Gabor's signal expansion, in which case the kernel $S(t, \omega)$ has the form of a discrete set of Dirac functions in the time-frequency domain, as we will see in section 5. The representation (1.2.4), i.e., choosing the kernel $S(t, \omega)$ in relation (1.2.5) equal to the sliding-window spectrum $S_w(t, \omega)$, is the best possible one in the sense that for this choice the $L^2$-norm of $S(t, \omega)$ takes its minimum value. To see this we multiply both sides of Eqs. (1.2.4) and (1.2.5) by $\varphi^*(t')$, integrate over $t'$, and conclude from the equivalence of the right-hand sides of the resulting equations that $S_w(t, \omega)$ and $S(t, \omega) - S_w(t, \omega)$ are orthogonal in the sense

$$\frac{1}{2\pi} \iint S_w(t, \omega)[S(t, \omega) - S_w(t, \omega)]^* dt d\omega = 0; \qquad (1.2.6)$$

hence, the relationship

$$\begin{aligned}
\frac{1}{2\pi} \iint |S(t, \omega)|^2 dt d\omega &= \frac{1}{2\pi} \iint |S_w(t, \omega)|^2 dt d\omega \\
&+ \frac{1}{2\pi} \iint |S(t, \omega) - S_w(t, \omega)|^2 dt d\omega
\end{aligned}$$

$$(1.2.7)$$

holds. It will be clear from Eq. (1.2.7) that the $L^2$-norm of $S(t, \omega)$ takes its minimum value if $S(t, \omega) - S_w(t, \omega) = 0$, i.e., if we choose the kernel $S(t, \omega)$ equal to the sliding-window spectrum $S_w(t, \omega)$.

## 1.2.2  SPACE AND FREQUENCY SHIFT

Let $S_w(t, \omega)$ be the sliding-window spectrum of the signal $\varphi(t)$; the sliding-window spectrum of the shifted and modulated signal $\varphi(t - t_o)e^{j\omega_o t}$ then takes the form $S_w(t - t_o, \omega - \omega_o)e^{-j(\omega - \omega_o)t_o}$. In particular, the squared modulus of the sliding-window spectrum, which is also known as the *physical spectrum*, or the *spectrogram*, has the following property: a time or frequency shift of the signal yields the same time or frequency shift for the squared modulus of the sliding-window spectrum.

## 1.2.3  SOME INTEGRALS CONCERNING THE SLIDING-WINDOW SPECTRUM

The integral of the squared modulus of the sliding-window spectrum over the frequency variable $\omega$,

$$\frac{1}{2\pi} \int |S_w(t, \omega)|^2 d\omega = \int |\varphi(t')|^2 |w(t' - t)|^2 dt'. \qquad (1.2.8)$$

can be interpreted as a weighted version of the intensity $|\varphi(t)|^2$, whereas the integral over the time variable $t$,

$$\int |S_w(t, \omega)|^2 dt = \frac{1}{2\pi} \int |\bar{\varphi}(\omega')|^2 |\bar{w}(\omega' - \omega)|^2 d\omega', \qquad (1.2.9)$$

can be considered as a weighted version of $|\bar{\varphi}(\omega)|^2$. The integral of the squared modulus over the entire time-frequency domain,

$$\frac{1}{2\pi} \int\int |S_w(t, \omega)|^2 dt d\omega = \left( \int |\varphi(t)|^2 dt \right) \left( \int |w(t)|^2 dt \right). \qquad (1.2.10)$$

is equal to the product of the total energy of the signal and the total energy of the window function.

## 1.2.4  DISCRETE-TIME SIGNALS

The concept of a sliding-window spectrum can easily be extended to discrete-time signals. Let $x[n]$ $(n = \ldots, -1.0.1.\ldots)$ denote such a discrete-time signal and let $w[n]$ represent a window sequence. Analogous to definition (1.2.1), the sliding-window spectrum is then defined as [cf. [746]. Eq. (6.1)]

$$S_w(n, \vartheta) = \sum_{n'} x[n']w^*[n' - n]e^{-j\vartheta n'}. \qquad (1.2.11)$$

Unlike Eq. (6.1) in [746], the definition (1.2.11) uses a complex conjugated version of the window; moreover, the window has not been time-reversed. The only reason for doing this is to get more elegant formulas in the remainder of this chapter. The sliding-window spectrum for a discrete-time signal is a function of the two variables $n$ and $\vartheta$: the time index $n$ is *discrete* and represents the position of the window, and the frequency variable $\vartheta$ is *continuous*. Of course, as in the case of normal Fourier transforms of discrete-time signals, the sliding-window spectrum $S_w(n, \vartheta)$ is *periodic* in $\vartheta$ with period $2\pi$.

The inversion formula (1.2.3) has its counterpart in the discrete-time case and then reads

$$x[n']w[n] = \frac{1}{2\pi}\int_{2\pi} S_w(n' - n, \vartheta)e^{j\vartheta n'}d\vartheta, \qquad (1.2.12)$$

where $\int_{2\pi} d\vartheta$ represents integration over one period $2\pi$: the counterpart of the inversion formula (1.2.4) takes the form

$$x[n']\sum_n |w[n]|^2 = \sum_n \frac{1}{2\pi}\int_{2\pi} S_w(n, \vartheta)w[n' - n]e^{j\vartheta n'}d\vartheta. \qquad (1.2.13)$$

The other properties of the sliding-window spectrum have their counterparts in the discrete-time case, as well, which counterparts can easily be derived.

## 1.3  Sampling Theorem for the Sliding-Window Spectrum

We can reconstruct the signal from the sliding-window spectrum via the inversion formula (1.2.3) or (1.2.4). However, in order to reconstruct the signal, we need not know the entire sliding-window spectrum: it suffices to know its values at the points of the *Gabor lattice* $(t = mT, \omega = k\Omega)$ with $\Omega T = 2\pi$, where $m$ and $k$ take all integer values [313]. In quantum mechanics this lattice is known as the Von Neumann lattice [678, 38]. Note that the Gabor lattice is rectangular, and that the rectangular cells occupy an area of $2\pi$ in the time-frequency domain (see Fig. 1.1).

Let the values of the sliding-window spectrum at the sampling points $(t = mT, \omega = k\Omega)$ be called $s_{mk}$. We thus have the relation

$$s_{mk} = S_w(mT, k\Omega) = \int \varphi(t)w^*(t - mT)e^{-jk\Omega t}dt. \qquad (1.3.1)$$

We shall now demonstrate how the signal can be found when we know the values $s_{mk}$ of the *sampled* sliding-window spectrum.

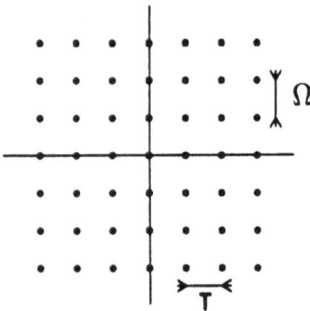

FIGURE 1.1: The Gabor lattice.

We first define the function $\tilde{s}(t,\omega)$ by a Fourier series with Fourier expansion coefficients $s_{mk}$:

$$\tilde{s}(t,\omega) = \sum_m \sum_k s_{mk} e^{-j(m\omega T - k\Omega t)}; \qquad (1.3.2)$$

note that the function $\tilde{s}(t,\omega)$ is periodic in $t$ and $\omega$, with periods $T$ and $\Omega$, respectively. The inverse relationship has the form

$$s_{mk} = \frac{1}{2\pi} \int_T \int_\Omega \tilde{s}(t,\omega) e^{j(m\omega T - k\Omega t)} dt d\omega, \qquad (1.3.3)$$

where the integrations extend over one period $T$ and one period $\Omega$, respectively. We remark that Parseval's *energy theorem* leads to the relationship

$$\frac{1}{2\pi} \int_T \int_\Omega |\tilde{s}(t,\omega)|^2 dt d\omega = \sum_m \sum_k |s_{mk}|^2. \qquad (1.3.4)$$

Furthermore, from the signal $\varphi(t)$ we define the two-dimensional function $\tilde{\varphi}(t,\omega)$ by

$$\tilde{\varphi}(t,\omega) = \sum_m \varphi(t + mT) e^{-jm\omega T}; \qquad (1.3.5)$$

the equivalent definition in terms of the frequency spectrum $\bar{\varphi}(\omega)$ reads

$$\tilde{\varphi}(t,\omega) = \frac{e^{j\omega t}}{T} \sum_k \bar{\varphi}(\omega + k\Omega) e^{jk\Omega t}. \qquad (1.3.6)$$

The slight asymmetry between these two definitions could be removed, if desired, by multiplying the right-hand sides of these definitions by $\sqrt{T} e^{-j\omega t/2}$. [Note that the equivalence of the two definitions (1.3.5) and (1.3.6) implies an interesting relationship between a function and its Fourier transform, viz.,

$$T \sum_m \varphi(t + mT)e^{-jm\omega T} = e^{j\omega t} \sum_k \tilde{\varphi}(\omega + k\Omega)e^{jk\Omega t} \quad \text{(with } \Omega T = 2\pi),$$

which is, in fact, a generalized form of Poisson's sum formula known in Fourier theory!] We remark that the function $\tilde{\varphi}(t, \omega)$ is periodic in the frequency variable $\omega$ with period $\Omega$, and *quasi*-periodic in the time variable $t$, with quasi-period $T$:

$$\tilde{\varphi}(t + mT, \omega + k\Omega) = \tilde{\varphi}(t, \omega)e^{j\omega mT}. \tag{1.3.7}$$

The inverse relationship of Eq. (1.3.5) has the form

$$\varphi(t + mT) = \frac{1}{\Omega} \int_\Omega \tilde{\varphi}(t, \omega)e^{jm\omega T}d\omega. \tag{1.3.8}$$

and a similar relationship exists for the definition (1.3.6); it will be clear that the variable $t$ in relation (1.3.8) can be restricted to an interval of length $T$, with $m$ taking on all integer values. Parseval's energy theorem now leads to the relationship

$$\frac{1}{2\pi} \int_T \int_\Omega |\tilde{\varphi}(t, \omega)|^2 dt d\omega = \frac{1}{T} \int |\varphi(t)|^2 dt. \tag{1.3.9}$$

Relation (1.3.5) provides a means to represent a one-dimensional time function by a two-dimensional time-frequency function on a rectangle with *finite area* $\Omega T = 2\pi$. The two-dimensional function $\tilde{\varphi}(t, \omega)$ associated to the one-dimensional function $\varphi(t)$, according to definition (1.3.5), is known as the *Zak transform* because Zak was the first who systematically studied this transformation in connection with solid state physics [992, 993, 994]. Some of its properties were known long before Zak's work, however. The same transform is called the Weil-Brezin map and it is claimed that the transform was already known to Gauss [798]. It was also used by Gel'fand (see, for instance, [756], Chapter XIII); Zak seems, however, to have been the first to recognize it as the versatile tool it is. The Zak transform has many interesting properties and also interesting applications to signal analysis, for which we refer to [428, 429].

With the help of the functions $\tilde{s}(t, \omega)$, $\tilde{\varphi}(t, \omega)$ and a similar function $\tilde{w}(t, \omega)$ (a Zak transform, again) associated with the window function $w(t)$, relation (1.3.1) can be transformed into

$$\tilde{s}(t, \omega) = T\tilde{\varphi}(t, \omega)\tilde{w}^*(t, \omega). \tag{1.3.10}$$

The transition from (1.3.1) to (1.3.10) goes as follows. We first write down the definition (1.3.2) of the function $\tilde{s}(t, \omega)$

$$\tilde{s}(t, \omega) = \sum_m \sum_k s_{mk} e^{-j(m\omega T - k\Omega t)}$$

and substitute the sample values $s_{mk}$ from relation (1.3.1):

$$\tilde{s}(t,\omega) = \sum_m \sum_k \left( \int \varphi(t')w^*(t'-mT)e^{-jk\Omega t'}dt' \right) e^{-j(m\omega T - k\Omega t)}.$$

We rearrange factors

$$\tilde{s}(t,\omega) = \sum_m \left[ \int \varphi(t')w^*(t'-mT)\left(\sum_k e^{-jk\Omega(t'-t)}\right) dt' \right] e^{-jm\omega T}$$

and replace the sum of exponentials by a sum of Dirac functions

$$\tilde{s}(t.\omega) = \sum_m \left[ \int \varphi(t')w^*(t'-mT)\left(T\sum_k \delta(t'-t-kT)\right) dt' \right]$$
$$\times e^{-jm\omega T}.$$

We rearrange factors again,

$$\tilde{s}(t.\omega) = T\sum_m \left(\sum_k \int \varphi(t')w^*(t'-mT)\delta(t'-t-kT)dt'\right)$$
$$\times e^{-jm\omega T}.$$

and evaluate the integral

$$\tilde{s}(t.\omega) = T\sum_m \left(\sum_k \varphi(t+kT)w^*(t+kT-mT)\right) e^{-jm\omega T}.$$

After a final rearranging of factors we find

$$\tilde{s}(t.\omega) = T\sum_k \varphi(t+kT)e^{-jk\omega T}$$
$$\times \left(\sum_m w(t+[k-m]T)e^{-j(k-m)\omega T}\right)^*,$$

in which expression we recognize the definitions for the functions $\tilde{\varphi}(t,\omega)$ and $\tilde{w}(t,\omega)$ [cf. definition (1.3.5)]; hence

$$\tilde{s}(t,\omega) = T\tilde{\varphi}(t,\omega)\tilde{w}^*(t,\omega).$$

In fact, we have now solved the problem of reconstructing the signal from its sampled sliding-window spectrum:

- from the sample values $s_{mk}$ we determine the function $\tilde{s}(t,\omega)$ via definition (1.3.2);

- from the window function $w(t)$ we derive the associated function $\tilde{w}(t, \omega)$ via definition (1.3.5);

- under the assumption that division by $\tilde{w}^*(t, \omega)$ is allowed, the function $\tilde{\varphi}(t, \omega)$ can be found with the help of relation (1.3.10);

- finally, the signal follows from $\tilde{\varphi}(t, \omega)$ by means of the inversion formula (1.3.8).

A simpler reconstruction method, however, becomes apparent in section 5, when we have studied Gabor's signal expansion.

Problems may arise in the case that $\tilde{w}(t, \omega)$ has zeros. In that case *homogeneous solutions* [55] $\tilde{z}(t, \omega)$, say, may occur, for which the relation

$$T\tilde{z}(t, \omega)\tilde{w}^*(t, \omega) = 0 \qquad (1.3.11)$$

holds. Relation (1.3.11), which is similar to relation (1.3.10) with $\tilde{s}(t, \omega) = 0$, can be transformed into the relation

$$\int z(t)w^*(t - mT)e^{-jk\Omega t}dt = 0, \qquad (1.3.12)$$

which is similar to relation (1.3.1) with $s_{mk} = 0$, and which shows that the sliding-window spectrum of a homogeneous solution $z(t)$ vanishes at the Gabor lattice [425]. We conclude that the existence of homogeneous solutions makes the reconstruction of the signal from its sampled sliding-window spectrum *non-unique*: if $\varphi(t)$ is a possible reconstruction, then $\varphi(t) + z(t)$ is a possible reconstruction, too.

### 1.3.1  DISCRETE-TIME SIGNALS

We can extend the concept of sampling of the sliding-window spectrum to the discrete-time case, as well. Let $N$ be a positive integer and let $\Theta$ be defined by $\Theta = 2\pi/N$. In the discrete-time case the Gabor lattice can then be described by $(n = mN, \vartheta = k\Theta)$. Let the values of the sliding-window spectrum at these sampling points be denoted by $s_{mk}$: we thus have the relation [cf. relation (1.3.1)]

$$s_{mk} = S_w(mN, k\Theta) = \sum_n \varphi[n]w^*[n - mN]e^{-jk\Theta n}. \qquad (1.3.13)$$

Of course, due to the periodicity of $S_w(n, \vartheta)$ in the frequency variable $\vartheta$ with period $2\pi$, the array of coefficients $s_{mk}$ is periodic in $k$ with period $N$.

Analogous to the continuous-time case, reconstruction of the discrete-time signal $\varphi[n]$ from the sample values $s_{mk}$ requires the function $\tilde{s}(n, \vartheta)$ defined by its Fourier series coefficients $s_{mk}$ [cf. definition (1.3.2)],

$$\tilde{s}(n, \vartheta) = \sum_m \sum_{k=<N>} s_{mk}e^{-j(m\vartheta N - k\Theta n)}. \qquad (1.3.14)$$

where $\sum_{k=<N>}$ represents summation over one period $N$. Note that the function $\tilde{s}(n, \vartheta)$ is periodic in the time index $n$ and the frequency variable $\vartheta$, with periods $N$ and $\Theta$, respectively. The inverse relationship has the form [cf. relation (1.3.3)]

$$s_{mk} = \frac{1}{N} \sum_{n=<N>} \frac{1}{\Theta} \int_{\Theta} \tilde{s}(n, \vartheta) e^{j(m\vartheta N - k\Theta n)} d\vartheta. \tag{1.3.15}$$

2urthermore, we need the Zak transform of the signal $\varphi[n]$, which is defined by [cf. definition (1.3.5)]

$$\tilde{\varphi}(n, \vartheta) = \sum_m \varphi[n + mN] e^{-jm\vartheta N}. \tag{1.3.16}$$

The Zak transform is now periodic in the frequency variable $\vartheta$ with period $\Theta$, and quasi-periodic in the time index $n$ with quasi-period $N$ [cf. relation (1.3.7)]:

$$\tilde{\varphi}(n + mN, \vartheta + k\Theta) = \tilde{\varphi}(n, \vartheta) e^{jm\vartheta N}. \tag{1.3.17}$$

The inverse relationship reads [cf. relation (1.3.8)]

$$\varphi[n + mN] = \frac{1}{\Theta} \int_{\Theta} \tilde{\varphi}(n, \vartheta) e^{jm\vartheta N} d\vartheta, \tag{1.3.18}$$

where now the time index $n$ can be restricted to an interval of length $N$, with $m$ taking on all integer values.

The case $N = 1$ and, consequently, $\Theta = 2\pi$ deserves special attention: in that case there is maximum overlap between the window sequence $w[n]$ and its direct neighbors $w[n \pm N]$. The formulas can then be simplified. Without losing any information, we can take $k \equiv 0$ in relation (1.3.13), and $n \equiv 0$ in relations (1.3.14)-(1.3.18), for instance. Relation (1.3.13) then reduces to a simple correlation

$$s_{m0} = \sum_n \varphi[n] w^*[n - m]: \tag{1.3.19}$$

note that, moreover, the values $s_{m0}$ become real when the signal $\varphi[n]$ and the window sequence $w[n]$ are real. Furthermore, relation (1.3.14) and (1.3.15) then constitute a normal Fourier transformation pair, and so do relations (1.3.16) and (1.3.18).

## 1.4   Examples of Window Functions

We shall consider some examples of window functions $w(t)$, and we shall determine their associated two-dimensional functions $\tilde{w}(t, \omega)$, confining ourselves throughout to the fundamental interval $(-\frac{1}{2}T < t \leq \frac{1}{2}T, -\frac{1}{2}\Omega < \omega \leq \frac{1}{2}\Omega)$; for other combinations of $t$ and $\omega$, we can use the (quasi-)periodicity property (1.3.7).

## Rect Window Function

As a first example we consider a rectangular window function whose width equals $T$ (see Fig. 1.2):

$$w(t) = \text{rect}\left(\frac{t}{T}\right) = \begin{cases} 1 & \text{for } -\frac{1}{2}T < t \le \frac{1}{2}T \\ 0 & \text{elsewhere.} \end{cases} \qquad (1.4.1)$$

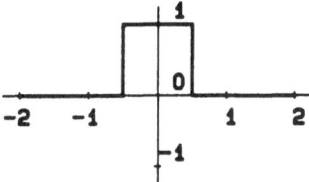

FIGURE 1.2: A rectangular window function, cf. Eq. (1.4.1).

The associated two-dimensional function $\tilde{w}(t, \omega)$ follows readily via definition (1.3.5); in the fundamental interval it reads

$$\tilde{w}(t, \omega) = 1. \qquad (1.4.2)$$

This example can easily be generalized to an arbitrary window function $w(t)$ that is limited to the interval $-\frac{1}{2}T < t \le \frac{1}{2}T$; in the fundamental interval the associated function $\tilde{w}(t, \omega)$ reads

$$\tilde{w}(t, \omega) = w(t). \qquad (1.4.3)$$

## Sinc Window Function

Our second example is the band-limited function (see Fig. 1.3)

$$w(t) = \text{sinc}\left(\frac{t}{T}\right) = \frac{\sin(\pi \frac{t}{T})}{\pi \frac{t}{T}}. \qquad (1.4.4)$$

This function and the rectangular window function of the first example are *dual* to each other, i.e., the Fourier transform of one function has the same form as the other function. The Fourier transform of the sinc window function reads

$$\bar{w}(\omega) = T \, \text{rect}\left(\frac{\omega}{\Omega}\right), \qquad (1.4.5)$$

and its associated two-dimensional function $\tilde{w}(t, \omega)$, which can readily be derived using definition (1.3.6), takes the form

$$\tilde{w}(t, \omega) = e^{j\omega t} \qquad (1.4.6)$$

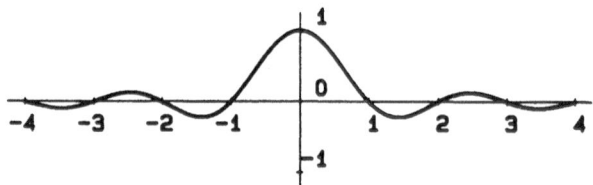

FIGURE 1.3: A sinc window function, cf. Eq. (1.4.4).

in the fundamental interval.

This example can easily be generalized to an arbitrary function $w(t)$ that is band-limited to the interval $-\frac{1}{2}\Omega < \omega \leq \frac{1}{2}\Omega$; in the fundamental interval the associated function $\tilde{w}(t,\omega)$ reads

$$\tilde{w}(t,\omega) = \frac{e^{j\omega t}}{T}\bar{w}(\omega). \qquad (1.4.7)$$

## 1.4.1  GAUSSIAN WINDOW FUNCTION

As our final continuous-time example we consider the Gaussian window function (see Fig. 1.4)

$$w(t) = \sqrt[4]{2}e^{-\pi(\frac{t}{T})^2}; \qquad (1.4.8)$$

the factor $\sqrt[4]{2}$ in this definition has been included to normalize $\int |w(t)|^2 dt$ to unity.

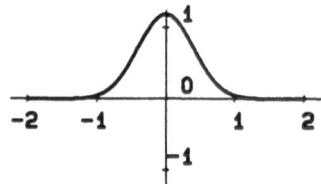

FIGURE 1.4: A Gaussian window function, cf. Eq. (1.4.8).

A Gaussian function has several advantages: its Fourier transform is again Gaussian, and the product of the effective width in the time domain and the one in the frequency domain takes the theoretical minimum value [704, 707]. The associated two-dimensional function $\tilde{w}(t,\omega)$ follows via definition (1.3.5); in the fundamental interval it takes the form

$$\tilde{w}(t,\omega) = \sqrt[4]{2}e^{-\pi(\frac{t}{T})^2}\theta_3(\pi\zeta^*). \qquad (1.4.9)$$

where $\theta_3(\zeta)$ is a *theta function* [950, 3] with *nome* $q = e^{-\pi}$, and where, for convenience, we have set $\zeta = \omega/\Omega + jt/T$.

Since $\tilde{w}(t, \omega)$ has a simple zero for $(t = \frac{1}{2}T, \omega = \frac{1}{2}\Omega)$ in this case, a homogeneous solution $\tilde{z}(t, \omega)$ may occur in the signal reconstruction process, reading in the fundamental interval (up to a constant factor)

$$\tilde{z}(t, \omega) = 2\pi \delta(t - \tfrac{1}{2}T)\delta(\omega - \tfrac{1}{2}\Omega) \tag{1.4.10}$$

and thus, with the help of the inversion formula (1.3.8)

$$z(t) = T \sum_{m} (-1)^m \delta(t - \tfrac{1}{2}T - mT). \tag{1.4.11}$$

This homogeneous solution $z(t)$ is depicted in Fig. 1.5.

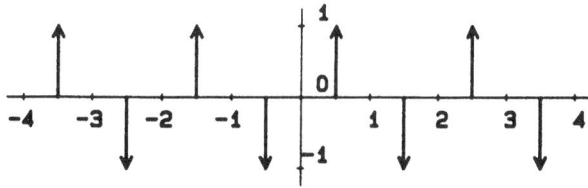

FIGURE 1.5: A homogeneous solution corresponding to a Gaussian window function, cf. Eq. (1.4.11).

## 1.4.2 DISCRETE-TIME SIGNALS

We consider two simple examples of discrete-time window sequences $w[n]$, and determine their associated two-dimensional functions $\tilde{w}(n, \vartheta)$ for different values of the shifting distance $N$. Our first discrete-time example is the symmetrical, three-point window sequence (see Fig. 1.6)

$$w[n] = \begin{cases} 1 & \text{for } n = 0 \\ \frac{1}{2}a & \text{for } n = \pm 1 \ (0 < a^2 < 1) \\ 0 & \text{elsewhere;} \end{cases} \tag{1.4.12}$$

note that for $a = 0.16$. we are dealing with a three-point Hamming window. For $N = 1$, the maximum-overlap case, we find

$$\tilde{w}(n, \vartheta) = (1 + a \cos \vartheta)e^{jn\vartheta}. \tag{1.4.13}$$

For $N = 2$ we find

$$\begin{cases} \tilde{w}(2m. \vartheta) &= e^{j2m\vartheta} \\ \tilde{w}(2m + 1. \vartheta) &= \frac{1}{2}a\left(1 + e^{j2\vartheta}\right)e^{j2m\vartheta}. \end{cases} \tag{1.4.14}$$

Note that in this case of partial overlap, the function $\tilde{w}(n, \vartheta)$ has zeros for $\vartheta = \frac{1}{2}\pi + r\pi \ (r = \dots, -1, 0, 1, \dots)$, and hence a homogeneous solution $z[n]$

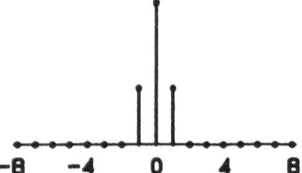

FIGURE 1.6: A symmetrical, three-point window sequence, cf. Eq. (1.4.12).

arises. Its associated function $\tilde{z}(n,\vartheta)$ is given by

$$\begin{cases} 2\tilde{z}(2m,\vartheta) &= 0 \\ 2\tilde{z}(2m+1,\vartheta) &= \pi(-1)^m z_1 \sum_r \delta\left(\vartheta - \tfrac{1}{2}\pi - r\pi\right). \end{cases} \qquad (1.4.15)$$

The homogeneous solution thus takes the form (see Fig. 1.7)

$$\begin{cases} z[2m] &= 0 \\ z[2m+1] &= (-1)^m z_1. \end{cases} \qquad (1.4.16)$$

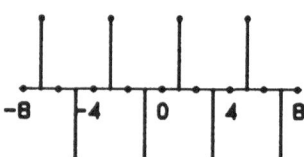

FIGURE 1.7: A homogeneous solution corresponding to a symmetrical, three-point window sequence, cf. Eq. (1.4.16).

Our second discrete-time example is the one-sided exponential window sequence (see Fig. 1.8)

$$w[n] = \begin{cases} e^{\alpha n} & \text{for } n \le 0 \ (\alpha > 0) \\ 0 & \text{for } n > 0. \end{cases} \qquad (1.4.17)$$

In the basic interval $-(N-1) \le n \le 0$, the associated function $\tilde{w}(n.\vartheta)$ takes the form

$$\tilde{w}(n.\vartheta) = e^{\alpha n}\frac{1}{1 - e^{-(\alpha - j\vartheta)N}}; \qquad (1.4.18)$$

the values of $\tilde{w}(n.\vartheta)$ outside this interval can be found by applying the quasi-periodicity relation (1.3.17).

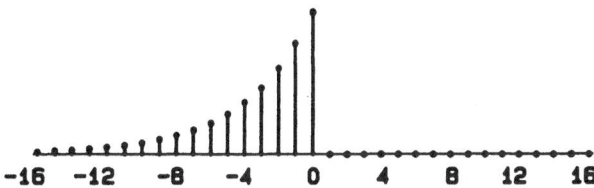

FIGURE 1.8: A one-sided exponential window sequence, cf. Eq. (1.4.17).

## 1.5  Gabor's Signal Expansion

In 1946, Gabor [313] suggested the expansion of a signal into a discrete set of Gaussian elementary signals [386, 48, 50, 55, 426]. Although Gabor restricted himself to an *elementary signal* that had a Gaussian shape, his signal expansion holds for rather arbitrarily shaped elementary signals [48, 50, 55]. With the help of *Gabor's signal expansion*, we can express the signal $\varphi(t)$ as a superposition of properly shifted and modulated versions of an elementary signal $g(t)$, say, yielding

$$\varphi(t) = \sum_m \sum_k a_{mk} g(t - mT) e^{jk\Omega t}, \tag{1.5.1}$$

where the time shift $T$ and the frequency shift $\Omega$ satisfy the relation $\Omega T = 2\pi$. Unlike the inversion formula (1.2.4), which represents the signal as a *continuum* of window functions, Gabor's signal expansion (1.5.1) represents the signal as a *discrete set* of elementary signals that are shifted over discrete distances $mT$ and that are modulated with discrete frequencies $k\Omega$.

In general, the discrete set of shifted and modulated elementary signals $g(t - mT) e^{jk\Omega t}$ need not be orthonormal, which implies that *Gabor's expansion coefficients* $a_{mk}$ cannot be determined in the usual way. In this section, however, we show how we can find a function $w(t)$, say, that is *bi-orthonormal* to the set of elementary signals in the sense

$$\int g(t) w^*(t - mT) e^{-jk\Omega t} dt = \delta_m \delta_k. \tag{1.5.2}$$

where $\delta_m$ is the Kronecker delta ($\delta_0 = 1, \delta_m = 0$ for $m \neq 0$): the choice of the symbol $w$ for this function — as if it was a window function — is intentional, as will become clear soon! With the help of this bi-orthonormal function $w(t)$, the expansion coefficients follow readily via

$$a_{mk} = \int \varphi(t) w^*(t - mT) e^{-jk\Omega t} dt. \tag{1.5.3}$$

The relationship between Gabor's signal expansion and the sliding-window spectrum becomes apparent by noting that the right-hand side of relation

(1.5.3) can be interpreted as a sampled sliding-window spectrum with window function $w(t)$ [see relation (1.3.1)].

To show the relationship between the Gabor expansion with the elementary signal $g(t)$ and expansion coefficients $a_{mk}$ on the one hand [cf. relation (1.5.1)], and the sliding-window spectrum with the window function $w(t)$ and the sample values $a_{mk}$ on the other hand [cf. relation (1.5.3)], we proceed as follows. We first derive a way to find the Gabor expansion coefficients $a_{mk}$ along the lines similar to the ones used in section 3. As we did for the array of sample values $s_{mk}$, we define from the array of Gabor coefficients $a_{mk}$ the function $\tilde{a}(t,\omega)$ according to definition (1.3.2). Furthermore, we introduce the function $\tilde{g}(t,\omega)$ associated to the elementary signal $g(t)$ according to definition (1.3.5) and also use the function $\tilde{\varphi}(t,\omega)$ associated to the signal $\varphi(t)$. We can then transform relation (1.5.1) into

$$\tilde{\varphi}(t,\omega) = \tilde{a}(t,\omega)\tilde{g}(t,\omega). \qquad (1.5.4)$$

The transition from (1.5.1) to (1.5.4) goes as follows. We first write down the definition (1.3.5) for the function $\tilde{\varphi}(t,\omega)$,

$$\tilde{\varphi}(t.\omega) = \sum_n \varphi(t + nT)e^{-jn\omega T}.$$

and substitute $\varphi(t + nT)$ from Gabor's signal expansion (1.5.1),

$$\tilde{\varphi}(t.\omega) = \sum_n \left( \sum_m \sum_k a_{mk} g(t + nT - mT)e^{jk\Omega t} \right) e^{-jn\omega T}.$$

We rearrange factors

$$\tilde{\varphi}(t,\omega) = \sum_m \sum_k a_{mk} e^{-j(m\omega T - k\Omega t)} \left( \sum_n g(t + [n - m]T)e^{-j(n-m)T} \right)$$

and recognize the definition (1.3.2) for the function $\tilde{a}(t,\omega)$ and the definition (1.3.5) for the function $\tilde{g}(t,\omega)$: hence

$$\tilde{\varphi}(t,\omega) = \tilde{a}(t,\omega)\tilde{g}(t,\omega).$$

In fact we have now solved the problem of finding Gabor's expansion coefficients, even in the case that the set of shifted and modulated elementary signals $g(t - mT)e^{jk\Omega t}$ is not orthogonal:

- from the signal $\varphi(t)$ and the elementary signal $g(t)$ we derive the associated functions $\tilde{\varphi}(t,\omega)$ and $\tilde{g}(t,\omega)$ via definition (1.3.5);

- under the assumption that division by $\tilde{g}(t,\omega)$ is allowed, the function $\tilde{a}(t,\omega)$ can be found by means of relation (1.5.4);

- finally, the expansion coefficients $a_{mk}$ follow from the function $\tilde{a}(t, \omega)$ with the help of the inversion formula (1.3.8).

We have just shown how Gabor's expansion coefficients could be determined when the signal $\varphi(t)$ and the elementary signal $g(t)$ are known; there is, however, a simpler way to find these expansion coefficients by means of relation (1.5.3). To prove this, we have to derive a function $w(t)$ that is bi-orthonormal to the elementary signal $g(t)$ in the sense of relation (1.5.2). Under the assumption, again, that division by $\tilde{g}(t, \omega)$ is allowed, we define the function $\tilde{w}(t, \omega)$ through the relation

$$T\tilde{g}(t, \omega)\tilde{w}^*(t, \omega) = 1. \tag{1.5.5}$$

Substitution of relation (1.5.5) into relation (1.5.4) yields

$$\tilde{a}(t, \omega) = T\tilde{\varphi}(t, \omega)\tilde{w}^*(t, \omega). \tag{1.5.6}$$

When we notice the resemblance between relation (1.5.6) and relation (1.3.10), it is not difficult to see that relation (1.5.6) can be transformed into relation (1.5.3), in the same way as relation (1.3.10) can be transformed into relation (1.3.1): the function $w(t)$ then follows from the function $\tilde{w}(t, \omega)$ by means of the inversion formula (1.3.8). Likewise, relation (1.5.5) can be transformed into the bi-orthonormality relation (1.5.2). We conclude that the expansion coefficients can be determined immediately by means of relation (1.5.3) when the signal $\varphi(t)$ and the function $w(t)$ are known. Note that the expansion coefficients $a_{mk}$ can be considered as the sample values of the sliding-window spectrum of the signal $\varphi(t)$ with window function $w(t)$.

Gabor's expansion coefficients may be non-unique in the case that $\tilde{g}(t, \omega)$ has zeros. In that case homogeneous solutions $\tilde{z}(t, \omega)$ may occur again [58], for which now the relation

$$\tilde{z}(t, \omega)\tilde{g}(t, \omega) = 0 \tag{1.5.7}$$

holds. Relation (1.5.7), which is similar to relation (1.5.4) with $\tilde{\varphi}(t, \omega) = 0$, can be transformed into the relation

$$\sum_m \sum_k z_{mk}g(t - mT)e^{jk\Omega t} = 0, \tag{1.5.8}$$

which is similar to relation (1.5.1) with $\varphi(t) = 0$. Relation (1.5.8) shows that certain arrays of non-zero coefficients in Gabor's signal expansion may yield a zero result. We thus conclude that Gabor's signal expansion may be *non-unique*: if the array of coefficients $a_{mk}$ yields the signal $\varphi(t)$, then the array $a_{mk} + z_{mk}$ yields the same signal.

The resemblance between relations (1.5.3) and (1.3.1) leads to another important conclusion. In section 3 it was shown how the signal could be reconstructed from the sampled sliding-window spectrum; we now conclude that there exists a simpler reconstruction method by means of Gabor's signal expansion (1.5.1), where we must identify the Gabor expansion coefficients $a_{mk}$ with the sample values $s_{mk}$ of the sliding-window spectrum: hence

$$\varphi(t) = \sum_m \sum_k s_{mk} g(t - mT) e^{jk\Omega t}. \qquad (1.5.9)$$

The bi-orthonormality of $w$ and $g$ as expressed by the bi-orthonormality relation (1.5.2) is, in fact, the reason why we can use the $w$-functions to find the coefficients of Gabor's expansion into $g$-functions [cf. relations (1.5.1) and (1.5.3)], and use the $g$-functions to reconstruct the signal from its sampled sliding-window spectrum with window function $w(t)$ [cf. relations (1.3.1) and (1.5.9)]. Zeros in either $\tilde{g}(t,\omega)$ or $\tilde{w}(t,\omega)$ may complicate matters, however. When we apply Parseval's energy theorem (1.3.9) to $w(t)$ or $g(t)$, and substitute from relation (1.5.5), we get the relationships

$$\frac{1}{T}\int |w(t)|^2 dt = \frac{1}{2\pi}\int_T \int_\Omega |\tilde{w}(t,\omega)|^2 dt d\omega$$

$$= \frac{1}{2\pi}\int_T \int_\Omega \left|\frac{1}{T\tilde{g}(t,\omega)}\right|^2 dt d\omega. \qquad (1.5.10)$$

$$\frac{1}{T}\int |g(t)|^2 dt = \frac{1}{2\pi}\int_T \int_\Omega |\tilde{g}(t,\omega)|^2 dt d\omega$$

$$= \frac{1}{2\pi}\int_T \int_\Omega \left|\frac{1}{T\tilde{w}(t,\omega)}\right|^2 dt d\omega. \qquad (1.5.11)$$

From these relationships we conclude that in the case that $\tilde{g}(t,\omega)$ or $\tilde{w}(t,\omega)$ has zeros, the required window function $w(t)$ or the required elementary signal $g(t)$ may not be quadratically summable. This consequence of the occurrence of zeros in $\tilde{g}(t,\omega)$ or $\tilde{w}(t,\omega)$ is even worse than the fact that Gabor's signal expansion is not unique or that the reconstruction from the sampled sliding-window spectrum is not unique due to homogeneous solutions; it may cause very bad convergence properties in the expansion or reconstruction method.

When we substitute from relation (1.5.9) into the definition (1.2.1) of the sliding-window spectrum, we obtain the relation

$$S_w(t,\omega) = \sum_m \sum_k s_{mk} \int g(t' - mT) e^{jk\Omega t'} w^*(t' - t) e^{-j\omega t'} dt'. \qquad (1.5.12)$$

Relation (1.5.12) enables us to express the sliding-window spectrum in terms of its sample values. We can write

$$S_w(t,\omega) = \sum_m \sum_k s_{mk} Q_w(t - mT, \omega - k\Omega) e^{-jm\omega T}, \qquad (1.5.13)$$

where we have used the shifting property of the sliding-window spectrum. and where we have introduced the *interpolation function*

$$Q_w(t,\omega) = \int g(t')w^*(t'-t)e^{-j\omega t'}\,dt'. \qquad (1.5.14)$$

Note that the interpolation function is, in fact, the sliding-window spectrum of the function $g(t)$ with window function $w(t)$, and that its property

$$Q_w(mT,k\Omega) = \delta_m\delta_k \qquad (1.5.15)$$

is equivalent to the bi-orthonormality property (1.5.2). By interchanging $g$ and $w$ in relation (1.5.14), we get

$$Q_g(t,\omega) = Q_w^*(-t,-\omega)e^{-j\omega t}, \qquad (1.5.16)$$

which is, in fact, the sliding-window spectrum of the function $w(t)$ with window function $g(t)$.

### 1.5.1 DISCRETE-TIME SIGNALS

For discrete-time signals, Gabor's expansion takes the form [cf. relation (1.5.1)]

$$\varphi[n] = \sum_m \sum_k a_{mk}g[n-mN]e^{jk\Theta n}: \qquad (1.5.17)$$

it will not be difficult to express all the other relationships in this section for discrete-time signals, as well. The interpolation property (1.5.13), for instance, then takes the form

$$S_w(n,\vartheta) = \sum_m \sum_{k=<N>} s_{mk}Q_w(n-mN,\vartheta-k\Theta)e^{-jm\vartheta N}, \qquad (1.5.18)$$

where, again, $Q_w(n,\vartheta)$ is the sliding-window spectrum of the elementary signal $g[n]$ with window sequence $w[n]$.

   The case $N = 1$, with maximum overlap between the window sequence $w[n]$ and its direct neighbors $w[n \pm N]$, deserves special attention, again, and the formulas can be simplified. Without losing any information, we can now take $k \equiv 0$ in relation (1.5.17), for instance, which then reduces to a simple convolution

$$x[n] = \sum_m a_{m0}g[n-m]. \qquad (1.5.19)$$

## 1.6  Examples of Elementary Signals

For some elementary signals $g(t)$, which we have considered already as window functions in section 4. we shall determine the corresponding window functions $w(t)$ and the interpolation functions $Q_w(t,\omega)$.

### 1.6.1   Rect Elementary Signal

For the *rect* elementary signal $g(t) = \text{rect}(t/T)$, cf. Fig. 1.2, we readily find

$$T\tilde{w}(t,\omega) = 1 \tag{1.6.1}$$

in the fundamental interval and hence

$$Tw(t) = \text{rect}\left(\frac{t}{T}\right). \tag{1.6.2}$$

The interpolation function $Q_w(t,\omega)$ that corresponds to a rectangular window function reads

$$Q_w(t,\omega) = e^{-j\omega t/2} \operatorname{sinc}\left[\frac{\omega}{\Omega}\left(1 - \left|\frac{t}{T}\right|\right)\right]\left(1 - \left|\frac{t}{T}\right|\right)\text{rect}\left(\frac{t}{2T}\right). \tag{1.6.3}$$

Gabor's signal expansion using a rectangular elementary signal represents, in fact, a well-known way of expanding a signal: we simply consider the signal in successive intervals of length $T$ and describe the signal in each interval by means of a Fourier series. Note that $w(t)$ is proportional to the elementary signal $g(t)$ since, in this case, the set of shifted and modulated elementary signals is orthogonal.

### 1.6.2   Sinc Elementary Signal

For the *sinc* elementary signal $g(t) = \operatorname{sinc}(t/T)$, cf. Fig. 1.3, we have

$$T\tilde{w}(t,\omega) = e^{j\omega t} \tag{1.6.4}$$

in the fundamental interval and hence

$$Tw(t) = \operatorname{sinc}\left(\frac{t}{T}\right); \tag{1.6.5}$$

the interpolation function corresponding to the sinc window function reads

$$Q_w(t,\omega) = e^{-j\omega t/2} \operatorname{sinc}\left[\frac{t}{T}\left(1 - \left|\frac{\omega}{\Omega}\right|\right)\right]\left(1 - \left|\frac{\omega}{\Omega}\right|\right)\text{rect}\left(\frac{\omega}{2\Omega}\right). \tag{1.6.6}$$

This example is simply the dual of the previous one. It will be clear that for a signal which is band-limited to the frequency interval $|\omega| < \frac{1}{2}\Omega$, Gabor's signal expansion represents the well-known ordinary sampling theorem.

### 1.6.3   Gaussian Elementary Signal

In the case of the Gaussian elementary signal $g(t) = \sqrt[4]{2}e^{-\pi(\frac{t}{T})^2}$, cf. Fig. 1.4, we have in the fundamental interval

$$\tilde{g}(t,\omega) = \sqrt[4]{2}e^{-\pi(\frac{t}{T})^2}\theta_3(\pi\zeta^*), \tag{1.6.7}$$

and thus

$$T\tilde{w}(t,\omega) = \frac{1}{\sqrt[4]{2}} e^{\pi(\frac{t}{T})^2} \frac{1}{\theta_3(\pi\zeta)}, \tag{1.6.8}$$

in which expressions we have set again $\zeta = \omega/\Omega + jt/T$. The function $1/\theta_3(\pi\zeta)$ can be expressed as

$$\frac{1}{\theta_3(\pi\zeta)} = \left(\frac{K_0}{\pi}\right)^{-3/2} \left(c_0 + 2\sum_{m=1}^{\infty}(-1)^m c_m \cos(2\pi m\zeta)\right), \tag{1.6.9}$$

where

$$c_m = \sum_{n=0}^{\infty}(-1)^n e^{-\pi(n+\frac{1}{2})(2m+n+\frac{1}{2})} \tag{1.6.10}$$

(see, for instance, [950], p. 489, Example 14); the constant $K_0 = 1.85407468$ is the *complete elliptic integral* for the modulus $\frac{1}{2}\sqrt{2}$ (see. for instance, [950], p. 524). It is now easy to determine $w(t)$ via the inversion formula (1.3.8), yielding

$$Tw(t + mT) = \frac{1}{\sqrt[4]{2}} \left(\frac{K_0}{\pi}\right)^{-3/2} e^{\pi(\frac{t}{T})^2}(-1)^m c_m e^{2\pi m\frac{t}{T}} \tag{1.6.11}$$

with $-\frac{1}{2}T < t \le \frac{1}{2}T$, and hence

$$Tw(t) = \frac{1}{\sqrt[4]{2}} \left(\frac{K_0}{\pi}\right)^{-3/2} e^{\pi(\frac{t}{T})^2} \sum_{n+\frac{1}{2}\ge|\frac{t}{T}|}(-1)^n e^{-\pi(n+\frac{1}{2})^2}. \tag{1.6.12}$$

This window function $w(t)$, which corresponds to the Gaussian elementary signal, is depicted in Fig. 1.9. A practical way to represent $w(t)$ is in the

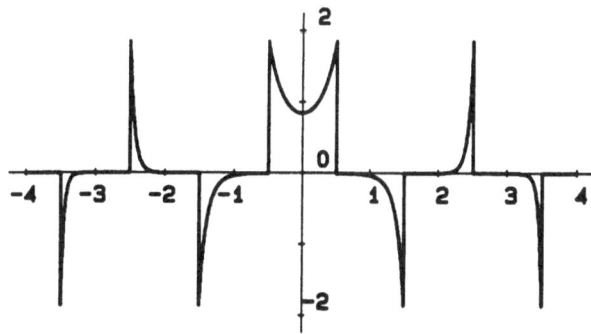

FIGURE 1.9: The window function that corresponds to a Gaussian elementary signal, cf. Eq. (1.6.12).

form [508]

$$Tw(t) = \frac{1}{\sqrt[4]{2}} \left(\frac{K_0}{\pi}\right)^{-3/2} (-1)^m C_m e^{\pi[(\frac{t}{T})^2 - (m+\frac{1}{2})^2]}, \qquad (1.6.13)$$

where $m$ is the non-negative integer defined by $(m - \frac{1}{2})T \leq |t| < (m + \frac{1}{2})T$
and where

$$C_m = \sum_{n=m}^{\infty} (-1)^{n-m} e^{-\pi[(n+\frac{1}{2})^2 - (m+\frac{1}{2})^2]}. \qquad (1.6.14)$$

Since $C_m$ is close to unity ($C_0 = 0.998133$, $C_1 = 0.999997$, ..., $C_\infty = 1$),
this representation leads to the approximation

$$Tw(t) \simeq \frac{1}{\sqrt[4]{2}} \left(\frac{K_0}{\pi}\right)^{-3/2} (-1)^m e^{\pi[(\frac{t}{T})^2 - (m+\frac{1}{2})^2]} \qquad (1.6.15)$$

with $(m - \frac{1}{2})T \leq |t| < (m + \frac{1}{2})T$.

Without proof, we mention some properties of this function $w(t)$. As
is also the case for the Gaussian function, the Fourier transform of $w(t)$
has the same form as $w(t)$ itself. Moreover, the function $w(t)$ satisfies the
differential equation

$$\frac{dw}{dt} = \frac{2\pi t}{T^2} w(t) - \frac{1}{\sqrt[4]{2}} \left(\frac{K_0}{\pi}\right)^{-3/2} \sum_m (-1)^m \delta(t - [m + \frac{1}{2}]T). \qquad (1.6.16)$$

More properties of this special $w(t)$ can be found elsewhere [428].

The interpolation function $Q_w(t,\omega)$ that corresponds to this window
function $w(t)$ takes the form

$$Q_w(t,\omega) = \frac{1}{2} \left(\frac{K_0}{\pi}\right)^{-3/2} e^{-\pi(\frac{t}{T})^2} \frac{\theta_1(\pi\zeta)}{\pi\zeta}. \qquad (1.6.17)$$

where, again, we have set $\zeta = \omega/\Omega + jt/T$ and where $\theta_1(\zeta)$ is again a theta
function [950, 3] with nome $q = e^{-\pi}$. Relation (1.6.17) can be expressed in
a more symmetrical form using Weierstrass' *sigma function* [950, 3] (with
$\omega' = j\omega = jK_0$; see [3], Sect. 18.14, Lemniscatic case), and then reads

$$Q_w(t,\omega) = e^{-j\omega t/2} e^{-\pi|\zeta|^2/2} \frac{\sigma(2K_0\zeta)}{2K_0\zeta}. \qquad (1.6.18)$$

From relation (1.6.18) we conclude that there seems to be a connection
with certain classical interpolation theorems [949, 428]. With the help of
relation (1.5.16) we find the interpolation function that corresponds to the
Gaussian window function $w(t) = \sqrt[4]{2} e^{-\pi(\frac{t}{T})^2}$:

$$Q_w(t,\omega) = e^{j\omega t/2} e^{-\pi|\zeta|^2/2} \frac{\sigma(2K_0\zeta^*)}{2K_0\zeta^*}. \qquad (1.6.19)$$

In the case of a Gaussian elementary signal $g(t)$, its associated function $\tilde{g}(t, \omega)$ has a simple zero for $(t = \frac{1}{2}T. \omega = \frac{1}{2}\Omega)$. In this case a homogeneous solution $\tilde{z}(t, \omega)$ may thus occur in the determination of the Gabor coefficients, reading in the fundamental interval (up to a constant factor)

$$\tilde{z}(t. \omega) = 2\pi\delta(t - \tfrac{1}{2}T)\delta(\omega - \tfrac{1}{2}\Omega), \tag{1.6.20}$$

[cf. relation (1.4.10)] and thus, with the help of the inversion formula (1.3.3),

$$z_{mk} = (-1)^{m+k}. \tag{1.6.21}$$

### 1.6.4 Discrete-Time Signals

We consider the symmetrical, three-point elementary signal $g[n]$

$$g[n] = \begin{cases} 1 & \text{for } n = 0 \\ \frac{1}{2}a & \text{for } n = \pm 1 \; (0 < a^2 < 1) \\ 0 & \text{elsewhere.} \end{cases} \tag{1.6.22}$$

which we considered already in section 4 as a window sequence. cf. Fig. 1.6. For the maximum-overlap case $(N = 1)$, we find

$$\tilde{w}(n. \vartheta) = \frac{e^{jn\vartheta}}{1 + a\cos\vartheta} \tag{1.6.23}$$

and the corresponding window sequence $w[n]$ thus takes the form (see Fig. 1.10)

$$w[n] = \frac{1}{\sqrt{1-a^2}} \left( \frac{\sqrt{1-a^2} - 1}{a} \right)^{|n|}. \tag{1.6.24}$$

In the case of partial overlap $(N = 2)$, we find

$$\begin{cases} \tilde{w}(2m. \vartheta) & = & e^{j2m\vartheta} \\ \tilde{w}(2m+1, \vartheta) & = & \dfrac{2}{a} \dfrac{e^{j2m\vartheta}}{1 + e^{-j2\vartheta}} \end{cases} \tag{1.6.25}$$

and hence (see Fig. 1.11)

$$\begin{cases} w[2m] & = & \begin{cases} \frac{1}{2} & \text{for } m = 0 \\ 0 & \text{for } m \neq 0, \end{cases} \\ w[2m+1] & = & \begin{cases} \frac{1}{2a}(-1)^m & \text{for } m \geq 0 \\ -\frac{1}{2a}(-1)^m & \text{for } m < 0. \end{cases} \end{cases} \tag{1.6.26}$$

As our final example we consider the one-sided exponential elementary signal, cf. Fig. 1.8,

$$g[n] = \begin{cases} e^{\alpha n} & \text{for } n \leq 0 \; (\alpha > 0) \\ 0 & \text{for } n > 0. \end{cases} \tag{1.6.27}$$

Martin J. Bastiaans

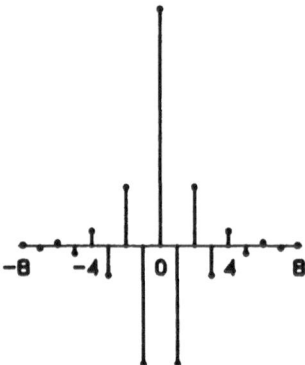

FIGURE 1.10: The window function corresponding to a symmetrical, three-point elementary signal, in the case of maximum overlap, cf. Eq. (1.6.24).

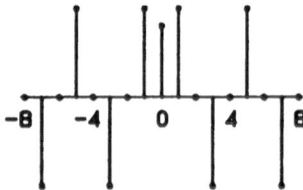

FIGURE 1.11: The window function corresponding to the symmetrical, three-point elementary signal, in the case of partial overlap, cf. Eq. (1.6.26).

In the basic interval $-(N-1) \leq n \leq 0$, the associated function $\tilde{g}(n,\vartheta)$ takes the form

$$\tilde{g}(n,\vartheta) = \frac{e^{\alpha n}}{1 - e^{-(\alpha - j\vartheta)N}}; \qquad (1.6.28)$$

the values of $\tilde{g}(n,\vartheta)$ outside this interval can be found by applying the quasi-periodicity relation (1.3.17). The function $\tilde{w}(n,\vartheta)$ now takes the form

$$\tilde{w}(n,\vartheta) = e^{-\alpha n}\left(1 - e^{-(\alpha + j\vartheta)N}\right) \qquad (1.6.29)$$

inside the basic interval. The corresponding window sequence $w[n]$ then reads (see Fig. 1.12)

$$w[n] = \begin{cases} \frac{1}{N}e^{-\alpha n} & \text{for } -(N-1) \leq n \leq 0 \\ -\frac{1}{N}e^{-\alpha n} & \text{for } 1 \leq n \leq N \\ 0 & \text{elsewhere.} \end{cases} \qquad (1.6.30)$$

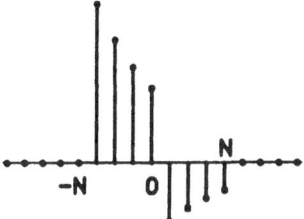

FIGURE 1.12: The window function corresponding to the one-sided exponential elementary signal, cf. Eq. (1.6.30).

We use this example to show once more the possible non-uniqueness of Gabor's signal expansion. In the limiting case $\alpha = 0$. the function $\tilde{g}(n, \vartheta)$ has zeros for $\vartheta = r(2\pi/N)$ $(r = \dots, -1, 0, 1, \dots)$, and an array of coefficients $z_{mk}$ arises whose associated function $\tilde{z}(n, \vartheta)$ in the basic interval, say, is given by

$$\tilde{z}(n, \vartheta) = \frac{2\pi}{N} z_n \sum_r \delta \left( \vartheta - r\frac{2\pi}{N} \right). \qquad (1.6.31)$$

The array $z_{mk}$ thus takes the form

$$z_{mk} = \frac{1}{N} \sum_{n=-(N-1)}^{0} z_n e^{-jk2\pi n/N} = z_{0k}. \qquad (1.6.32)$$

and yields a zero result when substituted in Gabor's signal representation.

## 1.7 Degrees of Freedom of a Signal

Gabor's signal expansion as introduced in section 5 is related to the *degrees of freedom* of a signal: each expansion coefficient $a_{mk}$ represents one complex degree of freedom. If a signal is, roughly, limited to the space interval $|t| < \frac{1}{2}a$ and to the frequency interval $|\omega| < \frac{1}{2}b$, the number of complex degrees of freedom equals the number of Gabor coefficients in the space-frequency rectangle with area $ab$, this number being about equal to the *time-bandwidth product* $ab/2\pi$. We shall consider this point in more detail in this section.

We first consider the propagation of Gabor's expansion coefficients through a *linear* (not necessarily time-invariant) *system*. A linear system that transforms an input signal $\varphi_i$ into an output signal $\varphi_o$ can be described in four different ways. depending on whether we describe the input and the output signal in the time or in the frequency domain. We thus have four equivalent input-output relationships,

$$\varphi_o(t_o) = \int h_{tt}(t_o, t_i)\varphi_i(t_i)dt_i, \tag{1.7.1}$$

$$\bar{\varphi}_o(\omega_o) = \int h_{\omega t}(\omega_o, t_i)\varphi_i(t_i)dt_i, \tag{1.7.2}$$

$$\varphi_o(t_o) = \frac{1}{2\pi}\int h_{t\omega}(t_o, \omega_i)\bar{\varphi}_i(\omega_i)d\omega_i, \tag{1.7.3}$$

$$\bar{\varphi}_o(\omega_o) = \frac{1}{2\pi}\int h_{\omega\omega}(\omega_o, \omega_i)\bar{\varphi}_i(\omega_i)d\omega_i, \tag{1.7.4}$$

in which the four *system functions* $h_{tt}$, $h_{\omega t}$, $h_{t\omega}$, and $h_{\omega\omega}$ are completely determined by the system. Relation (1.7.1) is the usual system representation in the time domain by means of the *impulse response* $h_{tt}(t_o, t_i)$; the function $h_{tt}(t, t_i)$ is the time domain response of the system at time $t$ due to the input impulse signal $\varphi_i(t) = \delta(t - t_i)$. Relation (1.7.4) is a similar system representation in the frequency domain; the function $h_{\omega\omega}(\omega, \omega_i)$ is the frequency domain response of the system at frequency $\omega$ due to the input $\bar{\varphi}_i(\omega) = 2\pi\delta(\omega - \omega_i)$, which is the Fourier transform of the harmonic input signal $\varphi_i(t) = e^{j\omega_i t}$. Relations (1.7.2) and (1.7.3) are hybrid system representations, since the input and the output signal are described in different domains.

Unlike the *four* system representations (1.7.1)-(1.7.4), there is only *one* system representation when we describe the input and the output signal by their Gabor coefficients (or by any other time-frequency representation). Let us describe the input signal $\varphi_i(t)$ and the output signal $\varphi_o(t)$ of a linear system by their Gabor expansions with expansion coefficients $a_{mk}^i$ and $a_{mk}^o$ and elementary signals $g_i(t)$ and $g_o(t)$ [with associated window functions $w_i(t)$ and $w_o(t)$], respectively; note that we have chosen *different* elementary signals for the input and the output signal. The input and output expansion coefficients are then related to each other by the relationship

$$a_{mk}^o = \sum_{m'}\sum_{k'} c_{mkm'k'}a_{m'k'}^i. \tag{1.7.5}$$

The coefficients $c_{mkm'k'}$ in this relationship are completely determined by the system and the elementary signals; indeed, when we combine the Gabor expansions of the input and the output signal with the system representation (1.7.2), for instance, we find

$$c_{mkm'k'} = \frac{1}{2\pi}\int\int h_{\omega t}(\omega, t)\bar{w}_o^*(\omega - k\Omega)$$
$$\times g_i(t - m'T)e^{j(m\omega T + k'\Omega t)}dtd\omega. \tag{1.7.6}$$

and similar relations for the other system functions.

As an example we consider the basic system. where the input signal $\varphi_i(t)$ is first truncated to the time interval $|t| < \frac{1}{2}a$ and then Fourier transformed; the resulting Fourier transform is truncated to the frequency interval $|\omega| < \frac{1}{2}b$ and then inverse Fourier transformed to yield the output signal $\varphi_o(t)$. Such a system can readily be described by a system function $h_{\omega t}(\omega, t)$, which in this case takes the form

$$h_{\omega t}(\omega. t) = \text{rect}\left(\frac{t}{a}\right) \text{rect}\left(\frac{\omega}{b}\right) e^{-j\omega t}. \tag{1.7.7}$$

For convenience, we shall choose the widths of the apertures for the input signal and for its Fourier transform equal to an odd multiple of the time and the frequency shift $T$ and $\Omega$. respectively: thus

$$a = (2M + 1)T \text{ and } b = (2K + 1)\Omega, \tag{1.7.8}$$

with $M$ and $K$ integers. When we substitute from relations (1.7.7) and (1.7.8) into relation (1.7.6). we conclude that the array of coefficients $c_{mkm'k'}$ can be expressed as a four-dimensional *discrete convolution* of two arrays $d_{mkm'k'}$ and $e_{mkm'k'}$, where the coefficients $d_{mkm'k'}$ are defined by

$$d_{mkm'k'} = \begin{cases} \delta_{m-m'}\delta_{k-k'} & \text{for } |m| \le M \text{ and } |k| \le K \\ 0 & \text{elsewhere}. \end{cases} \tag{1.7.9}$$

and the coefficients $e_{mkm'k'}$ are given by

$$e_{mkm'k'} = \frac{1}{2\pi} \int\int \text{rect}\left(\frac{t}{T}\right) \text{rect}\left(\frac{\omega}{\Omega}\right) e^{-j\omega t}\bar{\omega}_o^*(\omega - k\Omega)$$
$$\times g_i(t - m'T)e^{j(m\omega T + k'\Omega t)} dt d\omega. \tag{1.7.10}$$

A system whose Gabor coefficients $c_{mkm'k'}$ would have the form (1.7.9) is *ideal* in the sense that the Gabor coefficients of the output signal vanish outside the time-frequency rectangle with area $ab$. Hence. whereas the input signal of such an ideal system may have an infinite number of *degrees of freedom*, the number of degrees of freedom of the output signal. i.e.. the number of non-vanishing Gabor coefficients, is equal to the *time-bandwidth product* $ab/2\pi$. However. our system under consideration is not ideal: to find its Gabor coefficients $c_{mkm'k'}$, the ideal array $d_{mkm'k'}$ must be "smeared out" by convolving it with the array $e_{mkm'k'}$. The latter array is, in fact, the array of Gabor coefficients of the elementary system described by the system function (1.7.7). with the special choice $a = T$ and $b = \Omega$. i.e., $M = K = 0$.

Depending on the choice of the elementary signals for the input and the output signal, the array of coefficients $e_{mkm'k'}$ can be strongly concentrated. To show this we choose a *rect* elementary signal to describe the input signal and a *sinc* elementary signal to describe the output signal; thus

$$g_i(t) = \text{rect}\left(\frac{t}{T}\right) \tag{1.7.11}$$

and

$$\bar{w}_o(\omega) = \operatorname{rect}\left(\frac{\omega}{\Omega}\right). \tag{1.7.12}$$

We then find $e_{0000} = 0.873$, and the strong concentration becomes apparent by noting that

$$\sum_m \sum_k \sum_{m'} \sum_{k'} |e_{mkm'k'}|^2 = 1.$$

In general the value of $e_{0000}$ for this elementary system is given by

$$e_{0000} = \frac{1}{2\pi} \iint \operatorname{rect}\left(\frac{t}{T}\right) \operatorname{rect}\left(\frac{\omega}{\Omega}\right) e^{-j\omega t} \bar{w}_o^*(\omega) g_i(t) dt d\omega. \tag{1.7.13}$$

Furthermore, the identities

$$\sum_m \sum_k \sum_{m'} \sum_{k'} e_{mkm'k'} = \left(\sum_k \bar{w}_o(k\Omega)\right)^* \left(\sum_m g_i(mT)\right) \tag{1.7.14}$$

and

$$\sum_m \sum_k \sum_{m'} \sum_{k'} |e_{mkm'k'}|^2 = \left(\frac{1}{2\pi} \int |\bar{w}_o(\omega)|^2 d\omega\right) \left(\int |g_i(t)|^2 dt\right) \tag{1.7.15}$$

can be derived in a straightforward way, using the basic relation

$$\sum_n e^{jn\Omega t} = T \sum_n \delta(t - nT). \tag{1.7.16}$$

The ratio

$$\frac{|e_{0000}|^2}{\sum_m \sum_k \sum_{m'} \sum_{k'} |e_{mkm'k'}|^2} \tag{1.7.17}$$

can be considered as a *degree of concentration* of the array $e_{mkm'k'}$ around the coefficient $e_{0000}$. By applying a variational principle to the expression (1.7.17), it is not difficult to show that the degree of concentration has a stationary value when $g_i(t)$ and $\bar{w}_o(\omega)$ are chosen according to

$$g_i(t) = \Psi_{2m}\left(\frac{t}{T}\right) \operatorname{rect}\left(\frac{t}{T}\right) \tag{1.7.18}$$

and

$$\bar{w}_o(\omega) = \Psi_{2m}\left(\frac{\omega}{\Omega}\right) \operatorname{rect}\left(\frac{\omega}{\Omega}\right). \tag{1.7.19}$$

where the functions $\Psi_n(\xi)$ are the *prolate spheroidal wave functions* (see, for instance, [830]) defined by the eigenfunction equation

$$\int \Psi_n(\xi) \operatorname{rect}(\xi) e^{-2\pi j \xi \eta} d\xi = j^n \sqrt{\lambda_n} \Psi_n(\eta) \quad (n = 0, 1, \dots). \tag{1.7.20}$$

If we choose the elementary functions as in relations (1.7.18) and (1.7.19), the corresponding stationary value of the degree of concentration is equal to $\lambda_{2m}$. An optimum value is attained for $m = 0$, for which the degree of concentration takes the value $\lambda_0 = 0.783$. This is a slightly better result than choosing the elementary signals as in relations (1.7.11) and (1.7.12), in which case the degree of concentration takes the value 0.762.

We conclude that for a proper choice of the elementary signals the array $e_{mkm'k'}$ can be strongly concentrated. Since the Gabor coefficients $c_{mkm'k'}$ of the basic system under consideration can be found by convolving the ideal array $d_{mkm'k'}$ with the strongly concentrated array $e_{mkm'k'}$, the array of system coefficients $c_{mkm'k'}$ is very similar to the array $d_{mkm'k'}$. Hence, the number of degrees of freedom of the output signal of this system is about equal to the time-bandwidth product $ab/2\pi$. We remark that the way in which we have proved this has a clear physical interpretation. Roughly speaking, with the Gabor expansion of the input signal in mind, only those shifted and modulated versions of the elementary signal that can pass both the input (time) aperture and the Fourier (frequency) aperture will contribute to the output signal.

A slightly more general system than the one described by relation (1.7.7) is the one whose kernel $h_{\omega t}(\omega, t)$ takes the form

$$h_{\omega t}(\omega, t) = \sum_{m}\sum_{k} h_{mk} \operatorname{rect}\left(\frac{t}{T} - m\right) \operatorname{rect}\left(\frac{\omega}{\Omega} - k\right) e^{-j\omega t}. \qquad (1.7.21)$$

The array of system coefficients $c_{mkm'k'}$ can now be expressed as a four-dimensional convolution of the arrays $h_{mk}\delta_{m-m'}\delta_{k-k'}$ and $e_{mkm'k'}$. In the case that the array $e_{mkm'k'}$ is again strongly concentrated around the element $e_{0000}$, the Gabor coefficients of the input and the output signal are related by the simple relation

$$a^{o}_{mk} \simeq h_{mk}a^{i}_{mk}. \qquad (1.7.22)$$

For the special system described by relation (1.7.7), we easily find that the array $h_{mk}$ equals unity in the interval $(|m| \leq M, |k| \leq K)$ and vanishes outside that interval.

## 1.8 Optical Generation of Gabor's Expansion Coefficients for Rastered Signals

In this section we will describe an optical arrangement which is able to generate Gabor's expansion coefficients of a one-dimensional signal by optical means [54]. An important feature of the optical arrangement is that it accepts the one-dimensional signal on a *raster format*; hence, the two-dimensional nature of the optical processing system is fully utilized.

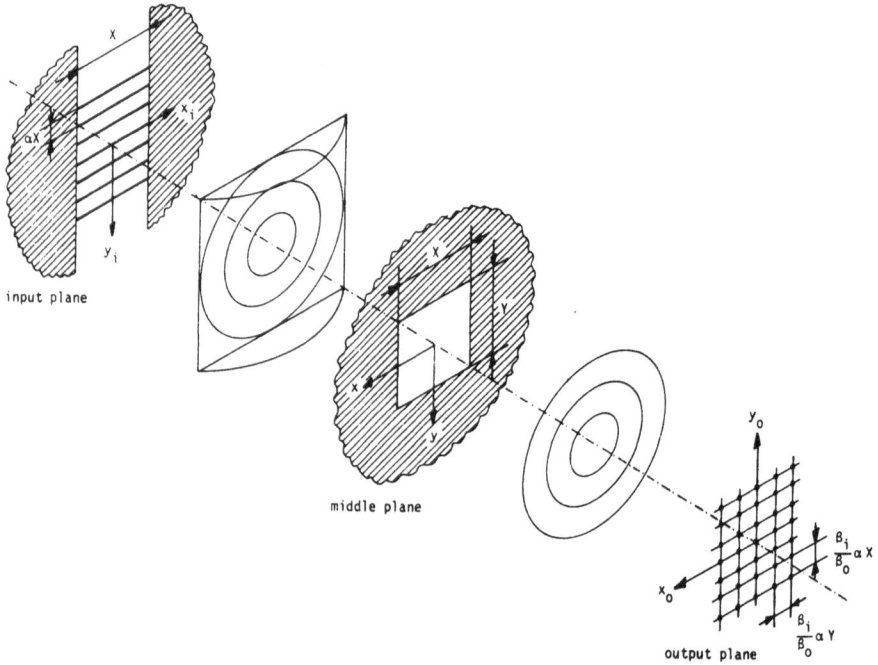

FIGURE 1.13: Optical setup for generation of Gabor's expansion coefficients.

Let us consider the optical arrangement depicted in Fig. 1.13. A plane wave of monochromatic laser light is normally incident on a transparency situated in the input plane. The transparency contains the signal $\varphi(x)$, say, in a rastered format. With $X = 2\pi/U$ being the width of this raster and $\alpha X$ being the spacing between the raster lines, the light amplitude $\varphi_i(x_i, y_i)$ just behind the transparency reads

$$\varphi_i(x_i, y_i) = \text{rect}\left(\frac{x_i}{X}\right) \sum_m \varphi(x_i + mX)\delta(y_i - m\alpha X).  \qquad (1.8.1)$$

An astigmatic optical system between the input plane and the middle plane performs a Fourier transformation in the $y$-direction and an ideal imaging (with inversion) in the $x$-direction. Such an astigmatic system can be realized as shown, for instance, using a combination of a spherical and a cylindrical lens. The astigmatic operation results in the light amplitude

$$
\begin{aligned}
\varphi_1(x, y) &= \iint e^{-j\beta_i y y_i} \delta(x - x_i)\varphi_i(x_i, y_i)dy_i \\
&= \text{rect}\left(\frac{x}{X}\right) \tilde{\varphi}(x, \alpha\beta_i y)  \qquad (1.8.2)
\end{aligned}
$$

just in front of the middle plane: the parameter $\beta_i$ contains the effect of the wavelength of the laser light and the focal length of the spherical lens. In Eq. (1.8.2) we have introduced the associated function $\tilde{\varphi}(x, u)$, the Zak transform of $\varphi(x)$, defined by equation (1.3.5).

A transparency with amplitude transmittance

$$m(x, y) = \text{rect}\left(\frac{x}{X}\right) \text{rect}\left(\frac{y}{Y}\right) X\tilde{w}^*(x, \alpha\beta_i y). \qquad (1.8.3)$$

where $Y = U/\alpha\beta_i = (2\pi/X)/\alpha\beta_i$, is situated in the middle plane. Just behind this transparency, the light amplitude takes the form

$$\varphi_2(x, y) = m(x, y)\varphi_1(x, y) = \text{rect}\left(\frac{x}{X}\right) \text{rect}\left(\frac{y}{Y}\right) \tilde{a}(x, \alpha\beta_i y). \qquad (1.8.4)$$

where use has been made of relation (1.5.6).

Finally, a two-dimensional Fourier transformation is performed between the middle plane and the output plane. Such a Fourier transformation can be realized as shown, for instance, using a spherical lens. The light amplitude in the output plane then takes the form

$$
\begin{aligned}
\varphi_o(x_o, y_o) &= \frac{1}{XY} \iint e^{-j\beta_o(x_o x - y_o y)} \varphi_2(x, y)\,dx\,dy \\
&= \sum_m \sum_k a_{mk} \,\text{sinc}\left(\frac{\beta_o x_o}{\alpha\beta_i X} - k\right) \text{sinc}\left(\frac{\beta_o y_o}{\alpha\beta_i X} - m\right).
\end{aligned}
$$

$$(1.8.5)$$

where Gabor's expansion coefficients $a_{mk}$ have been introduced: the parameter $\beta_o$, again, contains the effects of the wavelength of the light and the focal length of the spherical lens. We conclude that Gabor's expansion coefficients appear on a rectangular lattice of points

$$a_{mk} = \varphi_o\left(k\frac{\alpha\beta_i Y}{\beta_o}, m\frac{\alpha\beta_i X}{\beta_o}\right) \qquad (1.8.6)$$

in the output plane.

We remark that it is not an essential requirement that the input transparency consists of Dirac functions. When we replace the practically unrealizable Dirac functions $\delta(y - m\alpha Y)$ by realizable functions $d(y - m\alpha Y)$, say, then Eq. (1.8.1) reads

$$\varphi_i(x_i, y_i) = \text{rect}\left(\frac{x_i}{X}\right) \sum_m \varphi(x_i + mX)d(y_i - m\alpha X), \qquad (1.8.7)$$

and the light amplitude $\varphi_1(x, y)$ just in front of the middle plane takes the form

$$\varphi_1(x, y) = \text{rect}\left(\frac{x}{X}\right) \tilde{\varphi}(x, \alpha\beta_i y)\bar{d}(\beta_i y). \qquad (1.8.8)$$

The additional factor $\bar{d}(\beta_i y)$ can easily be compensated for by means of a transparency in the middle plane. Note that the special case $d(y) = \mathrm{sinc}(y/\alpha X)$, and thus $\bar{d}(\beta_i y) = \mathrm{rect}(y/Y)$, has the advantage that all the light from the input plane will fall inside the rectangle $\mathrm{rect}(x/X)\,\mathrm{rect}(y/Y)$ in the middle plane.

This technique to generate Gabor's expansion coefficients fully utilizes the two-dimensional nature of the optical system, its parallel processing features, and the large space-bandwidth product possible in optical processing. The technique exhibits a resemblance to *folded spectrum* techniques [194], where space-bandwidth products in the order of $3 \times 10^5$ are reported (see, for instance, [194], Chap. 8.3). In the case of speech processing, where speech recognition and speaker identification are important problems (see, for instance, [41]), such a space-bandwidth product would allow us to process speech fragments of about 1 min.

## 1.9   Conclusion

In this chapter we have derived a sampling theorem (in the time-frequency domain) for the sliding-window spectrum of a one-dimensional time signal, and we have shown how the signal can be reconstructed from the sampling values of its sliding-window spectrum by means of the Zak transform.

We have related the sampling of the sliding-window spectrum to Gabor's expansion of a signal in a discrete set of shifted and modulated elementary signals, and we have shown on the one hand that the sliding-window spectrum provides an easy way to determine Gabor's expansion coefficients and on the other hand that Gabor's signal expansion can elegantly be used to reconstruct the signal from its sampled sliding-window spectrum.

The key solution was that the window function, which is used in the sliding-window spectrum, and the elementary signal, which is used in Gabor's signal expansion, form a bi-orthonormal pair of function sets when shifted and modulated according to the Gabor lattice.

The Gabor lattice played a key role in this contribution. It is the regular lattice $(t = mT. \omega = k\Omega)$ with $\Omega T = 2\pi$ in the time-frequency domain, in which each cell occupies an area of $2\pi$. The density of the Gabor lattice is thus equal to the *Nyquist density* $1/2\pi$, which, as is well-known in information theory, is the minimum time-frequency density needed for full transmission of information. Gabor's expansion coefficients can then be interpreted as degrees of freedom of the signal.

It may be clear that a coarser lattice, with cells whose areas are larger than $2\pi$, leads to undersampling: we do not have enough freedom to be able to represent all possible signals. On the other hand, a finer lattice, with a density that is higher than the Nyquist density, leads to oversampling: dependence between the Gabor coefficients arises, and we can no longer interpret them as (independent) degrees of freedom.

Unfortunately, the Gabor lattice with its critical density $1/2\pi$ may lead to numerically unattractive properties; therefore, one might prefer a denser lattice, with $\Omega T < 2\pi$. This situation has not been considered in this chapter; an excellent review of denser lattices can be found in [244].

We conclude this chapter by drawing attention to some related topics: the rather modern *wavelet transform* of a signal and the way of representing a signal as a discrete set of *wavelets*. There is some resemblance between these topics and the ones that are studied in this chapter. But, whereas the sliding-window spectrum leads to a *time-frequency* representation of the signal, the wavelet transform leads to a *time-scale* representation. And whereas the Gabor lattice is linear in both the time and the frequency coordinate, the lattice that is used in the wavelet representation is non-linear. An excellent review on the wavelet transform can be found, again, in [244].

# 2

# Sampling in Optics

Franco Gori

## 2.1 Introduction

Illuminate an object with laser light and look at it through a diffraction grating. You will see a set of mutually displaced copies of the object. If the lateral extent of the object is small enough, the various copies do not overlap. You can easily devise a method for selecting a single copy of the object. For example, you can replace your own optical system, i.e., your eye, by a converging lens and let the multiple images that were impressing your retina be produced on a screen. Then, a hole on the screen will suffice to isolate a single image. You can even dispense with the laser light and repeat the observation in a more domestic environment by looking at a distant street lamp through a piece of fine fabric. In this case, all of the object copies except the central one will appear iridescent, but the basic phenomena will be the same.

Elementary optical experiments of this kind vividly illustrate the replicating effect of sampling as well as the possibility of recovering a luminous signal from its sampled version. This is because in optics certain transform operations are performed by Nature. It is not so for other phenomena exploited for the transmission of information (e.g., time-dependent electrical signals) where the transform of a signal, even the Fourier transform, is an abstract alternative representation of the signal more than the description of something that the signal itself displays somewhere in space. We can, of course, "see" the spectrum of a signal on a spectrum analyzer in much the same way as we could see any sophisticated transform of the signal on the monitor of a suitably programmed computer, but this is rather removed from the physical evolution of the phenomenon.

With such favorable elements, sampling was set to become a key tool in optics. It was in fact so and nowadays sampling procedures and sampling theorems are almost second nature to the opticist to such a pass that concepts and properties connected with sampling theory are often used without explicit mention. This does not mean that the subject has settled. On the contrary, refinements, extensions and new forms of sampling keep appearing in the optical literature.

Any attempt to give an exhaustive account of all the occurrences and uses of sampling in optics would be hopeless. It may even be added that somewhat like the Cartesian ovals, which from time to time someone finds out a new [869], known results of sampling theory begin to be rediscovered, because the scattering of papers among a lot of scientific journals, conference proceedings and books covering four decades makes it virtually impossible for anybody to have complete information.

Being well aware of this, we shall focus attention on some selected areas in which the application of sampling techniques has been particularly stimulating. One of the basic applications of sampling theory in optics is the estimation of the number of degrees of freedom of a wavefield. The deceptively simple evaluation of this number has met with several objections whose analysis has produced a host of results. This will be a central theme of the present contribution. We shall review the development of the main ideas on this subject from the beginning up to recent achievements. We shall refer, as far as possible, to simple, one-dimensional coherent cases where the important points can be appreciated in the neatest way. It will be seen that even when the mathematical techniques become slightly sophisticated, there is a clear connection to sampling. In this part, of course, the reference tool will be Fourier analysis.

Many optical phenomena are to be described by the Fresnel transform. Accordingly, we shall examine some fundamental properties of this transform and their optical significance. In particular, the use of sampling methods will be underlined. It will be seen that this can be used as a key for approaching the large body of techniques that use Fresnel phenomena for imaging and interferometry.

A tool of relevant interest for other optical phenomena is the Mellin transform. Here too, a sampling theory can be developed that leads to the so-called exponential sampling. Foundations and main features of this theory will be seen together with outlines of optical applications.

Although many optical fields of interest are coherent to a good approximation, most often we have to do with partially coherent fields. A complete description of them requires the use of coherence theory. Sampling plays a relevant role in this theory too and we shall give some hints to explain why this happens.

Several of the above quoted topics could be encompassed under the more general heading of the optical processing of information. This is a vast and rapidly expanding subject to which whole books are devoted and it could not possibly be reviewed in the present contribution. However, it can be useful to point out some of the reasons that make sampling theory so important for optical processing. This will be done in the final part.

Before entering the main themes, we shall devote some space to a short review of the history of sampling in optics.

Applications of sampling in optics are also described in previous review papers [440, 480].

## 2.2  Historical Background

There is a famous Molière's character who suddenly realizes that he has been speaking "in prose" for over forty years without even knowing. [1] This bears some resemblance to the situation of sampling in optics for which there is a sort of prehistoric period. This dates back to the beginning of our century when the thermodynamics of electromagnetic radiation was intensively investigated. In 1914. von Laue. who had already published some important papers on the entropy of radiation. identified (at least explicitly) the series coefficients with samples of the expanded function. Notice that this occurred one year before the result by Whittaker on the interpolatory or cardinal series [946].

The pioneering results by von Laue were later reviewed and extended by Landé who gave them the large audience of the *Handbuch der Physik* [522]. In spite of this. they lay somewhat dormant in optics. In that period in fact. problems connected to Degrees of Freedom (DOF) were of main concern to communication scientists dealing with electrical signals.

The full appreciation in optics of the concepts of DOF and sampling came much later when the ideas of the communication and information theories penetrated the field. A scientist whose contributions to both communication theory and optics were fundamental. namely. the Nobel laureate D. Gabor. led the way. In his paper on communication theory [313]. beside introducing the analytic signal (a common tool of modern optics). he proposed a discrete representation of a signal by means of Gaussian packets. He made this proposal on the grounds that Fourier analysis is at variance with some deeply rooted ideas of common sense. like the idea of instantaneous frequency of an acoustical signal [501]. Stressing analogies with quantum mechanics, he discussed possible representations of a time-dependent signal in a time-frequency plane (a phase space, in fact) and showed that the Gaussian packets occupied the minimum area permitted by the uncertainty relation. Because of the lack of orthogonality of the Gaussian packets, the evaluation of the expansion coefficients of a signal is not a trivial problem. Gabor suggested an approximate solution and an exact solution was found only much later by Bastiaans [48] [2]. Gabor's analysis was carried out in the realm of time-dependent signals but its relevance for optics was set to

---

[1] "... il y a plus de quarante ans que je dis de la prose sans que j'en susse rien." Molière: *Le Burgeois Gentilhomme.*

[2] See chapter 1 by Bastiaans in this volume.

become apparent shortly after. Indeed, it is not by chance that in 1948 Gabor announced the principle of holography [314]. In addition, in 1946, the first edition appeared of the famous book by Duffieux of the applications to optics of the Fourier integral [269]. In 1951, Gabor delivered a celebrated Ritchie lecture at the University of Edinburgh about optics and information whose content was replicated much later in the first volume of *Progress in Optics* [315]. Among many other points, some of whom will be reviewed later, he discussed the spatial version of the Gaussian packets that became widely known for their importance in laser optics [108, 109, 494]. Let us add that the expansion of a function by means of Gaussian packets became customary in the quantum theory of optical coherence through the use of coherent states [333, 568, 386] and that the uncertainty relations of Fourier analysis discussed in Gabor's paper of 1946 were later to play an important role in partial coherence theory [938, 574, 193, 961, 306, 59] as well as in optical processing [45, 41] through the use, e.g., of the Wigner function [953].

In the meantime, the masterful work of Shannon had appeared [818, 819]. Within a few years the first applications of information theory to optics were presented [562, 483, 484, 278, 907, 908, 546, 686, 317, 318, 563, 547, 964]. By the end of the 1950s and at the beginning of the 1960s, the sampling theorem began to be taught in optics textbooks [909, 687] and the general role of information theory in science was discussed [115]. The use of information theory spread out further in the optical literature [688, 29, 30, 35, 319, 381, 307, 552, 12]. In the same period, the optical processing of information began to be popularized [238, 913] and the connection between holography and communication theory was clarified [532] by the invention of off-axis holography. Finally, the publication of textbooks stressing the role of mathematical transforms and sampling [339, 704] marked the beginning of the full maturity period.

## 2.3   The von Laue Analysis

It is fair to begin by a short account of the 1914 paper by von Laue. Let us first define the number of DOF of a space-time field distribution as the number of parameters needed to specify the field. One can refer either to real or to complex DOF.

Von Laue considers a beam of monochromatic, linearly polarized light falling on a $\xi, \eta$ plane within a solid angle $\Omega$ and evaluates the number of DOF of the radiation illuminating a square region $|\xi| \leq \Xi$. He first expresses the field contributed at a typical point $(x, y, z)$ by a cone of radiation with vertex at the point $(\xi, \eta, 0)$. Using the Debye integral [106] he writes this contribution as follows:

$$\int_{-A}^{A} \int_{-A}^{A} e^{ik[\alpha(x-\xi)+\beta(y-\eta)+\gamma z]} \frac{d\alpha d\beta}{\gamma}, \tag{2.3.1}$$

where $k$ is the wave number of the field ($k = 2\pi/\lambda$, $\lambda$ being the wavelength). This is a superposition of plane waves whose wave vectors have direction cosines $\alpha, \beta$ and $\gamma$ and are contained in the solid angle $\Omega$ specified by the conditions $|\alpha| \leq A, |\beta| \leq A$. To obtain the complete field, say $V(x, y, z)$, von Laue multiplies expression (2.3.1) by a complex function $f(\xi, \eta)$ and integrates across the square $|\xi| \leq \Xi, |\eta| \leq \Xi$. He further assumes $A \ll 1$ so that $\gamma \cong 1$. The result is

$$
\begin{aligned}
V(x, y, z) &= \int_{-\Xi}^{\Xi} \int_{-\Xi}^{\Xi} f(\xi, \eta) \, d\xi \, d\eta \\
&\times \int_{-A}^{A} \int_{-A}^{A} e^{ik[\alpha(x-\xi)+\beta(y-\eta)+\gamma z]} \, d\alpha \, d\beta. \tag{2.3.2}
\end{aligned}
$$

On expanding $f(\xi, \eta)$ into a Fourier series and interchanging integrals. von Laue obtains the following expression:

$$
\begin{aligned}
V(x, y, z) &= 4 \, \Xi^2 e^{ikz} \int_{-A}^{A} \int_{-A}^{A} e^{ik(\alpha x + \beta y)} \tag{2.3.3} \\
&\times \sum_{m=-\infty}^{\infty} \sum_{n=-\infty}^{\infty} K_{mn} \, \operatorname{sinc}\left(\frac{2\,\Xi}{\lambda}\alpha - m\right) \\
&\times \operatorname{sinc}\left(\frac{2\,\Xi}{\lambda}\beta - n\right) \, d\alpha \, d\beta,
\end{aligned}
$$

where, as usual, we put $\operatorname{sinc}(t) = \sin(\pi t)/(\pi t)$. As a next step, von Laue gives an estimate of the integrals

$$\int_{-A}^{A} e^{ik\alpha x} \operatorname{sinc}\left(\frac{2\,\Xi}{\lambda}\alpha - m\right) \, d\alpha; \quad \int_{-A}^{A} e^{ik\beta y} \operatorname{sinc}\left(\frac{2\,\Xi}{\lambda}\beta - n\right) \, d\beta, \tag{2.3.4}$$

appearing in (2.3.4). Under the hypothesis $kA\Xi \gg 1$, he reaches the conclusion that these integrals are negligibly small unless the following conditions are met:

$$|m| < \frac{2\,\Xi\,A}{\lambda}: \qquad |n| < \frac{2\,\Xi\,A}{\lambda}. \tag{2.3.5}$$

The number of distinct pairs satisfying these conditions equals $(4\,\Xi\,A/\lambda)^2$ and this is also the number of complex coefficients $K_{mn}$ needed to specify the field, i.e., the number of complex DOF.

As far as the temporal DOF are concerned, the von Laue derivation is as follows. Let us consider a spectral interval $d\nu$ and a finite time span $T$. Expand the field into a time Fourier series within the interval $(0, T)$. The harmonics belonging to the interval $d\nu$ are the ones whose frequencies $n/T$ satisfy the inequalities

$$\nu < \frac{n}{T} < \nu + d\nu. \tag{2.3.6}$$

Their number amounts to $Td\nu$. Combining the spatial and temporal DOF, von Laue concludes that a beam of light with spectral width $d\nu$ and contained in a solid angle $\Omega$ possesses

$$\frac{4\,\Xi^2\Omega T\,d\nu}{\lambda^2} \tag{2.3.7}$$

complex DOF when it illuminates an area $4\,\Xi^2$ for a time $T$. Actually, von Laue refers to real DOF because he argues that no DOF are to be attributed to the phases.

We will not further discuss the von Laue paper but we cannot help remarking how pioneering it is. Note in fact that the sinc series appearing in (2.3.4) is nothing but the sampling expansion of what we now call the plane wave spectrum or angular spectrum of the radiation field [106, 339]. Such a spectrum can be expanded into a sampling series because its (inverse) Fourier transform (shorthand: FT) has a finite support (the square region $|\,\xi\,| \leq \Xi.\ |\,\eta\,| \leq \Xi$). Although von Laue does not notice explicitly that the coefficients $K_{mn}$ are samples of the angular spectrum, the substance of the sampling theorem is already present.

## 2.4   Degrees of Freedom of an Image

This section might well be subtitled "The struggle for superresolution." There is in fact an intimate connection between the concept of degrees of freedom of an image and that of resolving power, and although the degrees of freedom are used for many other purposes, one of the subjects that has turned out to be most intriguing is the possibility of overcoming the classical resolution limit.

In the following, we will not stick too strictly to the history of the subject. Instead, we will try to evidentiate some of the main ideas that developed in the field.

Let us recall that under certain simplifying assumptions [339, 575, 325, 106] an optical system can be thought of as a linear shift-invariant system characterized by its impulse response. Accordingly, the image function is the convolution between the object function and the impulse response. This holds true for any state of coherence of the light radiated by the object [106, 342], provided that a suitable meaning is given to the words object function and image function. For quasi-monochromatic spatially coherent light, the functions describing the object and the image represent disturbances, i.e., scalar field distributions: whereas they represent optical intensity distributions if the light radiated from the object is quasi-monochromatic and spatially incoherent. The impulse response to be used in these two cases is of course different: that pertaining to incoherent objects being proportional to the squared modulus of the coherent one. In the more general case of partially coherent illumination, both the object and the image are to be described by means of cross-spectral densities [569], i.e., by functions of pairs of points. The corresponding impulse response has the form of the product of the coherent responses evaluated at two distinct points.

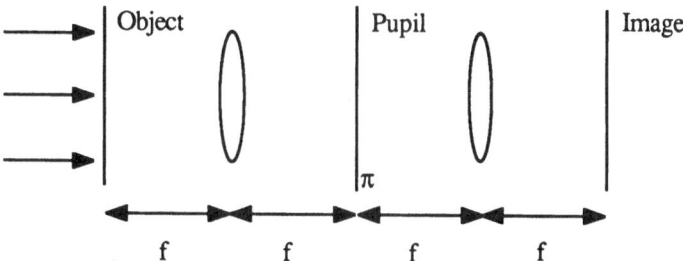

FIGURE 2.1: A unit magnification system.

Therefore, the fundamental function is the coherent impulse response. Using a one-dimensional notation, we shall denote it by $S(x)$. Its FT $\tilde{S}(p)$, namely, the coherent transfer function, is often called the pupil function because it can be thought of as the transmission function of the pupil of the optical system. The basic paradigm for coherent imaging is exemplified by the unit magnification system of Fig. 2.1. The object field distribution, say $f(x)$, propagating up to the pupil plane undergoes a Fourier transformation. With a suitable definition of the coordinates, the pupil plane $\pi$ can be considered as a plane of spatial frequencies where the object spectrum $\tilde{f}(p)$ is displayed. Such a spectrum is modified through multiplication by the pupil function. After that, light proceeds toward the image plane. This step can be described as an inverse FT if the coordinates axes in the object and image planes are opposedly oriented. In formulas, the image field distribution, say $g(x)$, can be expressed as the convolution

$$g(x) = \int f(y)S(x-y)\ dy, \tag{2.4.1}$$

where the integration region is (possibly) the whole $y$-axis. In equivalent terms, the spectra $\tilde{g}(p)$ and $\tilde{f}(p)$ of the image and the object, respectively, are related by

$$\tilde{g}(p) = \tilde{f}(p)\tilde{S}(p). \tag{2.4.2}$$

The venerable concept of resolving power can be introduced in a simple manner. A single object point gives rise to a light patch in the image. When the object is made up of two points, two light patches appear in the image. If the distance between the two object points is progressively reduced, the two patches overlap more and more until the overall image becomes practically indistinguishable from the one that would be produced by a single object point. The two object points are no longer "resolved." To quantify this statement, the coherence properties of the light radiated by the object points are to be taken into account [896]. We will not go into details. Suffice it to say that for any state of coherence one can define a resolution limit.

The resolution limit was thought of as an ultimate barrier for a very long time. Presumably, after the advent of quantum mechanics, the possibility of overcoming that limit also seemed to be prevented by the Heisenberg uncertainty principle. Yet, in the 1940s, superdirective or supergain antennas were discovered [797] and their optical counterparts, namely, superresolving pupils, were later proposed [906]. In principle, the impulse response of an optical system could be narrowed at will. Although the actual production of superresolving pupils presented formidable difficulties, it was clear that the resolving power concept lacked a solid foundation. The transfer of information from the object to the image was to be reexamined by different means. A possibility was offered by the sampling theorem.

## 2.4.1   USE OF THE SAMPLING THEOREM

In order to outline this approach, we shall refer to the simplest case. A one-dimensional coherent object field distribution $f(x)$ is imaged through an optical system whose pupil is a perfect low-pass filter extending from $-p_M$ to $p_M$ on an axis of spatial frequencies. The image field distribution $g(x)$ is then given by the convolution

$$g(x) = 2p_M \int_{-\infty}^{\infty} f(y)\,\mathrm{sinc}[2p_M(x-y)]\ dy. \tag{2.4.3}$$

Due to its band-limited nature, the image can be expressed through a sampling expansion of the form

$$g(x) = \sum_{n=-\infty}^{\infty} g\left(x_0 + \frac{n}{2p_M}\right)\mathrm{sinc}[2p_M(x-x_0)-n], \tag{2.4.4}$$

where $x_0$ is an arbitrary shift. In other words, the image is completely determined by a set of its samples taken at the Nyquist rate $1/2p_M$.[3]

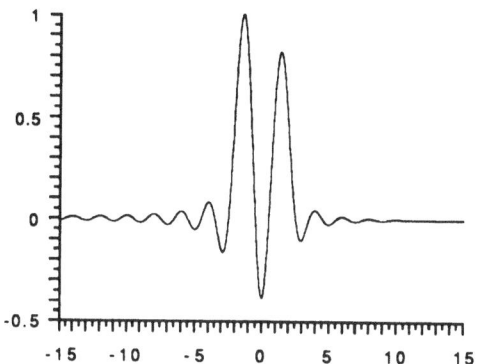

FIGURE 2.2: An image field distribution $g(x)$.

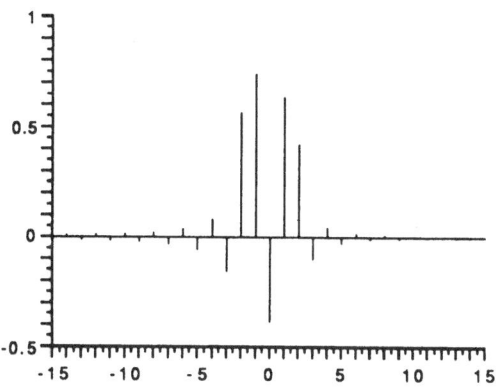

FIGURE 2.3: Set of samples of $g(x)$ obtained with $x_0 = 0$.

Let us consider the image distribution of Fig. 2.2, where, for the sake of simplicity, $g(x)$ is real and $2p_M = 1$. The set of samples obtained with $x_0 = 0$ is shown in Fig. 2.3. The vertical segments can be thought of as the positions and amplitudes of a set of point-like coherent sources that would give rise to our image. The important point to be made is that this is only one possible object out of infinitely many objects producing the same image. As an example, let us consider the samples obtained when $x_0 = 0.5$. These are drawn in Fig. 2.4. It is seen that only two samples are different from zero. [As a matter of fact, Fig. 2.2 was obtained as a plot of the function $\mathrm{sinc}(x - 1.5) + 0.8\,\mathrm{sinc}(x + 1.5)$]. It is tempting to conclude

---

[3]Sampling expansions suitable for space-variant systems also exist [593].

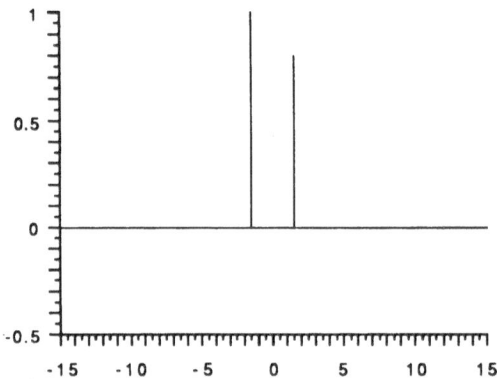

FIGURE 2.4: Set of samples of $g(x)$ obtained with $x_0 = 0.5$.

that the true object is a pair of well-resolved point-like sources. However, unless we have sufficient *a priori* information, no such conclusion can be drawn [907].

The classical formulation of the resolving power concept is based on the assumption that the observer has only to decide whether a single object point or two object points are present. In a similar hypothesis, the observer only needs one bit of information. Although this may be the case in some astronomical or spectroscopic investigation, the general situation is much more complicated than this.

The previous remarks point out the role of prior knowledge in scientific observations. There is an early optical example of the importance of this factor at the very beginning of modern science. This occurred when Galileo aimed his telescope at Saturn. He saw a surprising image and decided to communicate the results of his observations to Kepler. According to the style of those old days, he sent Kepler an anagram in Latin language. The decoded message meant "I observed a very high planet composed of three bodies." He was wrong, of course, but how did that occur? We can easily guess that the image was not very sharp and Galileo had no previous experience of planets surrounded by annuli. The most commonly known shape for celestial bodies was the sphere. As a consequence, he interpreted the image as produced by a central body with two much smaller bodies at its sides.

Let us come back to sampling and ask what changes are to be made in our arguments in the presence of prior knowledge. One of the simplest cases is perhaps when we know that the object has a finite extent, say from $-x_M$ to $x_M$. The image formation law is still given by (2.4.3) with the only difference that the infinite limits of integration are replaced by $\pm x_M$. We want to give an estimate of the number of DOF of the image. To this end, we note that the impulse response $\mathrm{sinc}(2p_M x)$ of our system has a width, roughly speaking, of about $1/(2p_M)$. Let us suppose that the object extent is much larger than this, i.e., that $4x_M p_M \gg 1$. Then, we can say that the image resulting from the convolution between $f(x)$ and $\mathrm{sinc}(2p_M x)$ has an overall extent slightly greater than $2x_M$. This, of course, is what common experience suggests. For example, no one looking through the viewfinder of a photographic camera expects to see an image much larger than the one predicted by geometrical optics. Within the practically finite extent $2x_M$ of the image we find a finite number $N$ of samples, namely,

$$N = 4x_M p_M, \qquad (2.4.5)$$

and this is also the number of complex DOF of the image. It is easily shown that the result is of the same type obtained in the von Laue analysis. For obvious reasons, the quantity $4x_M p_M$ is called the *space-bandwidth product* [553]. The term Shannon number [910] is also used.

## 2.4.2 SOME OBJECTIONS

The previous derivation of the number of DOF is admittedly crude and several objections can be raised against it. First, we have assumed that the only non-negligible samples are those falling within the geometrical image. This is based more on physical intuition than on mathematically sound arguments. As a matter of fact, it is not difficult to find examples in which many relevant samples are outside the geometrical image [715]. This may be seen, e.g., in Fig. 2.5 that gives the image produced by the following object. A group of 501 coherent point-like sources whose amplitudes are alternatively $+1$ or $-1$ are aligned between $-x_M$ and $x_M$, with a mutual spacing of $0.1/2p_M$. The condition $4x_M p_M \gg 1$ is satisfied and yet significant parts of the image are outside the interval $[-x_M, x_M]$. The horizontal unit in Fig. 2.5 equals $1/(2p_M)$. We note in passing that the same object can be used to evidentiate how minute changes of the object can produce large changes in the image. Suppose in fact that we pass from 501 to 500 point-like sources. The resulting image, shown in Fig. 2.6, exhibits several changes and, in particular, a change of phase of $\pi$ of one of the extreme peaks.

As a second objection, it may be observed that non-uniform samplings exist [495, 980, 544, 704, 351, 391, 956, 111]. For example, a band-limited function can be specified by a set of samples grouped in periodically repeated bunches provided that the average sample density satisfies the Nyquist condition. Within a bunch, the mutual distance between adjacent samples can be much smaller than $1/(2p_M)$. Suppose now that the image is specified by a non-uniform sampling of this type with one of the bunches falling within $[-x_M, x_M]$. Clearly enough, we can choose the sampling in such a way that the number of samples contained in the geometrical image is substantially greater or smaller than the value of $N$ given by Eq. (2.4.5).

A third objection is the following. It may be shown that the FT of a

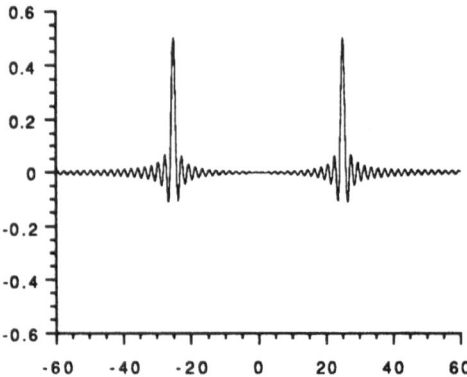

FIGURE 2.5: Image distribution of *501* coherent point-like sources with alternative ±1 amplitudes

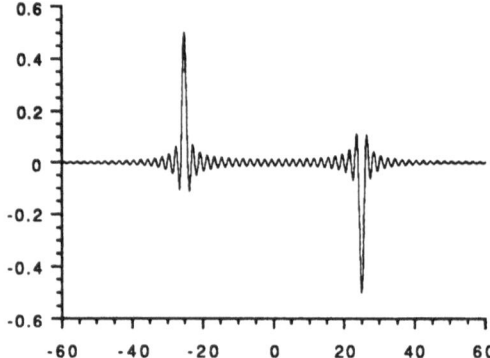

FIGURE 2.6: Image distribution of *500* coherent point-like sources with alternative ±1 amplitudes

function with finite support is analytic [965], [381, 987, 988]. This applies to the spectrum of the object field distribution. Such a spectrum is available in the interval $[-p_M, p_M]$ and, because of its analyticity properties, we can extrapolate it along the whole frequency axis. Therefore, a perfect recovery of the object should be possible. As the spectrum in $[-p_M, p_M]$ has the whole image as its own (inverse) FT, it may appear that we need the knowledge of the image field from $-\infty$ to $\infty$, but we can push our argument a little further. In fact, the image, being the (inverse) FT of a spectrum with finite support $[-p_M, p_M]$, is itself analytic and can be extrapolated starting from its knowledge in a finite interval, for example, $[x_M, x_M]$ or even in a smaller interval. In principle, any tiny piece of the image should be enough to reconstruct the object perfectly.

This conclusion, of course, sounds paradoxical. By the same token, when

listening to a music record, we could claim to be able to predict the future
melodies after few notes, on the grounds that the piece has a finite duration
and is reproduced through an amplifier with finite bandwidth. Having said
that, it remains to make clear where the above argument fails. The trap, of
course, is that when we descend from the platonic heaven of perfect Fourier
transforms and analytic functions to our laboratory, we find a world fraught
with noise and measurement errors [830]. We cannot assess the actual value
of the above arguments unless we find some procedure to implement the
reconstruction so that we can examine the effect of noise (including in it
measurement errors). This is where the ingenuity of researchers has produced the richest harvest of results. We shall now go briefly through some
of the most relevant approaches.

### 2.4.3 THE EIGENFUNCTION TECHNIQUE

We have a coherent object with finite support $[x_M, x_M]$ imaged through
the low-pass pupil $\text{rect}[p/(2p_M)]$, where $\text{rect}(t) = 1$ for $\mid t \mid \leq 1/2$ and is
otherwise zero. By a simple change of variables, the imaging equation can
be written

$$g(x) = \int_{-c/2}^{c/2} f(y)\text{sinc}(x - y) \ dy \quad (\mid x \mid \ \leq c/2),  \qquad (2.4.6)$$

where, to avoid multiplication of symbols, we made the substitutions

$$g\left(\frac{x}{2p_M}\right) \rightarrow g(x); \qquad f\left(\frac{x}{2p_M}\right) \rightarrow f(x),  \qquad (2.4.7)$$

and where

$$c = 4x_M p_M  \qquad (2.4.8)$$

is the space-bandwidth product. In optical terms, Eq. (2.4.6) describes the
imaging of a coherent object with support $[-c/2, c/2]$ through a low-pass
pupil extending from $-1/2$ to $1/2$. Only the region of the geometrical image
($|x| \leq c/2$) is considered. For the moment, noise is ignored. We have to
solve Eq. (2.4.6). Let $L_c^2$ be the space of square-integrable functions defined
in $[-c/2, c/2]$. We assume $f(y)$ to belong to $L_c^2$. Eq. (2.4.6) is a Fredholm
integral equation in the first kind, whose convolution kernel $\text{sinc}(x - y)$
is easily proved to be positive definite. A unique solution exists. It can
be found through the eigenfunction technique that we shall now sketch.
Consider the homogeneous Fredholm integral equation of the second kind

$$\int_{-c/2}^{c/2} \Phi(y)\,\text{sinc}(x - y) \ dy = \mu\Phi(x) \quad (\mid x \mid \leq \ c/2).  \qquad (2.4.9)$$

Because of the nature of the kernel, there exists a discrete set of real orthogonal eigenfunctions $\Phi_n(x)$ corresponding to positive and less than unity
eigenvalues $\mu_n$ ($n = 0, 1, \ldots$). The ordering is for decreasing eigenvalues

$$1 > \mu_0 > \mu_1 > \cdots \ .$$
(2.4.10)

The $\Phi_n(x)$ are the prolate spheroidal wavefunctions (shorthand: PSWF) [832, 519, 520]. A complete notation for them would be $\Phi_n(c; x)$ because there is a different family of PSWF for any value of $c$. Similarly, we should write $\mu_n(c)$. However, we shall drop such an explicit dependence on $c$.

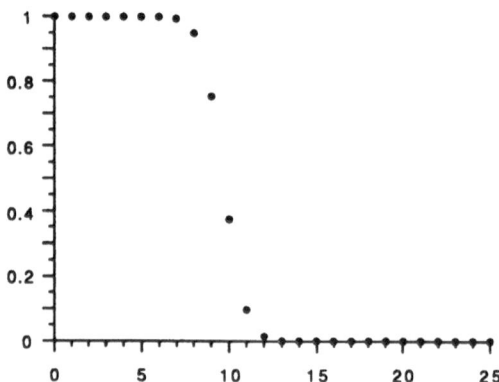

FIGURE 2.7: Eigenvalues versus $n$ for $c = 32/\pi$.

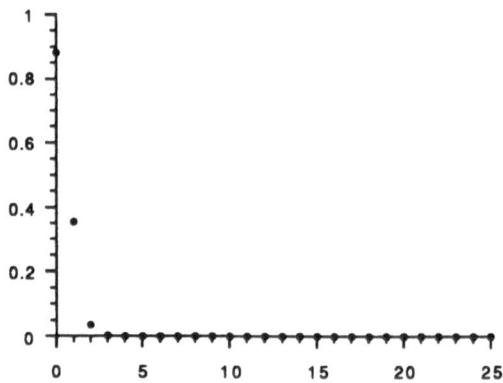

FIGURE 2.8: Eigenvalues versus $n$ for $c = 4/n$.

A few words about the PSWF are in order. Within the basic interval $[-c/2, c/2]$ they have a number of zeros equal to their order index, so that they become more and more rapidly oscillating when $n$ increases. Defining an equivalent spatial frequency as the inverse of twice the mean distance between adjacent zeros. we can say that, in $[-c/2, c/2]$, $\Phi_n(x)$ has an equivalent spatial frequency of $n/(2c)$. Observe now that. in optical terms. Eq. (2.4.9) defines a set of objects ($\Phi_n$) that, when imaged through the low-pass pupil. reproduce themselves within a constant ($\mu_n$). They can be called the eigenobjects of the optical system, in the sense that for any of them the image is a perfect replica of the object except for an amplitude factor. The surprising result is that there exist objects of arbitrarily large equivalent frequency that are perfectly imaged by the system. There is, however, a multiplying factor and we can expect it do drop to low values once the equivalent frequency exceeds the cutoff frequency (1/2) of the low-pass filter. This is exactly what happens as soon as $c$ exceeds a few units. As an example, the eigenvalues versus $n$ are represented in Fig. 2.7 for $c = 32/\pi$ [833]. The eigenvalues $\nu_n$ are almost unity for index values less than $c$ and then become extremely small. The transition from high to low eigenvalues occurs in a range of indices that grows proportionally to $\ln(c)$. Such a step function behavior is no longer valid when $c$ approaches 1. In this case, even the first eigenvalues are appreciably lower than one, as can be seen, e.g., from Fig. 2.8 referring to $c = 4/pi$. As we shall see, this has important consequences.

In optical terms, we would explain the small values of the $\mu_n$ by saying that the energy conveyed by the corresponding $\Phi_n$ goes mostly outside the pupil.

In order to clarify this point, we have to extend the definition of the PSWF outside $[-c/2, c/2]$. This can be done with the aid of Eq. (2.4.9). As the left-hand side makes sense for any $x$, we can remove the limitation $|x| > c/2$ and let $\Phi_n$ be defined in the outer region, $|x| > c/2$, through the convolution of the inner part with $\text{sinc}(x)$. We shall refer to $\Phi_n(x)$ defined everywhere as the extended $\Phi_n$ Note that, by its very definition, this is a band-limited function. The inner part of $\Phi_n$ will be termed the truncated $\Phi_n$ Now, if we assume the $\Phi_n$ to be normalized in the basic interval

$$\int_{-c/2}^{c/2} \Phi_n^2(x)\ dx = 1 \qquad (\forall n), \qquad (2.4.11)$$

then it can be proved that

$$\int_{-\infty}^{\infty} \Phi_n^2(x)\ dx = \frac{1}{\mu_n} \qquad (\forall n). \qquad (2.4.12)$$

We can read the left-hand side integrals in Eqs. (2.4.11) and (2.4.12) as a sort of measure of energy and we can say that the PSWF have their energy mostly inside or outside the basic interval depending on whether the order index is smaller or greater than $c$.

The behavior of the PSWF can be clarified a bit further by discussing an even more fundamental property of them. The PSWF are self-reproducing under finite FT. If we take the FT of the truncated $\Phi_n$, we obtain a function with the same shape as the extended $\Phi_n$ itself. There is a scale factor as well as an amplitude factor to be taken into account. The complete relation is

$$\int_{-c/2}^{c/2} \Phi_n(y) e^{-2\pi i p y} \, dy = i^{-n} \sqrt{c\mu_n} \; \Phi_n(cp). \qquad (2.4.13)$$

Conversely, the FT of the extended $\Phi_n$ is proportional to the following truncated version of it:

$$\int_{-\infty}^{\infty} \Phi_n(y) e^{-2\pi i p y} \, dy = i^{-n} \sqrt{c/\mu_n} \; \Phi_n(cp) \, \mathrm{rect}(p). \qquad (2.4.14)$$

Eq. (2.4.14), of course, is consistent with the fact that the extended $\Phi_n$ is band-limited.

It is useful to observe that there is a familiar example of functions that are self-reproducing under Fourier transformation. It is the set of the Hermite-Gauss functions. To facilitate comparison, we insert a parameter $c$ into their definition as follows:

$$G_n(x) = \frac{(2/c)^{\frac{1}{4}}}{\sqrt{2^n n!}} H_n\left(x\sqrt{\frac{2\pi}{c}}\right) e^{-\pi x^2/c}, \qquad (2.4.15)$$

where $H_n$ is the $n$th Hermite polynomial. The (infinite) FT of $G_n(x)$ is

$$\int_{-\infty}^{\infty} G_n(y) e^{-2\pi i p y} \, dy = i^{-n} \sqrt{c} \; G_n(cp). \qquad (2.4.16)$$

On comparing Eqs. (2.4.13) and (2.4.16), we can expect that for large $c$ the PSWF with index smaller than $c$ become similar to the Hermite-Gauss functions. In fact, this is the case. Incidentally, it is for this reason that we can assimilate the modes of a laser cavity with spherical mirrors to Hermite-Gauss beams [108].

The property expressed by Eq. (2.4.9) can be thought of as the result of two steps of the form (2.4.13). First, $\Phi_n$ is truncated to the basic interval and Fourier transformed. According to Eq. (2.4.13), this produces the extended $\Phi_n$ up to a change of scale and a complex factor. Note that the scale change is such that we find within the pupil the same inner part of $\Phi_n$ that was included in the object extent. After truncation by the pupil, an inverse FT gives the image. This implies again Eq. (2.4.13) or, more properly, its complex conjugate. The final result is the attenuated version of $\Phi_n$.

In a more formal way, we truncate to $[-1/2, 1/2]$ and make an inverse FT of both sides of Eq. (2.4.13). Interchanging order of integration on the left and changing variables on the right, we obtain

$$\int_{-c/2}^{c/2} \Phi_n(y) \, dy \int_{-1/2}^{1/2} e^{2\pi i p(x-y)} \, dp$$

$$= i^{-n} \sqrt{\frac{\mu_n}{c}} \int_{-c/2}^{c/2} \Phi_n(v) e^{2\pi i v x/c} \, dv. \qquad (2.4.17)$$

Using the complex conjugate of (2.4.13) we find that (2.4.17) coincides with (2.4.9).

We shall now use the PSWF to give a formal solution of Eq. (2.4.9). To this aim, we expand both $f(x)$ and $g(x)$ into a series of PSWF:

$$f(x) = \sum_{n=0}^{\infty} f_n \Phi_n(x). \qquad \left\{ f_n = \int_{-c/2}^{c/2} f(x) \Phi_n(x) \ dx \right\}. \qquad (2.4.18)$$

$$g(x) = \sum_{n=0}^{\infty} g_n \Phi_n(x). \qquad \left\{ g_n = \int_{-c/2}^{c/2} g(x) \Phi_n(x) \ dx \right\}. \qquad (2.4.19)$$

On inserting from Eq. (2.4.18) and (2.4.19) into Eq. (2.4.6) and taking into account Eq. (2.4.9) we find

$$g_n = \mu_n f_n \qquad (n = 0, 1, \ldots). \qquad (2.4.20)$$

Hence. on dividing the image coefficients $g_n$ by the corresponding eigenvalues $\mu_n$, we obtain the object coefficients $f_n$. i.e., the solution of our problem (see Eq. (2.4.18)).

What about the number of DOF of the image ? In the noiseless case considered so far, all of the (infinitely many) DOF of the object represented by the set of coefficients $f_n$ are transferred to the image in the set of the $g_n$. However, in the passage from the object to the image each coefficient is multiplied by the corresponding eigenvalue. In particular, the object coefficients with index exceeding $c$ by more than the width of the transition region are multiplied by very small numbers.

We have now to see how the recovery process is influenced by noise [142, 311, 937, 782, 769. 770, 910, 73. 786]. For noisy images, $g_n$ is affected by an error. Let

$$\bar{g}_n = g_n + \epsilon_n \qquad (n = 0, 1, \ldots) \qquad (2.4.21)$$

be the noisy value of $g_n$ with an error term $\epsilon_n$. The estimated value of $f_n$. say $\bar{f}_n$, is obtained from (2.4.21) through division by $\mu_n$:

$$\bar{f}_n = f_n + \frac{\epsilon}{\mu_n} \qquad (n = 0, 1, \ldots). \qquad (2.4.22)$$

It is seen that the error term is amplified $1/\mu_n$ times by this process. To give a feeling of this effect, we consider an example. Let $c = 32/\pi$ and suppose we want to evaluate $\bar{f}_n$ up to $n = 20$ (which corresponds to twice the number of DOF furnished by Eq. (2.4.5)). From the tables of [833] it turns out that when we try to recover $\bar{f}_{20}$, the error term is multiplied by the frightening figure of $10^{12}$. Unless the errors on the image coefficients decrease as fast as the $\mu_n$, this would induce a disaster. Unfortunately. in most realistic situations, the causes of error are likely to behave as a sort of white noise, thus producing a mean squared value of $\epsilon_n$ independent from $n$. This prevents the recovery of object coefficients with index significantly greater than $c$. The enormous amplification of data errors occurring in the evaluation of the solution is a symptom of a typical pathology of inverse problems, namely, ill-posedness [899, 791, 74]. The methods for treating ill-posed problems in such a way as to obtain sensible solutions, i.e, methods for regularizing the problem have had a large development in recent years. In particular, those methods have been applied to the present problem of image recovery [34, 928, 78, 79]. Here we shall limit ourselves to very simple considerations.

The simplest way to cope with the noise problem is to use a truncated series expansion in the object recovery. This means that we use Eq. (2.4.22) up to a value of $n$ determined as follows. Let $\phi$ and $\eta$ be the root mean square values of $f_n$ and $\epsilon_n$, respectively, in the hypothesis that both the object and the noise are white processes. To prevent noise from overcoming the signal, we stop the series at the maximum value of $n$ such that the following inequality is still satisfied:

$$\mu_n > \frac{\eta}{\phi},\qquad\qquad (2.4.23)$$

Denote by $N'$ this value of $n$. Then. $N'$ is the number of DOF of the image. It depends, of course, on the ratio $\eta/\phi$, let us say the noise to signal ratio. Nevertheless, due to the sharp fall-off of the eigenvalues. for any realistic case, the value of $N'$ will not exceed $c$ very much. More exactly, $N'$ is determined by the width of the transition region so that the estimate of the number of DOF furnished by Eq. (2.4.5) is to be corrected by adding a term growing like $\ln(c)$. Let us see an example. For $c = 50/\pi \simeq 15.9$ and $\eta/\phi = 10^{-3}$ (which means rather good experimental data) the tables of [833] give $N' = 19$. Now, passing from $N = 16$ to $N' = 19$ does not imply a very significant increase of resolution. So, after all, the sampling based estimate is asymptotically correct (for large $c$).

The substantial agreement between the sampling and the eigenfunction approaches can be expressed by saying that, for large $c$, the eigenvalues behave approximately like samples of the pupil taken at the Nyquist rate $1/(2x_M)$. They are very near to one up to $n \leq c$ and very near to zero for greater indices. This pictorial remark can be converted into a rigorous statement through the Szegö's theorem [464, 516. 517]. Under much more general conditions, e.g., for differently shaped pupils, the theorem asserts that in a well-defined sense the eigenvalues are asymptotically approximated by the pupil samples.

If $c$ is not large. say for $c$ of the order of unity. things are different. In this case. even the recovery of a couple of object coefficients beyond the Shannon limit is significant in that it corresponds to obtaining a resolution two or three times greater that the classical one. A value of $c$ near to one means an object extent approximately equal to the width of the impulse response. At first sight, it may appear that this would seldom occur. Yet, there is an important class of instruments in which this is the case. It is the class of scanning instruments [958] where the object is illuminated by a tiny spot of light moving across it. At any time, the effective object reduces to the illuminated region and the space-bandwidth product is then small. We can say that the trick is to use the temporal DOF of the optical channel that are normally unexploited. Obviously, the illuminating spot can hardly have a rectangular profile, so that our analysis is to be somewhat extended. A hint on this will be given later.

It is useful to observe that the recovery process up to the index $N'$ can be described as the result of the application to the image of a certain integral operator [40]. Referring to the noisy case, we write the truncated series estimate of the object. say $f_{N'}(x)$, as

$$f_{N'}(x) = \sum_{n=0}^{N'} \bar{f}_n \Phi_n(x).$$ (2.4.24)

It is easily seen that $f_{N'}(x)$ is obtained from the noisy image, say $\bar{g}(y)$, in the following way:

$$f_{N'}(x) = \int_{-c/2}^{c/2} \bar{g}(y) R_{N'}(x,y) \; dy.$$ (2.4.25)

where

$$R_{N'}(x,y) = \sum_{n=0}^{N'} \frac{1}{\mu_n} \Phi_n(x) \Phi_n(y).$$ (2.4.26)

The virtue of this approach is to evidentiate that the kernel $R_{N'}(x,y)$ is independent from the particular image to which the recovery process is to be applied. Note that $R_{N'}(x,y)$ is a shift-variant kernel.

## 2.4.4 THE GERCHBERG METHOD

Elegant as they are mathematically, the PSWF are not simple to use for numerical evaluations [829, 833. 309]. The situation is even worse when the eigenfunction technique is to be used for more general imaging processes in which the eigenfunctions are seldom known analytically. Therefore, methods of image processing not requiring the explicit knowledge of the eigenfunctions are welcomed. One of these is an iterative method proposed in [329] and originally explained through energy considerations in the domains of the image and its FT.

An analysis of the Gerchberg method in terms of PSWF was independently given in [346. 256, 705]. Here, we shall use the approach given in [346] because it affords an easy extension of the method to more general imaging situations.

First, let us give Eq. (2.4.6) a more compact form by introducing the symbol $K$ for the integral operator as follows

$$g(x) = (Kf)(x) \equiv \int_{-c/2}^{c/2} f(y)\operatorname{sinc}(x-y)\,dy \qquad (|x| \le c/2). \quad (2.4.27)$$

The basic idea is to transform this Fredholm integral equation of the first kind into one of the second kind and solve the last by a Neumann series. This is done by introducing a sort of complementary operator

$$\bar{K} = U - K, \tag{2.4.28}$$

where $U$ is the identity operator. Using $\bar{K}$. Eq. (2.4.27) can be written

$$f(x) = g(x) + (\bar{K}f)(x). \tag{2.4.29}$$

which is a Fredholm equation of the second kind. Following the usual procedure we can try to solve Eq. (2.4.29) by iteration. Successive approximations of the solution are

$$
\begin{aligned}
f^{(1)}(x) &= g(x). \\
f^{(2)}(x) &= g(x) + (\bar{K}f^{(1)})(x) = g(x) + (\bar{K}g)(x). \\
f^{(3)}(x) &= g(x) + (\bar{K}f^{(2)})(x) = g(x) + (\bar{K}g)(x) + (\bar{K}^2 g)(x). \\
&\vdots \qquad \vdots \\
f^{(M)}(x) &= g(x) + (\bar{K}f^{(M-1)})(x) \\
&= g(x) + (\bar{K}g)(x) + (\bar{K}^2 g)(x) + \cdots + (\bar{K}^{M-1}g)(x).
\end{aligned}
\tag{2.4.30}
$$

The effect of successive iterations can be studied with the aid of the PSWF. Using Eqs. (2.4.28) and (2.4.9). we see that the action of $\bar{K}$ on $\Phi_n(x)$ is expressed by the relation

$$(\bar{K}\Phi_n)(x) = (1 - \mu_n)\Phi_n(x). \tag{2.4.31}$$

On inserting from Eq. (2.4.19) into the last of Eqs. (2.4.31) and taking into account Eq. (2.4.31), we easily obtain

$$f^{(M)}(x) = \sum_{n=0}^{\infty} g_n \frac{1 - (1 - \mu_n)^M}{\mu_n}\Phi_n(x) = \sum_{n=0}^{\infty} p_n^{(M)} f_n \Phi_n(x). \tag{2.4.32}$$

where we used Eq. (2.4.20) and we defined the quantities

$$p_m^{(M)} = 1 - (1 - \mu_n)^M \qquad (n = 0.1.\dots;\ M = 1.2.\dots). \tag{2.4.33}$$

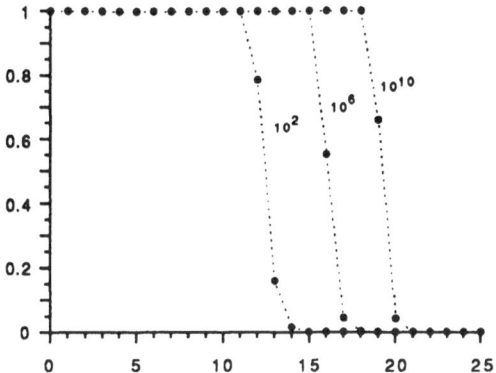

FIGURE 2.9: Plot of $p_n^{(M)}$ vs. $n$ for $c = 32/\pi$ and $M = 10^2, 10^6, 10^{10}$.

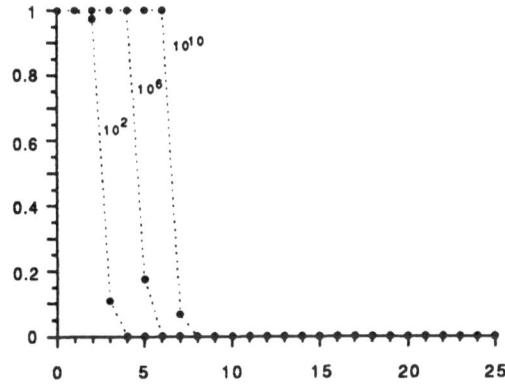

FIGURE 2.10: Plot of $p_n^{(M)}$ vs. $n$ for $c = 4/\pi$

Taking into account that $\mu_n < 1$, we see that the coefficients $p_n^{(M)}$ tend to unity for $M \rightarrow \infty$. In the same limit, $f^{(M)}$ tends to the solution of Eq. (2.4.29). On comparing Eq. (2.4.32) to Eq. (2.4.19) and taking into account Eq. (2.4.20), we note that for any finite $M$ the net effect of the iterations is that the eigenvalues $\mu_n$ are replaced by the quantities $p_n^{(M)}$. These quantities are drawn in Figs. 2.9 and 2.10, for $c = 32/\pi$ and $c = 4/\pi$, respectively, for a few values of $M$ (indicated near each curve). The behavior of the $p_n^{(M)}$ with respect to $n$ is similar to the one exhibited by the $\mu_n$ (see Figs. 2.7 and 2.8) except that the transition from high to low values occurs at a progressively higher index for increasing $M$. In a sense, this is equivalent to recovering the image that would be produced by a progressively wider pupil.

For small eigenvalues $\mu_n$, the number of iterations required to bring $p_n^{(M)}$ near to one grows approximately like $1/\mu_n$ as can be easily deduced from

Eq. (2.4.33). If $c$ is near to one, as in Fig. 2.10, a rather small number of iterations (e.g., $M = 100$) produces an equivalent bandwidth two or three times larger than the original one. To obtain the same result when $c$ is greater than some units, as in Fig. 2.9, we would need an impractically high number of iterations. One the other hand, two remarks are to be made. First, variations of the method can give faster convergence and even closed form implementation [784, 565, 185, 597, 599]. Second, the recovery of object coefficients corresponding to too small eigenvalues is actually precluded by noise problems. From this point of view, the Gerchberg method suffers from the same limitations as the direct eigenfunction technique.

The virtues of the Gerchberg methods are: a) it allows one to implement the eigenfunction technique without requiring their explicit knowledge; b) it works with imaging kernels of many types. In the simplest form of the method, the iterations are stopped when the scientist realizes that noise is beginning to dominate the reconstruction. This is practically done by using some plausibility criterion for the reconstructed object. Alternatively, the prior knowledge about the object can be incorporated into a regularized form of the method [200, 1].

The Gerchberg method implies repeated convolutions and truncations. This is usually done by switching back and forth between the two domains where the object and its Fourier transform are defined. Numerically, this is obtained by using the discrete FT with fast algorithms [114]. It is curious to observe that on approximating continuous with discrete FT, one makes a sampling of both a function and its FT, assuming that the replica effects produced by sampling do not cause overlapping. This amounts to assuming that both the function and its FT can have finite support and this is prohibited, strictly speaking, by the same analyticity properties that are at the root of the recovery processes [295].

It is seen from Eqs. (2.4.31) that the output of each iteration is fed back to the input of a system described by the operator $\bar{K}$. This remark suggests that the Gerchberg method could be implemented by analogical systems with feedback. This can be in fact demonstrated [255, 580]. It is to be recalled that the Gerchberg method has been preceded by a similar method originally proposed in [920] and later reconsidered and extended by several authors [431, 310, 395, 893, 414]. The main difference [365] is that the original van Cittert method does not use the *a priori* information about the finite extent of the object so that it essentially performs an ordinary inverse Fourier filtering [310]. The analytical technique underlying the Gerchberg method was actually known in mathematics after the work in [523]. Also, it must be mentioned that a somewhat similar iterative method was used by Landau in a technique relating to companded signals [514, 518].

The Gerchberg method is a prototype of constrained iterative algorithms and can be extended and generalized in several ways [745, 982, 955, 184, 781, 423, 539, 581, 795, 834, 867, 564, 777, 472, 231].

## 2.5 Superresolving Pupils

We have seen methods for giving approximate solutions to the recovery problem using, either explicitly or implicitly, the PSWF. The result can be described as an image with increased resolution with respect to the image given by the original pupil.

In principle, an attractive alternative is to replace the so-called clear pupil, i.e., the pupil with transmission function rect($p$), by a superresolving pupil (shorthand: SRP) [906]. This would give an increased resolution in real time without any post-processing of the image. Here, we want to see something more about such pupils and to establish a link between them and the previous approaches.

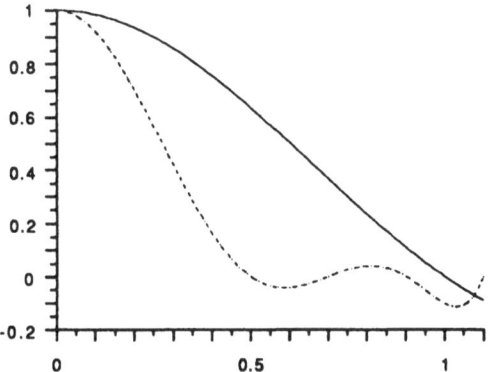

FIGURE 2.11: A plot of $S(x)$ (dotted line) and sinc($x$) (solid line).

Let us first outline the most elementary way to design a SRP. Denote by $S(x)$ the impulse response of the pupil. As the pupil function vanishes outside $[-1/2, 1/2]$, $S(x)$ can be expressed through the sampling expansion

$$S(x) = \sum_{n=-\infty}^{\infty} S(n) \operatorname{sinc}(x - n). \qquad (2.5.1)$$

The goal is to find the values to be given to the samples $S(n)$ in order to produce an impulse response that is narrower than sinc($x$) on a limited interval. Once this is done, the FT of Eq. (2.5.1) gives the Fourier series expansion of the required pupil function. The values $S(n)$ can be found by trial and error as follows. Suppose that only a finite number of them, say $2M + 1$, are different from zero. We now require $S(x)$ to be one at the origin and to vanish at $2M$ selected points $x_n$ ($n = \pm 1, \pm 2, \ldots, \pm M$). Then, Eq. (2.5.1) furnishes a set of $2M + 1$ linear equations in the $2M + 1$ unknown quantities $S(n)$. The points $x_n$ are chosen in such a way as to force $S(x)$ to be narrower than sinc($x$) and to remain at low values on a certain interval. An example helps to visualize this approach. Let $M = 4$ and choose $x_{\pm 1} = \pm 0.5$, $x_{\pm 2} = \pm 0.7$, $x_{\pm 3} = \pm 0.9$, $x_{\pm 4} = \pm 1.1$. The function $S(x)$ found on solving the problem is drawn in Fig. 2.11 where sinc($x$) is also given for the sake of comparison. It is seen that the resolution is increased by a factor of 2. The values of $S(n)$ ($n = 0.1.....4$) turn out be $S(0) = 1$; $S(\pm 1) = -0.101692$; $S(\pm 2) = 390.289$; $S(\pm 3) = 4522.14$: $S(\pm 4) = 6530.01$. It is clear from the high values of $S(n)$ for $\mid n \mid > 1$ that the impulse response drastically increases outside the interval $[-1.1, 1.1]$. These external sidelobes are the price to be paid for obtaining superresolution. To prevent the sidelobes from blurring the image, the object must have a finite extent equal to half the interval on which $S(x)$ has been kept under control (this is to say an extent 1.1 in our example). Therefore, the use of SRP requires the prior knowledge that the object vanishes outside a certain interval. As an empirical result, one finds that when the useful interval is increased and/or the width of the central core of $S(x)$ is decreased, the outer sidelobes become higher and higher. The design of SRP belongs to the general problem of approximating, on a given interval, an arbitrary function by means of a band-limited function [891]. A systematic approach can be based on the use of the PSWF [308, 309]. Suppose that we want $S(x)$ to have a prescribed form in $[-c/2. c/2]$. In such an interval, $S(x)$ can be expanded into the PSWF series

$$S(x) = \sum_{n=0}^{\infty} s_n \Phi(x) \qquad \left\{ s_n = \int_{-c/2}^{c/2} S(x) \Phi_n(x) \ dx \right\}. \qquad (2.5.2)$$

Replace now the truncated $\Phi_n$ by their extended version. This gives rise to a function that coincides with $S(x)$ for $\mid x \mid \leq c/2$ and is band-limited. Using Eq. (2.4.14), we immediately find that the required pupil function, say $\tilde{S}(p)$, is

$$\tilde{S}(p) = \sqrt{c} \ \text{rect}(p) \sum_{n=0}^{\infty} \frac{i^{-n}}{\sqrt{\mu_n}} s_n \Phi_n(cp). \qquad (2.5.3)$$

Therefore, we can choose at will $S(x)$ for $\mid x \mid \leq c/2$. Equation (2.5.3) gives us the pupil that produces the wanted response. Of course, we have no control about the behavior of $S(x)$ for $\mid x \mid > c/2$ and this is where again high sidelobes make their appearance. Note that the useful interval for the object to be correctly imaged is $[-c/4. c/4]$. We add that a similar approach can be used for a closely related problem in antenna theory, namely, to find the line source that best approximates a specified radiation pattern [759]. Furthermore. the PSWF offer an analytical solution [828] to the complementary optical problem of apodization, i.e.. suppression of the sidelobes of the impulse response.

In spite of their conceptual simplicity. SRP are difficult to use except for few cases [234. 110, 384]. The main reasons for this are the following. First. the pupil must be realized with high precision because even small errors can divert light from the outer sidelobes into the central region, thus disturbing the image. Second, the pupil is inefficient from the energetic point of view because much of the energy collected by the pupil is thrown into the outer regions of the image. Third, the pupil lacks flexibility. It would be useful in fact to be able to change the amount of superresolution and the consequent height of the outer sidelobes when different levels of noise are to be faced. This, of course, is impossible for a fixed pupil. An alternative possibility is to simulate the effect of the SRP by postprocessing the image obtained with a clear pupil. We can easily understand that in order to simulate the effect of the SRP we can merely convolve the image given by the clear pupil with the superresolved impulse response. In the Fourier domain, in fact, this amounts to multiplying the original pupil rect($p$) by $\tilde{S}(p)$. i.e.. to simulate the presence of the SRP in the image formation process. In this way, the impulse response of the SRP can be changed at will [257. 360].

Comparing this type of postprocessing with the one seen in subsection 2.4.3, we note that in both cases the image is acted on by an integral operator, the main difference being that $S(x)$ acts as a convolution kernel. whereas $R_{N'}(x,y)$ is shift-variant.

## 2.5.1 SINGULAR VALUE ANALYSIS

As we shall see later, the eigenfunction technique exemplified by the use of the PSWF can be extended in several directions. Nonetheless, it cannot be applied when the input and the output of our linear system belong to different functional spaces. In this case, a generalization is given by the singular value analysis that we shall briefly sketch here [9, 80, 75, 76]. Let us consider the following linear operator $A$:

$$(Af)(x) = \int f(y)H(x,y)dy, \qquad (2.5.4)$$

where the integration limits and the definition range for $x$ are given when the kernel $H(x,y)$ is specified. We also consider the adjoint operator $A^+$ defined by

$$(A^+h)(x) = \int h(y)H^*(y,x)dy, \qquad (2.5.5)$$

where the asterisk denotes the complex conjugate. We assume $H(x,y)$ to satisfy the condition

$$\int\int |H(x,y)|^2 dxdy < \infty. \qquad (2.5.6)$$

Then, the operators $A^+A$ and $AA^+$ possess a discrete set of eigenfunctions $u_n$ and $v_n$. respectively, corresponding to the same set of non-negative eigenvalues $\alpha_n^2$. The eigensystems are to be found by solving the equations

$$
\begin{aligned}
(A^+ A u_n)(x) &= \alpha_n^2 u_n(x) \quad (n = 0, 1, \ldots), \\
(A A^+ v_n)(x) &= \alpha_n^2 v_n(x) \quad (n = 0, 1, \ldots).
\end{aligned}
\tag{2.5.7}
$$

The functions $u_n$, $v_n$ are the singular functions of the operator $A$ and the (real non-negative) numbers $\alpha_n$ are the corresponding singular values. The action of $A$ $(A^+)$ on $u_n$ $(v_n)$ is the following:

$$
\begin{aligned}
(A u_n)(x) &= \alpha_n v_n(x) \quad (n = 0, 1, \ldots). \tag{2.5.8} \\
(A^+ v_n)(x)(x) &= \alpha_n u_n(x) \quad (n = 0, 1, \ldots). \tag{2.5.9}
\end{aligned}
$$

Suppose now that the equation

$$
g(x) = (A f)x
\tag{2.5.10}
$$

is to be solved. In general, the following expansion holds:

$$
f(x) = \sum_{n=0}^{\infty} a_n u_n(x) + r(x).
\tag{2.5.11}
$$

where $r(x)$ belongs to the null space of $A$. In addition. $g(x)$ admits the expansion

$$
g(x) = \sum_{n=0}^{\infty} b_n v_n(x).
\tag{2.5.12}
$$

On inserting (2.5.11) into (2.5.10) and taking into account Eqs. (2.5.8) and (2.5.12) one finds

$$
b_n = \alpha_n a_n \quad (n = 0, 1, \ldots).
\tag{2.5.13}
$$

Equations (2.5.11)–(2.5.13) are similar to Eqs. (2.4.18)–(2.4.20). The eigenvalues are replaced by the singular values and two different sets of functions $u_n$, $v_n$ are used. In addition, depending on the properties of $A$. a part of $f(x)$, namely, $r(x)$, can be irretrievably lost.

We shall now give two examples of applications of this technique to the recovery problem discussed earlier. We saw that the image formation can be thought of as a two-step process. In each of them, a finite support function is Fourier transformed and the FT is given on a finite interval. Let us consider the problem of recovering the function starting from the knowledge of a portion of its FT. This occurs, for example, if we detect the object spectrum across the pupil, either directly or through inverse Fourier transformation of the field distribution all across the image axis. The task is equivalent to extrapolating the spectrum outside the pupil. Reversing roles. we can say that we want to extrapolate a function whose

FT is known to vanish outside a finite interval. This is why the present problem is also known as the extrapolation of a band-limited function. The operator $A$ describing this process is defined as follows:

$$(Af)(p) = \int_{-c/2}^{c/2} f(y)e^{-2\pi i p y}dy \qquad (|p| \leq 1/2), \qquad (2.5.14)$$

whose adjoint is written

$$(A^+q)(x) = \int_{-1/2}^{1/2} q(t)e^{2\pi i x t}dt \qquad (|x| \leq c/2). \qquad (2.5.15)$$

It is easily seen that the operators $A^+A$ and $AA^+$ are

$$(A^+Af)(x) = \int_{-c/2}^{c/2} f(y)\operatorname{sinc}(x-y)dy \quad (|x| \leq c/2), \qquad (2.5.16)$$

$$(AA^+q)(p) = \int_{-1/2}^{1/2} q(t)\operatorname{sinc}[c(p-t)]dt \quad (|p| \leq 1/2). \qquad (2.5.17)$$

Except for a scale factor, the two integral operators are identical to the one entering Eq. (2.4.9). As a consequence, both sets of singular functions reduce to the set of PSWF. The singular values $\alpha_n$ equal $\sqrt{\mu_n}$ (see Eqs. (2.4.13) and (2.5.8)). Because of the completeness properties of the PSWF, a perfect recovery would be possible in the noiseless case. The recovery of the $\alpha_n$ in the presence of noise could be examined along the same lines discussed with reference to Eq.( 2.4.20). It is clear, however, that the substitution of the $\mu_n$ with the $\alpha_n$ is an advantage because $\alpha_n > \mu_n$.

The same scheme can be applied to scanning instruments. Suppose that at a certain time the object is illuminated by a field distribution whose shape we denote by $P(x)$ and imaged by an optical system with impulse response $H(x,y)$. The field emerging from the object can be expressed as the product of the dimensionless illuminating profile $P(x)$ times the field $f(x)$ that the object would emit if it were illuminated by an orthogonal plane wave of unit amplitude. The image field $g(x)$ can be expressed through the action on $f(x)$ of a suitable operator $A$ defined as follows:

$$g(x) = (Af)(x) = \int_{-\infty}^{\infty} P(y)f(y)H(x,y)dy. \quad (-\infty < x < \infty) \qquad (2.5.18)$$

It is seen that Eq. (2.5.18) generalizes the previous case (see, e.g., Eq. (2.4.1)). In particular, when $P(y) = \operatorname{rect}(y/c)$ and $H(x,y) = \operatorname{sinc}(x-y)$ we come back to Eq. (2.4.6) except that now $x$ can vary on the whole axis.

In order to find what information about the object $f(x)$ can be obtained, some features of $A$ are to be studied. It may well happen that $A$ is not injective. In physical terms, this means that the object can be split into the sum of a transmittable part and one part that gives no image. The recovery problem then refers to the transmittable part only. A simple case is the one where both $H$ and $f$ are sinc functions with the same width. This occurs when the same low-pass optical system is used both to illuminate and to image the object [76]. The singular system can be found in closed form [359]. Once again, the sampling theorem appears in the analysis. In fact, it turns out that the transmittable part of the object can be obtained by sampling it and deleting certain samples. In addition [77], for confocal scanning microscopy, an analytic inversion formula can be found starting from a sampling expansion of the image.

## 2.5.2  INCOHERENT IMAGING

As we said at the beginning of the present section, when the object radiates quasi-monochromatic spatially incoherent light, the image formation law involves optical intensities. For the one-dimensional case with a clear pupil extending from $-p_M$ to $p_M$, the law can be written

$$g(x) = 2p_M \int_{-\infty}^{\infty} f(y) \operatorname{sinc}^2[2p_M(x-y)]dy, \qquad (2.5.19)$$

where $f(x)$ and $g(x)$ are the optical intensity distributions across the object and the image, respectively. The proportionality factor in front of the integral is such that a maximum value of one is attained by the incoherent transfer function (also called the optical transfer function [339]). The latter, say $H(p)$, is proportional to the autocorrelation of the actual pupil. In the present case, we have

$$H(p) = \left[1 - \frac{|p|}{2p_M}\right] \operatorname{rect}\left[\frac{p}{4p_M}\right]. \qquad (2.5.20)$$

The analysis carried out earlier could now be rephrased for incoherent imaging. However, there are certain differences that deserve attention. First, the bandwidth is twice the one pertaining to coherent imaging, although the comparison is to be made with some caution because now the transfer function refers to the spatial frequencies of the object intensity instead of the object field. In addition, the incoherent transfer function is not flat within the band of frequencies accepted by the system. This suggests that an image enhancement could be obtained by restoring to the correct weight the frequency components inside the bandwidth of the system. Let us give a simple example of this inverse filtering process. Figure 2.12 represents the image of two point-like (mutually incoherent) sources of equal weight centered at $\pm 0.41/(2p_M)$. The unit on the horizontal axis equals $1/(2p_M)$. Clearly, the two sources are not resolved. We now filter the spectrum of this image so as to produce an equivalent transfer function that is flat in

$(-2p_M, 2p_M)$.[4] This amounts to replacing the original $\mathrm{sinc}^2(2p_m x)$ impulse response by $\mathrm{sinc}(4p_M x)$. The resulting corrected image is given in Fig. 2.13. Two resolved sources can be recognized (at least, if we know that either one or two sources are present, see subsection 2.4.1). Yet, the image shows some artifacts consisting of spurious oscillations. The main remark is that negative sidelobes are present, a feature inconsistent with the physical meaning of the function as an optical intensity.

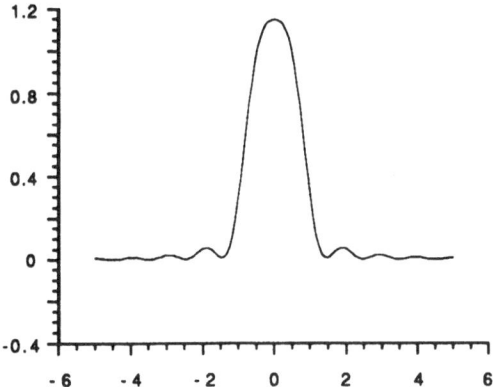

FIGURE 2.12: Image distribution of two point-like mutually incoherent sources.

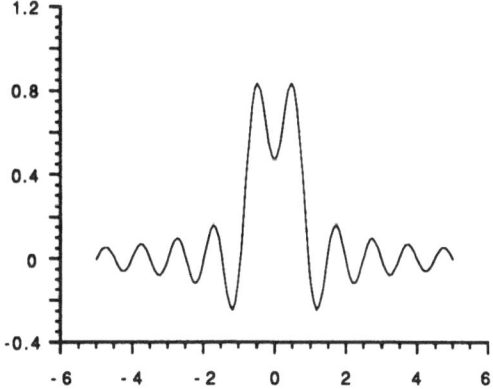

FIGURE 2.13: Image distribution of Fig. 2.12 after filtering.

Here, we encounter a basic constraint about the solutions of inverse problems relating to incoherent imaging. Being an intensity, the solution must

---

[4]Even in the noiseless case, the inverse filtering has to exclude the points $p = \pm 2p_M$ where $H(p)$ vanishes.

be non-negative. The role of this constraint also appears clearly when we apply the sampling theory to the present problem. The function $g(x)$ as expressed by Eq. (2.5.19) is band-limited and can be written as the sampling expansion

$$g(x) = \sum_{n=-\infty}^{\infty} g\left[x_0 + \frac{n}{4p_M}\right] \text{sinc}[4p_M(x - x_0) - n], \qquad (2.5.21)$$

where $x_0$ is an arbitrary shift. At first sight, this looks similar to Eq. (2.4.4) except for the doubling of the bandwidth. We can again say that the image is completely determined by a set of its samples taken at the Nyquist rate $1/(4p_M)$. Yet, the samples are no longer independent from one another because they have to produce a superposition where the negative lobes of the sinc functions are everywhere compensated by positive contributions. We add that the problem of sampling non-negative functions is related to the so-called phase problem [688, 936, 496, 32, 292, 260, 671].

The present imaging process could be examined through the eigenfunction technique. Although the eigenfunctions and the eigenvalues are not analytically known, they can be computed numerically [73, 345, 362]. In particular, it turns out that the eigenvalues, say $\mu_n$, are approximately given by $\mu_n = [1 - n/(8x_M p_M)]$, up to values of $n$ smaller than $8x_M p_M$. For greater values of $n$, the eigenvalues are extremely small. This again agrees with the Szegö's theorem which connects the sampling and the eigenfunction approaches.

Superresolution techniques quite similar to the ones already discussed for the coherent case can be applied to incoherent imaging. Furthermore, the positivity constraint can be included in the superresolving algorithms [310, 406, 408].

It is to be noted that the comparison between coherent and incoherent imaging is not an immediate one. In the first case, one deals primarily with field quantities, whereas in the second case intensities are the basic quantities [339, 342]. In addition, noise effects are different [205, 206].

## 2.5.3  SURVEY OF EXTENSIONS

Up to now, we limited ourselves to one-dimensional imaging without aberrations. The analysis can be extended in several directions. First, two-dimensional geometries can be considered. If a coherent object with a finite support possessing an area $S$ is imaged through a clear pupil covering a finite area $P$ in the spatial frequency plane, the number of complex DOF of the image equals SP. This essentially is the so-called Gabor's theorem [315] whose one-dimensional version leads to Eq. (2.4.5). As in the one-dimensional case, the validity of this result can be justified through different approaches using suitable forms of the sampling theorem [319, 94] or eigenfunction techniques [827, 515, 910, 356] or general properties of the solutions of the wave equation [656, 657]. Further extensions refer to the presence of aberrations [363, 259, 33] as well as to images produced by point-like element pupils [357, 358, 831]. The number of DOF for scattered fields can be evaluated in a similar way [364] and this leads to interesting analogies to the DOF in holography [556, 424, 739].

It is interesting to observe that superresolution can be linked to one of the newest approaches to data processing, namely, neural computing [589, 2].

We discussed several superresolution methods. Nevertheless, other methods exist [559, 22, 560. 725, 672. 724, 293, 726, 881, 407, 195].

## 2.6 Fresnel Sampling

We have seen the essential role played by the FT in the description of imaging processes. Nevertheless, most optical propagation phenomena are to be described by means of another transformation, namely, the Fresnel transform. This is a linear transformation depending on one real parameter $\alpha$. For a function $f(x)$, the $\alpha$-Fresnel transform to be denoted by $\mathcal{E}_\alpha\{f\}(x)$ or $\hat{f}_\alpha$, is [348]

$$\mathcal{E}_\alpha\{f\}(x) \equiv \hat{f}_\alpha(x) = \sqrt{-i\alpha} \int_{-\infty}^{\infty} f(\xi) e^{\pi i \alpha (x-\xi)^2} d\xi. \qquad (2.6.1)$$

The inversion formula reads

$$f(\xi) = \sqrt{i\alpha} \int_{-\infty}^{\infty} \hat{f}_\alpha(x) e^{-\pi i \alpha (x-\xi)^2} dx, \qquad (2.6.2)$$

so that inverse transform simply equals the Fresnel transform of parameter $-\alpha$.

The extension to the two-dimensional case is straightforward. The connection between the Fresnel transform and the Fresnel or paraxial approximation of physical optics is immediately seen from the diffraction integral [339, 106]

$$V_z(x, y) = -\frac{i e^{ikz}}{\lambda z} \int \int V_0(\xi. \eta) e^{i \frac{k}{2z} [(x-\xi)^2 + (y-\eta)^2]} d\xi d\eta. \qquad (2.6.3)$$

where $V_0$ and $V_z$ are the (coherent) field distributions across the planes $z = 0$ and $z = $ const. $> 0$, respectively. Except for the factor $\exp(ikz)$, we see that $V_z$ is the two-dimensional Fresnel transform of $V_0$ with parameter

$$\alpha = \frac{1}{\lambda z}, \qquad (2.6.4)$$

because $k = 2\pi/\lambda$.

A function $f(x)$ is said to be $\alpha$-Fresnel limited (shorthand: $\alpha$-Fl) if $\hat{f}_\alpha(x)$ vanishes outside some finite interval, say $[-x_0, x_0]$. Functions of this type possess several properties. For example, it can be proved than an $\alpha$-Fl function cannot be $\beta$-Fl also if $\alpha \neq \beta$, nor can it be band-limited in the Fourier sense. In addition, an $\alpha$-Fl function cannot be purely real. For $\alpha$-Fl functions the following sampling expansion holds [348]:

$$
f(x) = e^{-\pi i \alpha x^2} \sum_{n=-\infty}^{\infty} f\left(\frac{n}{2|\alpha|x_0}\right) e^{\pi i \alpha \left[\frac{n}{2\alpha x_0}\right]^2}
$$
$$
\times \operatorname{sinc}(2|\alpha|x_0 x - n). \tag{2.6.5}
$$

Conversely, if a function $f(x)$ vanishes outside a finite interval, say $[-x_M, x_M]$, then its $\alpha$-Fresnel transform can be expressed as follows:

$$
\hat{f}_\alpha(x) = e^{\pi i \alpha x^2} \sum_{n=-\infty}^{\infty} \hat{f}_\alpha\left(\frac{n}{2|\alpha|x_M}\right) e^{-\pi i \alpha \left[\frac{n}{2\alpha x_M}\right]^2}
$$
$$
\times \operatorname{sinc}(2|\alpha|x_M x - n). \tag{2.6.6}
$$

The symmetry between Eqs. (2.6.5) and (2.6.6) is due to the fact that the inverse $\alpha$-Fresnel transform is simply obtained by changing $\alpha$ into $-\alpha$.

Let us see the optical significance of these results, limiting ourselves to the one-dimensional case. We assume $f(x)$ to be a coherent field distribution with finite support $[-x_M, x_M]$ emerging from a plane $z = 0$. From the one-dimensional analog of Eq. (2.6.3) we see that the field distribution across a plane $z = \text{const.} > 0$ is given, up to a phase term $\exp(ikz)$, by $\hat{f}_\alpha(x)$ where $\alpha$ is specified by Eq. (2.6.4). By virtue of Eq. (2.6.6), the propagated field is completely determined by its samples spaced at a distance $\lambda z/(2x_M)$ from one another. Such a distance grows linearly with $z$ and this gives a measure of how the information density varies on free propagation.

In the case of FT, sampling a function produces an infinite set of replicas of the spectrum. In optical terms, if we cover with a grating a coherently radiating object put in the front focal plane of a converging lens, we obtain an array of copies of the object spectrum in the back focal plane. We can ask whether anything similar can be observed in Fresnel optics.

In order to answer the previous question it is useful to give an alternative representation of the samples of $\hat{f}_\alpha$. Let us introduce the function

$$
h(x) = f(x)e^{\pi i \alpha x^2}. \tag{2.6.7}
$$

whose Fourier series expansion in $[-x_M, x_M]$ reads

$$
h(x) = \sum_{m=-\infty}^{\infty} h_m e^{\pi i m \frac{x}{x_M}}. \tag{2.6.8}
$$

where

$$
h_m = \frac{1}{2x_M} \int_{-x_M}^{x_M} f(x)e^{\pi i \alpha x^2 - \pi i m \frac{x}{x_M}} \, dx.
$$

It is easily seen that the samples of $\hat{f}_\alpha$ can be given the form

$$\hat{f}_\alpha \left( \frac{n}{2|\alpha|x_M} \right) = 2x_M \sqrt{-i\alpha} e^{\pi i \alpha \left[ \frac{n}{2\alpha x_M} \right]^2} h_n. \qquad (2.6.9)$$

In particular, suppose that $\alpha$ satisfies the condition

$$\frac{1}{4\alpha x_M^2} = 2r \qquad (r : \text{integer}). \qquad (2.6.10)$$

In this case, the samples $\hat{f}_\alpha$ are proportional to the Fourier coefficients of $h(x)$. Because of Eq. (2.6.4), a set of planes exists

$$z_r = \frac{8rx_M^2}{\lambda} \qquad (r : \text{integer}), \qquad (2.6.11)$$

where Eq. (2.6.10) is satisfied. Let us put an ideal sampling mask in any of these planes. Disregarding the unessential factor $\exp(ikz)$, the field emerging from the mask can be approximated by a function, say $s(x)$, of the form

$$\begin{aligned} s(x) &= \sum_{n=-\infty}^{\infty} \hat{f}_\alpha \left( \frac{n}{2|\alpha|x_M} \right) \delta \left( x - \frac{n}{2|\alpha|x_M} \right) \\ &= 2x_M \sqrt{-i\alpha} \sum_{n=-\infty}^{\infty} h_n \delta \left( x - \frac{n}{2|\alpha|x_M} \right). \end{aligned} \qquad (2.6.12)$$

The field produced across a typical plane $z = \text{const.} > 0$ at a distance $D$ beyond the mask can be evaluated, up to a phase term $\exp(ikD)$, as the $\beta$-Fresnel transform of $s(x)$ with $\beta = 1/(\lambda D)$. We find without difficulty

$$\hat{s}_\beta(x) = 2x_m \sqrt{-\alpha\beta} e^{\pi i \beta x^2} \sum_{n=-\infty}^{\infty} h_n e^{\pi i \beta \left( \frac{n}{2\alpha x_M} \right)^2} e^{\pi i \beta x \frac{n}{|\alpha| x_M}}. \qquad (2.6.13)$$

It is seen that the series in Eq. (2.6.13) gives a periodic function. Now, it is possible to choose $D$ and hence $\beta$ in such a way that

$$\frac{\beta}{(2\alpha x_M)^2} = \frac{1}{\lambda D(2\alpha x_M)^2} = 2q \qquad (q : \text{integer}). \qquad (2.6.14)$$

On inserting Eqs. (2.6.10) and (2.6.14) into Eq. (2.6.13) we obtain

$$\hat{s}_\beta(x) = \frac{1}{4rx_M} \sqrt{-\frac{q}{r}} e^{\pi i \frac{q}{2r^2} \left( \frac{x}{2x_M} \right)^2} \sum_{n=-\infty}^{\infty} h_n e^{-2\pi i n \frac{x}{P}}, \qquad (2.6.15)$$

where

$$P = \frac{r}{q} 2x_M \qquad (r, q : \text{integers}). \qquad (2.6.16)$$

The series in (2.6.15) represents the periodic repetition of a magnified or demagnified version of $h(-x)$ depending on whether $r$ is greater or smaller than $q$. If we consider the intensity distribution, the phase factor in front of the series cancels out. Using Eq. (2.6.7), we conclude that multiple images are formed of the original object intensity $|f(x)|^2$.

The above phenomenon is somehow connected to the self-imaging of periodic objects, or Talbot effect [909, 659, 555, 140, 677, 499, 417]. Let $f(x)$ be periodic. To facilitate comparison with the previous case, we denote the period by $2x_M$. The Fourier series expansion of $f(x)$ is of the form

$$f(x) = \sum_{n=-\infty}^{\infty} f_n e^{\pi i n \frac{x}{x_M}},$$

$$f_n = \frac{1}{2x_M} \int_{-x_M}^{x_M} f(x) e^{\pi i n \frac{x}{x_M}} dx. \qquad (2.6.17)$$

The $\alpha$-Fresnel transform of $f(x)$ turns out to be

$$\hat{f}_\alpha(x) = \sum_{n=-\infty}^{\infty} f_n e^{-\pi i \frac{n^2}{4\alpha x_M^2}} e^{\pi i n \frac{x}{x_M}}. \qquad (2.6.18)$$

If condition (2.6.10) is satisfied, Eq. (2.6.18) becomes identical to the series expansion (2.6.17). Accordingly, at any plane satisfying Eq. (2.6.11), the same field distribution is found (except for the $\exp(ikz)$ term) as at the $z = 0$ plane.

The effects described so far are among the simplest of a rather large class of multiple image production and reproduction phenomena in Fresnel optics. In fact, one can observe a lot of similar phenomena, e.g., the Lau effect, passing from coherently to incoherently radiating objects [42, 421, 347, 884, 885, 887]. For a complete review the reader is referred to [716].

Finally, we note that an extension of the PSWF to Fresnel transform problems can be made [349]. This can be of help when dealing with Fresnel propagation phenomena.

## 2.7  Exponential Sampling

In several optical problems, the Mellin transform is of interest. Different definitions of this transform are used in the optical literature depending on the problem to be treated. We shall now dwell a bit on these definitions and on the related sampling theorems. Further, we shall point out the classes of problems where these notions come in handy.

A first definition of the Mellin transform [111, 194], say $f_M(P)$ or $[Mf(x)](p)$, of a function $f(x)$ defined in $[0, \infty)$ is the following:

$$f_M(p) = [Mf(x)](p) = \int_0^\infty f(x) x^{-2\pi i p} \frac{dx}{x}. \qquad (2.7.1)$$

provided that the integral converges. The change of variable $x = \exp(t)$ leads to the alternative expression

$$f_M(p) = \int_{-\infty}^{\infty} f(e^t)e^{-2\pi ipt}dt. \qquad (2.7.2)$$

This gives a clear operational meaning to the Mellin transform. We first pass from $f(x)$ to $f[\exp(t)]$. This operation, which expands the interval $[0.1]$ of the $x$-axis to the whole negative semiaxis $t < 0$, is equivalent to drawing the graph of $f(x)$ on a semilogarithmic plot, as exemplified in Fig. 2.14 and 2.15 where the triangle function $(1 - |x - 100|/50)\,\text{rect}[(x - 100)/100]$ is plotted with a linear and a logarithmic horizontal axis, respectively. Note that if we change $f(x)$ into $f(ax)$, i.e., if we expand $(a < 1)$ or compress

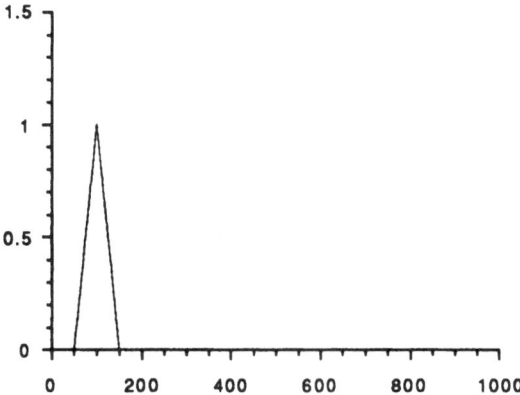

FIGURE 2.14: Plot of the triangle function with a linear horizontal axis.

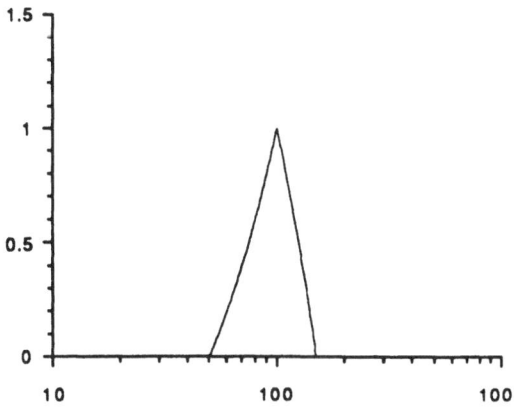

FIGURE 2.15: Plot of the triangle function with a logarithmic horizontal axis.

$(a > 1)$ $f(x)$ horizontally, the semilogarithmic plot is simply shifted to the right or to the left by a quantity proportional to $|\ln(a)|$, respectively. This is shown in Figs. 2.16 and 2.17 which refer to the same triangular function as Figs. 2.14 and 2.15, with a factor $a = 0.2$.

The second operation implied by Eq. (2.7.2) is simply a FT. In view of the properties of the first operation, the overall effect of a horizontal magnification of $f(x)$ on its Mellin transform is multiplication by a phase term. More precisely, we have

$$[Mf(ax)](p) = a^{2\pi ip}[Mf(x)](p). \tag{2.7.3}$$

This property is of interest in optical processing [194], for example for pattern recognition, when a certain shape is to be recognized regardless of the scale factor.

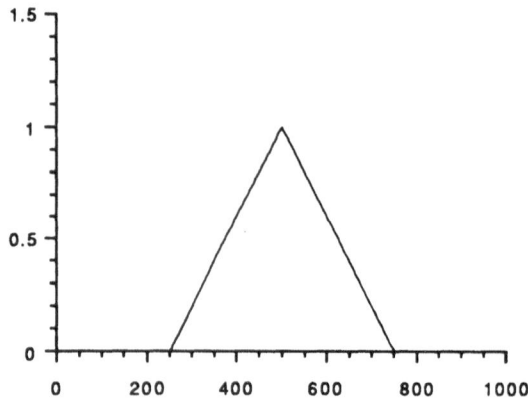

FIGURE 2.16: Plot of the triangle function in Fig. 2.14, scaled with a factor $a = 0.2$.

Using Eq. (2.7.2) together with the Fourier inversion formula, we obtain

$$f(x) = \int_{-\infty}^{\infty} f_M(p) x^{2\pi ip} dp, \tag{2.7.4}$$

as the Mellin inversion formula.

Suppose that a function $f(x)$ is Mellin band-limited, in the sense that $f_M(p)$ vanishes for $|p| \geq p_M$. Then, we see from Eq. (2.7.2) that $f[\exp(t)]$ is band-limited in the ordinary, Fourier sense. Accordingly, we can write an expansion of the form (2.4.4)

$$f(e^t) = \sum_{n=-\infty}^{\infty} f(e^{\frac{n}{2p_M}}) \operatorname{sinc}(2p_M t - n). \tag{2.7.5}$$

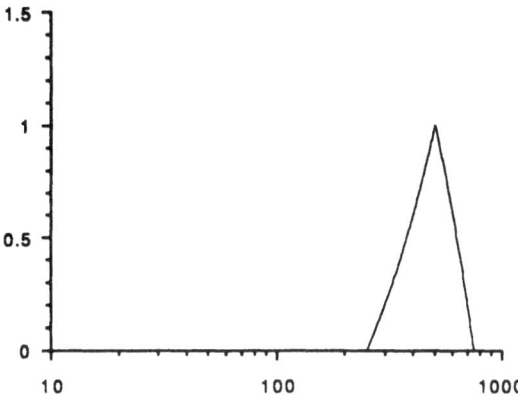

FIGURE 2.17: Plot of the triangle function in Fig. 2.15, scaled with a factor $a = 0.2$.

where, for the sake of simplicity, the origin shift has been omitted. Written in terms of the original variable, Eq. (2.7.5) becomes

$$f(x) = \sum_{n=-\infty}^{\infty} f(e^{\frac{n}{2p_M}}) \operatorname{sinc}(2p_M \ln(x) - n). \tag{2.7.6}$$

This constitutes the sampling expansion for Mellin band-limited functions. Note that the samples are not equally spaced. Instead, the ratio between the positions of adjacent samples is a constant and a resolution ratio $\exp[1/(2p_M)]$ can be introduced. Roughly speaking, we can say that in a Mellin band-limited function independent pieces of information tend to accumulate near the origin. To underline this peculiar distribution of samples one speaks of exponential sampling [697].

We shall now introduce a different definition of Mellin transform which is useful for many problems [81]. Let us denote by $\check{f}(p)$ this alternative form of the Mellin transform of $f(x)$:

$$\check{f}(p) = \int_0^\infty f(x) x^{-2\pi i p} \frac{dx}{\sqrt{x}}. \tag{2.7.7}$$

Using again the change of variable $x = \exp(t)$, Eq. (2.7.7) becomes

$$\check{f}(p) = \int_{-\infty}^\infty f(e^t) e^{t/2} e^{-2\pi i p t} \, dt. \tag{2.7.8}$$

In addition to the passage from $f(x)$ to $f[\exp(t)]$, the present definition of Mellin transform implies multiplication by $\exp(t/2)$. After that, a FT is performed. It is easily seen that the new definition preserves the property expressed by Eq. (2.7.3) except that now the right-hand side has to be divided by $\sqrt{a}$. The inversion formula now becomes

$$f(x) = \frac{1}{\sqrt{x}} \int_{-\infty}^{\infty} \check{f}(p) x^{2\pi i p} dp. \qquad (2.7.9)$$

For functions that are Mellin band-limited, i.e., for which Eq. (2.7.7) vanishes when $|p| > p_M$, a sampling theorem can be derived with a reasoning analogous to the one leading to Eq. (2.7.6). The sampling expansion now reads

$$f(x) = \frac{1}{\sqrt{x}} \sum_{n=-\infty}^{\infty} f(e^{\frac{n}{2p_M}}) e^{\frac{n}{4p_M}} \operatorname{sinc}[2p_M \ln(x) - n]. \qquad (2.7.10)$$

The Mellin transform is the basic tool for solving integral equations of the form

$$g(x) = \int_0^{\infty} K(xy) f(y) dy, \qquad (2.7.11)$$

where $f(x)$ is to be recovered starting from the knowledge of $K(xy)$ and $g(x)$. If we perform a Mellin transform of $g(x)$, we have from Eqs. (2.7.7) and (2.7.11)

$$\begin{aligned} \check{g}(p) &= \int_0^{\infty} g(x) x^{-2\pi i p} \frac{dx}{\sqrt{x}} \\ &= \int_0^{\infty} f(y) \, dy \int_0^{\infty} K(xy) x^{-2\pi i p} \frac{dx}{\sqrt{x}}, \end{aligned} \qquad (2.7.12)$$

where the integration order has been interchanged. With the change of variable $xy = t$, Eq. (2.7.12) becomes

$$\begin{aligned} \check{g}(p) &= \int_0^{\infty} f(y) y^{2\pi i p} \frac{dy}{\sqrt{y}} \int_0^{\infty} K(t) t^{-2\pi i p} \frac{dt}{\sqrt{t}} \\ &= \check{f}(-p) \check{K}(p). \end{aligned} \qquad (2.7.13)$$

We see from Eq. (2.7.13) that the action of the integral operator (2.7.11) on $f(x)$ is equivalent to a filtering of its Mellin transform. A comparison with Eq. (2.4.2) shows that for product kernels $K(xy)$ the Mellin transform plays the same type of role as the FT for convolution kernels.

As a simple example of the occurrence of (2.7.11) in optics, let us consider a transparent medium in which opaque spherical particles with various diameters are suspended. This system is illuminated by collimated coherent light and the far-field intensity distribution is observed. The contribution given by a single particle with radius $a$ can be approximately evaluated as the Fraunhofer diffraction pattern of a disk with the same radius. We assume that the particles are wandering through the medium and that the far-field intensity is averaged over a large time interval. Therefore, interference effects among the fields diffracted from different particles cancel out and the various diffraction patterns add to one another on an intensity basis. Because of Babinet's principle, at any point of the far-field except the origin the single contribution has the same shape as the intensity in the well-known Airy pattern [106]. i.e.,

$$\pi^2 a^4 \left[ \frac{2J_1(ka\vartheta)}{ka\vartheta} \right]^2, \qquad (2.7.14)$$

where $\vartheta$ is the (supposedly small) scattering angle and $J_1$ is the Bessel function of the first kind and order one. If the particle radii are distributed according to a certain (unknown) function $r(a)$, the total intensity is

$$I(\vartheta) = I_0 \int_0^\infty \left[ \frac{J_1^2(ka\vartheta)}{(ka\vartheta)^2} \right] a^4 r(a) \, da, \qquad (2.7.15)$$

where $I_0$ is a constant. Eq. (2.7.15) is of the form (2.7.11) with $f(y)$ and $g(x)$ replaced by $a^4 r(a)$ and $I(\vartheta)$, respectively. The kernel $K$ is given by the terms within square brackets multiplied by $I_0$. Equation (2.7.15) is to be solved in order to find the particle size distribution $r(a)$ [945]. Other examples include polydispersity analysis by photon correlation spectroscopy [697]. aerosol size distribution analysis [935] and laser velocimetry [237].

Depending on the problem at hand, the solution of Eq. (2.7.11) can be sought by various methods. The deceptively simplest method is, of course, inverse filtering leading to

$$\check{f}(p) = \frac{\check{g}(p)}{\check{K}(-p)}, \qquad (2.7.16)$$

for those values of $p$ where $\check{K}(-p)$ is different from zero. However. in most cases, $\check{K}(p)$ is likely to be extremely small for certain ranges of $p$ values thus producing error amplification when Eq. (2.7.16) is applied to noisy data. Furthermore. in many cases, $\check{K}(p)$ vanishes outside a finite interval $[-p_M, p_M]$. The solution of Eq. (2.7.11) then requires a discussion somewhat similar to the one connected to imaging through convolution kernels (see Sec. 2.4), including the use of sampling theorems and singular functions. For more information. the reader is referred to [81, 82].

Here and in the previous sections, we have seen examples of linear transforms applied to optical problems. Other transforms are of interest in optics, for example. Abel, Hankel. Hartley, Hilbert, Laplace and Radon transforms [111, 112, 387, 422, 704]. A general sampling theorem [582] applies to any linear transform whenever the function to be transformed is Fourier band-limited. say in $[-p_M, p_M]$. Let us express the linear operator in the

form (2.5.4) and insert under the integral symbol the sampling expansion
of $f(y)$. This gives

$$(Af)(x) = \sum_{n=-\infty}^{\infty} f\left(\frac{n}{2p_M}\right) H_\ell\left(x, \frac{n}{2p_M}\right), \qquad (2.7.17)$$

where $H_\ell$ is a low-pass filtered version of the kernel $H$ defined as follows:

$$H_\ell(x, t) = \int_{-\infty}^{\infty} H(x, y) \operatorname{sinc}[2p_M(t - y)] \, dy. \qquad (2.7.18)$$

It will be noted that Eq. (2.7.17) looks similar to a numerical integration
formula, e.g., trapezoidal rule, for evaluating $(Af)(x)$. Actually, it is an
exact formula. Examples of the use of Eq. (2.7.17) for numerical evaluations
are given in [582].

## 2.8   Partially Coherent Fields

The flow of information in a partially coherent field is to be described by
means of the correlation functions. The neatest way to do this is to work in
the space-frequency domain where, for each temporal frequency, the spatial
correlation properties of the field are accounted for by the cross-spectral
density [569]. It has been shown [962] that for virtually any source the cross-
spectral density $W(\mathbf{R}_1, \mathbf{R}_2, \nu)$ at two space points with radius vectors $\mathbf{R}_1$
and $\mathbf{R}_2$ can be expressed as the average

$$W(\mathbf{R}_1, \mathbf{R}_2, \nu) = \langle V^*(\mathbf{R}_1, \nu) V(\mathbf{R}_2, \nu) \rangle \qquad (2.8.1)$$

on an ensemble of realizations of monochromatic fields at the temporal
frequency $\nu$. Most features of the partially coherent field can then be traced
back to features of the underlying monochromatic fields. Therefore, it is not
surprising that sampling theory plays a role in coherence theory.

The best known partially coherent fields are the ones produced by spatially incoherent planar sources. Let us denote by $I_0(\rho, \nu)$ the optical intensity at frequency $\nu$ in a typical source point with radius vector $\rho$ (in
the source plane). The cross-spectral density $W_D(\mathbf{r}_1, \mathbf{r}_2, \nu)$ at two points
possessing radius vectors $\mathbf{r}_1$ and $\mathbf{r}_2$ in a plane parallel to the source plane
at a distance $D$ from it, is given by the van Cittert-Zernike theorem [106]

$$W_D(\mathbf{r}_1, \mathbf{r}_2, \nu) = \frac{e^{i\frac{k}{2D}(r_2^2 - r_1^2)}}{D^2} \int I_0(\rho, \nu) e^{-2\pi i \rho \frac{\mathbf{r}_2 - \mathbf{r}_1}{\lambda D}} \, d^2\rho. \qquad (2.8.2)$$

It is seen that, except for a geometrical factor (which is independent of
the source intensity distribution), $W_D$ equals the FT of $I_0(\rho, \nu)$. Suppose
now that $I_0(\rho, \nu)$ has a finite support. Then, $W_D$ is proportional, through

a known factor, to a band-limited function and it can be determined by sampling [776]. In addition, the integral in Eq. (2.8.2) depends on $r_1$ and $r_2$ only through their difference. This has given rise to synthesis techniques for sampling partially coherent fields [352, 354, 340, 783, 273, 99]. Let us give an elementary one-dimensional example assuming a source width $2x_M$. We can sample $W_D$ at $x_2 - x_1 = n\Delta x$ $(n = 0, 1, ...)$ where

$$\Delta x = \frac{\lambda D}{2x_M}. \tag{2.8.3}$$

Let us consider a linear array of four equally spaced points at a distance $\Delta x$ from each other, as in Fig. 2.18. Assuming that $W_D$ can be measured at any pair of points of the array, we obtain four samples corresponding to $n = 0, 1, 2, 3$. However, the array is a redundant one because the distance $x2 - x1 = \Delta x$ can be obtained with $4 - n$ different pairs $(n = 0, 1, 2, 3)$. The array of Fig 2.18 eliminates the redundancy (except, of course, for $n = 0$) and furnishes samples of $W_D$ up to $n = 6$ using the same number of elements. It is easy to imagine that in the two-dimensional case similar synthesis procedures can be applied leading to arrays where the number of distinct pairs afforded by $N$ point-like elements grows like $N^2$. Such techniques have found wide application in radioastronomy [334, 229].

$$\bullet \ \ \bullet \ \ \bullet \ \ \bullet \qquad\qquad \bullet \ \ \bullet \qquad\quad \bullet \qquad \bullet$$

FIGURE 2.18: Linear array of four equally spaced points at a distance $\Delta x$ from each other

We can ask whether there exist more general phenomena causing $W_D$ to be band-limited. One fundamental reason is the following. The angular spectrum of a monochromatic field generally comprises both homogeneous and inhomogeneous or evanescent waves [106, 339]. The latter are strongly damped on free propagation. At distances greater than a few wavelengths from any material surface the spectrum is well approximated by the homogeneous part only. Across any plane where this condition is met each member of the ensemble of monochromatic fields involved in the average of Eq. (2.8.1) has a spatial structure of the form

$$V(\mathbf{r}, \nu) = \int_c \tilde{V}(\mathbf{p}, \nu) e^{2\pi i \mathbf{p} \cdot \mathbf{r}} d^2 p, \tag{2.8.4}$$

where $\mathbf{r}$ is a radius vector in the chosen plane and $C$ is the circle of radius $1/\lambda$ in the spatial frequency plane. According to Eq. (2.8.1), the corresponding cross-spectral density is

$$W(\mathbf{r}_1, \mathbf{r}_2, \nu) = \int_c \int_c \langle \tilde{V}^*(\mathbf{p}_1, \nu) \tilde{V}^*(\mathbf{p}_2, \nu) \rangle e^{2\pi i (\mathbf{p}_2 \cdot \mathbf{r}_2 - \mathbf{p}_1 \cdot \mathbf{r}_1)} d^2 p_1 d^2 p_2. \tag{2.8.5}$$

Whatever the form of the average under the integral sign, $W$ is clearly band-limited with its spectrum included in the four-dimensional domain $C \times C$ [963]. Hence, $W$ can be determined through a suitable sampling.

An extensive treatment of partially coherent fields along the previous lines was given in [319]. Actually, Gamo's work referred to the mutual intensity but most of his results could be easily rephrased in terms of cross-spectral density.

The preceding remarks deal with spatial coherence. At any point in space, the temporal coherence depends on the (temporal) power spectrum at the same point [106] through the Wiener-Kintchine theorem. According to that theorem, the temporal autocorrelation function and the power spectrum form a Fourier pair. This is the basis of Fourier spectroscopy where sampling theory has an essential role [923, 605].

## 2.9   Optical Processing

The processing of information by optical means is a very wide subject ranging from old and well-established techniques (e.g., phase-contrast microscopy [106, 339]) to rather new and rapidly developing ones (e.g., optical implementation of neural networks). Indeed, several topics dealt with in previous sections can also be encompassed in optical processing.

Sampling theory is a tool of continuous use in this field. As an example, let us consider the production of multiple images of an object illuminated with coherent light [839, 897]. In principle, this is easily obtained with the system of Fig. 2.1 putting a suitable sampling mask in the pupil plane. Indeed, if the object has a finite support, the spectrum displayed in the pupil plane is a band-limited function and its sampling produces a multitude of replicas of the image in the output plane. Generally, one would like to produce replicas with one and the same intensity. If the sampling mask is a set of holes (or slits for one-dimensional cases) in an opaque background, replicas of different weights are produced because of the non-zero size of the holes. In fact, the mask behaves as an amplitude grating [106] whose diffraction orders are known to be differently weighted. A possible solution is to use a phase sampling mask like a Dammann grating [241, 916]. This is a grating whose transmission function jumps between $+1$ and $-1$ a number of times in a period. The positions of the transition points can be calculated so as to produce a finite number of diffraction orders of equal weight. As there is no absorption, the efficiency, i.e., the fraction of power directed in the wanted orders, can be very high. Note that a grating of this type can be used as a multiple beam splitter. In addition, phase gratings can be used for spatial filtering purposes [389]. Another application of sampling occurs in image subtraction with both coherent and partially coherent light [332].

Besides applications of this type, there are countless occurrences of sampling in optical processing. It is worthwhile to review some fundamental reasons for this. To begin with, suppose that we want to measure a certain optical intensity distribution across a plane region. Whatever type of detector we choose. the measurement implies some form of sampling [287, 412, 143, 411, 199]. As a typical example let us consider a one-dimensional array of detectors. Each detector has a certain width $\Delta x$ over which it integrates the optical intensity. Let the distance between centers of two adjacent detectors be $x_1$ and let $N_d$ be the number of detectors. The output of the array can be thought of as the result of two processes. First. the optical intensity is convolved with $\mathrm{rect}(x/\Delta x)$. Second. it undergoes a finite sampling. Assuming linearity of the detection process. the relationship between the output signal $s(x)$ and the optical intensity $I(x)$ can be written

$$s(x) \;=\; \mathrm{rect}\left(\frac{x}{N_d \Delta x}\right) \sum_{n=-\infty}^{\infty} \delta(x - nx_1)$$

$$\times \left\{ I(\xi) * \mathrm{rect}\left(\frac{\xi}{\Delta x}\right) \right\}(x), \qquad (2.9.1)$$

where proportionality factors have been omitted and the convolution operation is denoted by $*$. Let us briefly discuss the effect of this type of sampling. The convolution of $I(x)$ and $\mathrm{rect}(x/\Delta x)$ gives a low-pass filtering of the FT of $I(x)$. say $\tilde{I}(p)$. The equivalent transfer function is of the form $\mathrm{sinc}(\Delta x p)$ and gives severe attenuation of spatial frequencies above $1/\Delta x$. As a result, the function that we actually sample is not $I(x)$ but a blurred reproduction of it. Then. there is the sampling effect producing replicas of the filtered version of $\tilde{I}(p)$. spaced at a distance $1/x_1$ from one another. This will give rise to aliasing errors. Finally. there is the effect arising from the finite extent of the sampled region. Because of this effect, the set of replicas is convolved with a $\mathrm{sinc}(N_d \Delta x p)$ function. Taking into account these effects and using some prior knowledge about the intensity distribution to be measured. one chooses the above parameters so as to keep errors below an acceptable level.

A peculiar effect. namely. moiré fringes. may be produced by aliasing errors when the intensity distribution to be sampled, for example. a halftone screened image. has an underlying periodic structure. In this case, a double sampling at two different rates can help select frequency components less suffering from aliasing errors thus eliminating moiré fringes [684, 971]

Similar remarks apply when it comes to making a permanent record of the intensity distribution by means of a photographic emulsion. The detectors are now silver halide grains. Although they are somewhat different from each other and irregularly distributed within the emulsion, the same description approximately applies with reference to the mean values of grain width and grain distance.

Detection of coherent field distribution follows the same pattern because the lack of phase-sensitive detectors forces us to use as the measurable quantity the intensity obtained in the interference between the unknown field and a reference field.

Another operation in which sampling is almost invariably present is the spatial modulation of a light wave. Beyond reasons like the periodic repetition of cells with a certain area (as in many spatial light modulators) there is another fundamental motivation. If we want to impart a certain phase distribution to a wavefield, we can use a suitable phase object. On the other hand, this is generally impractical because of the difficulties of realizing arbitrary phase objects. It is much easier to code the phase distribution through an amplitude modulation. This is, of course, the very principle of holography. An early example of this in optical processing is the holographic matched filter of [925] where both the amplitude and the phase distributions of the signal to be processed are modified by the passage through a hologram. In order to isolate the term of interest in the diffracted field from other unwanted terms, off-axis holography is used. This implies an inclined reference beam and this is equivalent to translating the spectrum of the signal to be recorded on a carrier spatial frequency.

Although in some cases holographic filters can be produced in an analogical manner, most frequently they are synthesized with the aid of a computer [552, 554]. The basic layout of a computer-generated hologram is an array of cells. One (complex) value of the function (i.e., an amplitude and a phase value) is coded in each cell by a suitable control of the amplitude transmission function. Typically, a transparent dot is drawn on an opaque background in each cell. The size and the position of the dot account for the amplitude and the phase, respectively, of the sample to be synthesized. What is to be underlined is that this arrangement in a lattice of cells requires the use of sampling theory from the outset. Furthermore, extensions of the sampling theory can be usefully employed. For example, as the samples are approximated by dots of non-zero area, errors occur. They can be eliminated by a suitable predistortion of the function to be synthesized [44].

Computer-generated holography is today a rich topic with a lot of applications. For review papers, the reader is referred to [531, 141].

In ordinary holography, the use of sampling for high-density storage and interferometry was discussed in [197, 198, 242, 361. 914].

A frequently occurring task is the sampling of a two-dimensional function whose spectrum is limited by a circle. let us say a circle band-limited function. Of course, a sampling lattice with square cells can be used. However. this is a redundant sampling because the alias-free region in the frequency plane is a whole square circumscribed to the required circle. This implies that the sample density is $4/\pi$ times higher than it would be required by the circle area. Better results are obtained if the spectrum replicas produced by the sampling are packed according to a hexagonal geometry. This occurs when a sampling lattice with hexagonal cells is used [721, 648, 669]. The redundancy factor decreases from $4/\pi$ to $2\sqrt{3}/\pi$ but the problem remains of further reducing the redundancy.

A sampling theorem referring explicitly to circle band-limited functions exists [319, 94]. It implies a continuous sampling along suitable circles. Although this theorem is conceptually important for problems like the evaluation of the number of DOF, the practical implementation of circle sampling is not easy. A notable exception occurs when the circle band-limited function is also circularly symmetric. In this case, the problem is actually one-dimensional and we can use the sampling theorem based on the Fourier-Bessel series [704] or on the Dini series [888].

A different way to reduce redundancy has been found in [215] exploiting the mutual dependence of the samples in the redundant case [588]. We shall discuss the basic idea in a simple case.

Let us consider a two-dimensional sampling array with a square cell lattice. The side of the cell equals one. The lattice sites are depicted in Fig. 2.19 as the union of two sets represented by crosses and dots, respectively. The

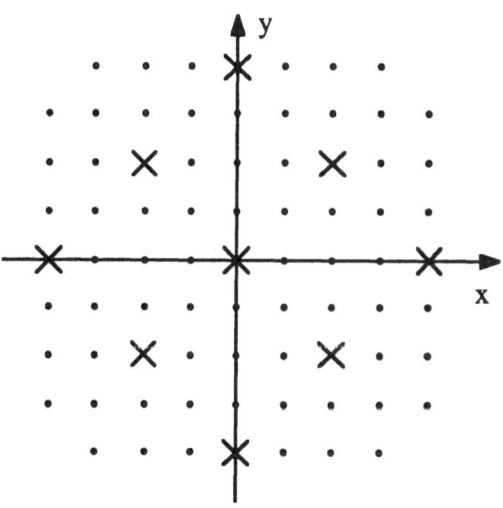

FIGURE 2.19: A two-dimensional sampling array with a square cell lattice.

first, called the cross set, is itself a square lattice with side $2\sqrt{2}$ rotated through $\pi/4$ with respect to the $x-$ and $y-$ axes. The second set will be called the dot set.

Suppose that the whole array is used to sample a band-limited function whose spectrum is confined within a circle of radius $1/2$ centered at the origin. This gives rise to a periodic repetition of the spectrum as schema-

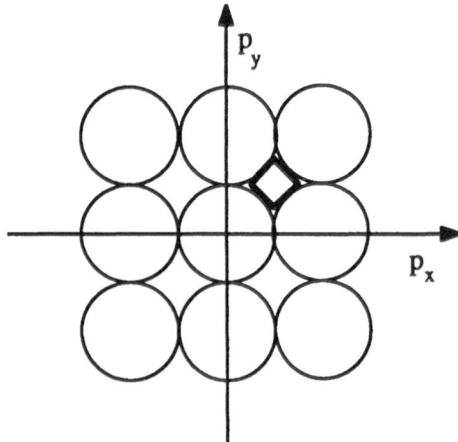

FIGURE 2.20: Spectrum of a band-limited function sampled with the array of Fig. 2.19.

tized in Fig. 2.20 where the replicas of the spectrum are projected onto the plane of the spatial frequencies $p_x, p_y$. Let us consider now the spectrum of the samples corresponding to the cross set. This is a (two-dimensional) periodic function whose period is a square with side $1/(2\sqrt{2})$ rotated through $\pi/4$ with respect to the $p_x, p_y-$ axes, (see Fig. 2.21). We focus our attention on the square evidentiated by heavy lines in Fig. 2.21. When drawn in Fig. 2.20, such a square falls in a free region where the spectrum of the whole set of samples vanishes. We conclude that, within the chosen square, the spectrum of the cross samples is opposite to that of the dot samples so that the first is immediately deduced from the second. Accordingly, the dot samples are sufficient to find also the cross samples. The latter are therefore redundant and can be eliminated, thus increasing the sampling efficiency. Building on this idea, more sophisticated and highly efficient sampling schemes can be devised [215]. The contribution of Cheung in this volume discusses other methods by which sampling density can be reduced. Another type of sampling that can be usefully employed for certain band-limited functions is polar sampling [858, 863]. The reader is referred to the contribution by Stark in this volume for such a subject. Finally, a sampling theorem that applies to three-dimensional optical microscopy has been established in [681].

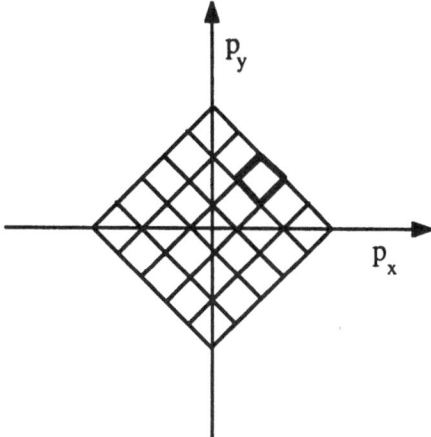

FIGURE 2.21: Spectrum of the samples corresponding to the cross set.

## 2.10   Conclusion

We have been wandering through a number of applications of the sampling theory to optics. Admittedly, the tour is unequally weighted and far from complete. The author hopes that a couple of points are underlined by the present chapter. First: sampling, in one form or another, appears recurrently in optical problems. Second: many new and extended forms of sampling continue to be discovered and deserve further research.

# 3

# A Multidimensional Extension of Papoulis' Generalized Sampling Expansion with the Application in Minimum Density Sampling

Kwan F. Cheung

This chapter is divided into two parts. In Part I, Papoulis' one-dimensional generalized sampling expansion (GSE) is extended to multidimensional (M–D) band-limited functions. In Part II, using sample decimation, we will utilize the M–D GSE formulation to reduce the sampling density of M–D band-limited functions. The ultimate reduction leads to the minimum sampling density, which is equal to the area of the support of the function's Fourier spectrum. This rate is analogous to the definition of the Nyquist rate for one-dimensional (1–D) band-limited functions.

## Part I
## A Multidimensional Extension of Papoulis' Generalized Sampling Expansion

## 3.1  Introduction

The generalized sampling expansion (GSE), initially formulated by Papoulis [707, 706], unifies a broad class of extensions generated from the Shannon sampling theorem [818] under a generalized setting. Consider a $2\sigma$–band-limited function $f(t)$ (i.e., $f(t)$ has finite energy and $F(\nu) = 0$ for $|\nu| \geq \sigma$ where $\sigma$ is measured in Hertz):

$$F(\nu) = \int_{-\infty}^{\infty} f(t)e^{-j2\pi\nu t}dt.$$

$$f(t) = \int_{-\sigma}^{\sigma} F(\nu)e^{j2\pi\nu t}d\nu.$$

According to the Shannon sampling theorem. the function is uniquely determined from the sample set $\{f(nT_g)\}$. where $T_g = 1/2\sigma$ is the Nyquist interval.

In a more general setting. the $m^{th}$ order ($m = 1,2,\ldots$ is the order of expansion) GSE allows $f(t)$ to be uniquely determined by $m$ sample sets: $\{\{g_k(nT)\}|k = 0 \text{ to } m-1\}$. The signal $g_k(t)$ is the response of a linear system, $h_k(t)$, with the input $f(t)$:

$$g_k(t) = h_k(t) * f(t). \quad k = 0 \text{ to } m-1.$$

The sampling interval $T$ is equal to $mT_g$, $m$ times the Nyquist interval. Thus, every $g_k(t)$ is sampled at $1/m^{th}$ the Nyquist rate. There are $m$ sample sets. The overall sampling rate is still equal to the Nyquist rate. If $m = 1$ and $H_1(\nu) = 1$, the conventional Shannon sampling theorem results.

Two extensions that fall under the GSE are interlaced (or bunched) sampling and signal-derivative sampling [111, 544]. Both can be formulated as second-order expansions of the 1–D GSE. Interlaced sampling has $H_1(\nu) = 1$ and $H_2(\nu) = e^{j2\pi\alpha}$ ($0 < \alpha < T$). Signal-derivative sampling has the same $H_1(\nu)$ with $H_2(\nu) = j2\pi\nu$.

Papoulis showed that there exist interpolation functions for each of the two-dimensional cases above. Let $y_k(t)$ be the interpolation function for the $k^{th}$ sample set so that $f(t)$ can be reconstructed. The interpolation formula for reconstructing $f(t)$ is

$$f(t) = \sum_{k=0}^{m-1} \sum_{n=-\infty}^{\infty} g_k(nT)y_k(t - nT).$$

The block diagram representing the $m^{th}$-order GSE is depicted in Fig. 3.1. A special contribution of this generalization is that the merit of different sampling extensions can be compared in a common analytical setting. By utilizing the GSE. Cheung and Marks [210, 212] demonstrated that there is a class of sampling theorems which are ill–posed [590]. In particular. given the samples are contaminated with noise. the interpolation noise variance is unbounded for these ill-posed sampling theorems.

We will first briefly review the formulation of 1–D GSE.

## 3.2   GSE Formulation

The $m^{th}$-order GSE follows from the partitioning of the spectrum $F(\nu)$. In particular, $F(\nu)$ is partitioned into $m$ equal portions. Each partition has an extension of $c$:

$$c = \frac{1}{T}. \tag{3.2.1}$$

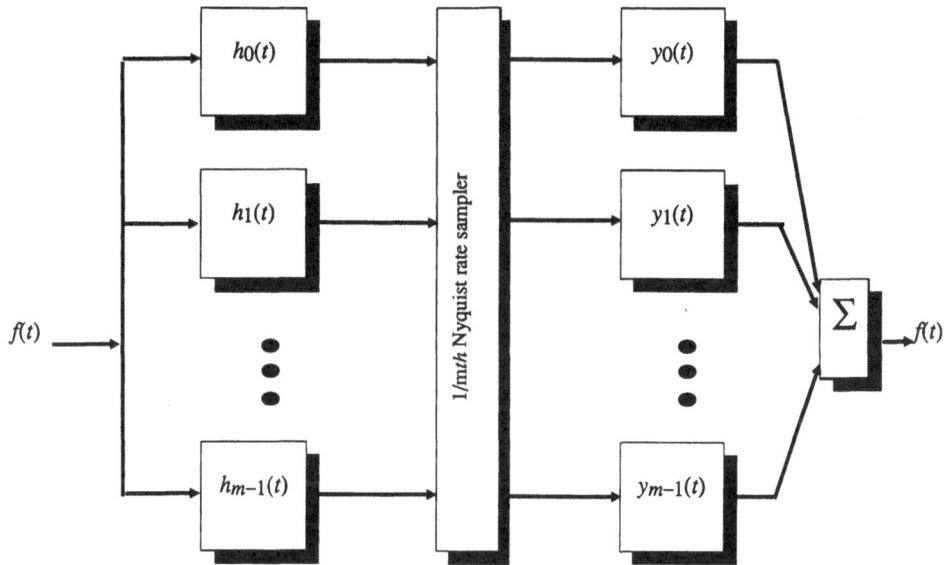

FIGURE 3.1: The architecture of an $m^{th}$-order generalized sampling expansion.

This partitioning of $F(\nu)$ is shown in Fig. 3.2. The support of the first partition, $F_0(\nu)$, lies in the region $(-\sigma, -\sigma + c)$. We denote this region as $\mathcal{A}$. Let $G_k(\nu)$ be the spectrum corresponding to the sample set $\{g_k(nT)\}$. Then for $k = 0$ to $m - 1$,

$$G_k(\nu) = c \sum_{i=0}^{m-1} H_k(\nu + ic)\hat{F}_i(\nu); \quad \nu \in \mathcal{A}. \tag{3.2.2}$$

where $\hat{F}_i(\nu)$ is the $i^{th}$ partition of $F(\nu)$ shifted to $\mathcal{A}$:

$$\hat{F}_i(\nu) \;=\; F_i(\nu + ic); \quad \nu \in \mathcal{A}.$$

Equation (3.2.2) can be put in matrix form as

$$\vec{G} = c\underline{H}\vec{F}, \tag{3.2.3}$$

where $\underline{H}$ is a square matrix of dimension $m$, $\vec{G}$ is the vector with entries $\{G_k(\nu)\}$ and $\vec{F}$ is the vector with entries $\{\hat{F}_i(\nu)\}$. Existence of a solution requires $\underline{H}$ to be non-singular in every point of $\mathcal{A}$. If this condition is satisfied. then

$$\vec{F} = \frac{1}{c}\underline{H}^{-1}\vec{G}. \tag{3.2.4}$$

The $m$ partitions of $F(\nu)$ are restored in $\mathcal{A}$ as a linear combination of the spectra of the $m$ sample sets. The $m$ restored partitions are then shifted to

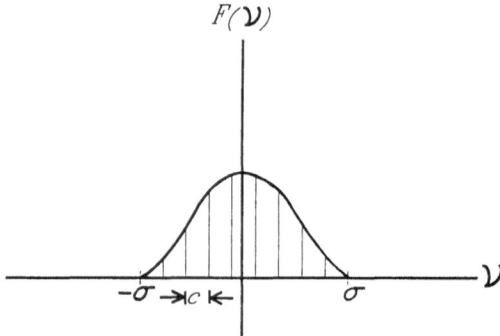

FIGURE 3.2: A 1–D band-limited function is partitioned into $m$ partitions in the 1–D GSE. The support of each partition has an extension of $c$.

their respective positions and the spectrum $F(\nu)$ is reconstructed. Shifting of these partitions in the $\nu$ domain corresponds to modulation in the $t$ domain. Specifically,

$$
\begin{aligned}
f(t) &= \int_{-\sigma}^{\sigma} F(\nu)e^{j2\pi\nu t}d\nu \\
&= \sum_{i=0}^{m-1} \int_{-\sigma+ic}^{-\sigma+(i+1)c} F_i(\nu)e^{j2\pi\nu t}d\nu \\
&= \int_{\mathcal{A}} \sum_{i=0}^{m-1} \left[ \hat{F}_i(\nu)e^{j2\pi ict} \right] e^{j2\pi\nu t}d\nu.
\end{aligned}
\tag{3.2.5}
$$

The square brackets in (3.2.5) enclose the modulation process which shifts the partitions to their respective positions. Let $\vec{E}$ be the *carrier* vector:

$$
\vec{E} = \begin{pmatrix} 1 & e^{j2\pi ct} & e^{j2\pi 2ct} & \cdots & e^{j2\pi(m-1)ct} \end{pmatrix}^T
\tag{3.2.6}
$$

(the superscript $T$ denotes transposition). Equation (3.2.5) can then be written in the following form:

$$
f(t) = \int_{\mathcal{A}} \vec{E}^T \vec{F} e^{j2\pi\nu t}d\nu.
\tag{3.2.7}
$$

By substituting (3.2.4) into (3.2.7), we obtain

$$
f(t) = \frac{1}{c} \int_{\mathcal{A}} \vec{E}^T \underline{H}^{-1} \vec{G} \, e^{j2\pi\nu t}d\nu.
\tag{3.2.8}
$$

The product $\vec{E}^T \underline{H}^{-1} \vec{G}$ in (3.2.8) generates the expression of the interpolation formula for restoring $f(t)$. In particular, let

$$\underline{H}^T \vec{Y} = \vec{E}, \tag{3.2.9}$$

where

$$\vec{Y} = \left( \begin{array}{cccc} Y_0(\nu, t) & Y_1(\nu, t) & \cdots & Y_{m-1}(\nu, t) \end{array} \right)^T.$$

Equation (3.2.8) becomes

$$f(t) = \frac{1}{c} \int_{\mathcal{A}} \vec{Y}^T \vec{G} e^{j2\pi\nu t} d\nu \tag{3.2.10}$$

from which the interpolation formula for restoring $f(t)$ results:

$$f(t) = \sum_{k=0}^{m-1} \sum_{n=-\infty}^{\infty} g_k(nT) y_k(t - nT). \tag{3.2.11}$$

where

$$y_i(t) = \frac{1}{c} \int_{\mathcal{A}} Y_i(\nu, t) e^{j2\pi\nu t} d\nu \tag{3.2.12}$$

is the interpolation function for the $i^{th}$ sample set. Existence of the solution requires the matrix $\underline{H}$ to be non-singular at every point in $\mathcal{A}$.

## 3.3  M–D Extension

We now proceed to extend the 1–D GSE to M–D band-limited functions. As will be demonstrated, the key of this extension lies in the manner by which a period of the samples' spectrum is partitioned. Before proceeding, we have a brief presentation of the M–D sampling theorem in order to maintain the notational uniformity. A more in-depth treatment is given by Marks [590].

### 3.3.1  M–D Sampling Theorem

The M–D sampling theorem, introduced by Petersen and Middleton [721], is a direct extension of the Shannon sampling theorem. Let $f(\vec{t})$ be a M–D $\mathcal{B}$–band-limited function, where $\vec{t} = (t_1, t_2, \ldots, t_N)^T$. Assume $F(\vec{\nu}) = 0$ for

$\vec{\nu} \notin \mathcal{B}$ where $\mathcal{B}$ is a region contained within an M–D hypersphere of finite radius B centered at the origin:

$$\mathcal{B} = \{\vec{\nu} \,|\, \vec{\nu}^T \vec{\nu} < B, B > 0\}$$

and $\vec{\nu} = (\nu_1, \nu_2, \dots, \nu_N)^T$. Such finite energy signals can be uniquely represented by their samples. Let the sampling lattice be denoted by the matrix–vector pair: $[\underline{V}, \vec{p}]$. The vector $\vec{p}$ is the offset vector of the sampling lattice from the origin of the $\vec{t}$ plane. The matrix

$$\underline{V} = \left[\vec{v}_1 \,|\, \vec{v}_2 \,|\, \cdots \,|\, \vec{v}_N\right]$$

is referred to as the *sampling matrix*. Each column vector, $\vec{v}_i$, indicates the direction of the $i^{th}$ sampling dimension and its norm is the sampling interval in that dimension. Thus, $\underline{V}$ governs the periodicity of the geometry of the sampling lattice. Let $\hat{f}(\vec{t})$ be the output function of the sampler:

$$\hat{f}(\vec{t}) = \sum_{\vec{n}} f(\underline{V}\vec{n} + \vec{p})\delta(\vec{t} - \underline{V}\vec{n} - \vec{p}),$$

where

$$\sum_{\vec{n}} = \sum_{n_1=-\infty}^{\infty} \sum_{n_2=-\infty}^{\infty} \cdots \sum_{n_N=-\infty}^{\infty}$$

and $\delta(\vec{t}) = \delta(t_1)\delta(t_2)\cdots\delta(t_N)$ is the $N$–D Dirac delta function. Without loss of generality, we let $\vec{p}$ be the zero vector. The spectrum of $\hat{f}(\vec{t})$ replicates periodically:

$$\begin{aligned}
\hat{F}(\vec{\nu}) &= \sum_{\vec{m}} f(\underline{V}\vec{n}) e^{-j2\pi\vec{\nu}^T\underline{V}\vec{n}} \\
&= D \sum_{\vec{m}} F(\vec{\nu} - \underline{U}\vec{m}),
\end{aligned} \tag{3.3.1}$$

where

$$\underline{U} = \left[\vec{u}_1 \,|\, \vec{u}_2 \,|\, \cdots \,|\, \vec{u}_N\right]$$

and $D = 1/|\underline{V}|$ is the sampling density. The matrix $\underline{U}$ is referred to as the *periodicity* matrix. Its $N$ column vectors, referred to as the *periodicity vectors*, govern the replication pattern of $\hat{F}(\vec{\nu})$. Clearly, $\underline{V}$ is required to be non-singular. In other words, the $N$ sampling vectors are required to be linearly independent. An illustration of a sampling geometry of a two-dimensional function and the corresponding replication pattern is illustrated in Fig. 3.3.

A $N$–D cell, denoted as $\mathcal{C}$, is defined as a period of $\hat{F}(\vec{\nu})$. For M–D functions, there are many possible cell shapes. The two cells in Fig. 3.4 are

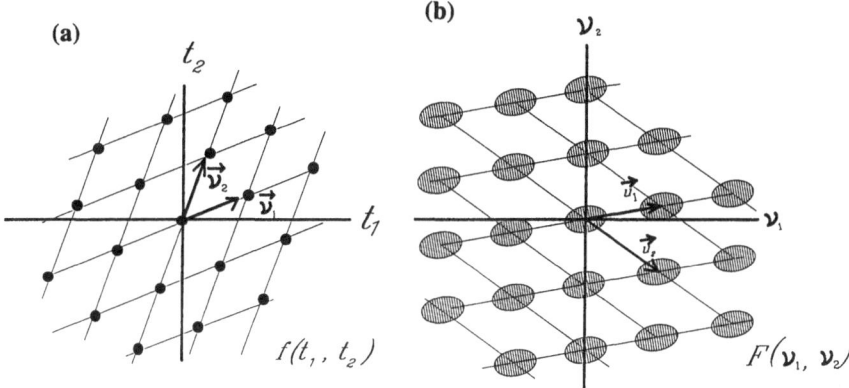

FIGURE 3.3: A 2-D illustration of a sampling geometry and the corresponding replication geometry.

the two possible cell shapes corresponding to a sampling matrix. Methods of constructing these cells are discussed by Dubois [267]. Of all possible cell shapes, the parallelopiped shape cell is the most straightforwardly constructed. Each pair of its parallel "legs" is constructed by a periodicity vector. For our running example. the parallelogram cell in Fig. 3.5 is constructed from the two periodicity vectors.

Given the samples, the signal is restored by the following interpolation formula:

$$f(\vec{t}) = \sum_{\vec{n}} f(\underline{V}\vec{n})\, h(\vec{t} - \underline{V}\vec{n}). \tag{3.3.2}$$

where

$$h(\vec{t}) = \frac{1}{D} \int_D e^{j2\pi\vec{v}^T\vec{t}} d\vec{v}$$

and

$$\int d\vec{v} = \int_{\nu_1} d\nu_1 \int_{\nu_2} d\nu_2 \cdots \int_{\nu_N} d\nu_N.$$

The region $D$, as shown in Fig. 3.5, is any region enclosing only the zeroth-order replication of $\hat{F}(\vec{v})$.

The function $f(\vec{t})$ is said to be sampled at the *Nyquist density* (whose 1–D counterpart is the Nyquist rate) if the corresponding spectral replications. $\hat{F}(\vec{v})$, is the most densely packed of all sampling geometries [588]. The sampling of a 2–D circularly band-limited function at the Nyquist density is illustrated in Fig. 3.6. The corresponding sampling geometry is hexagonal geometry.

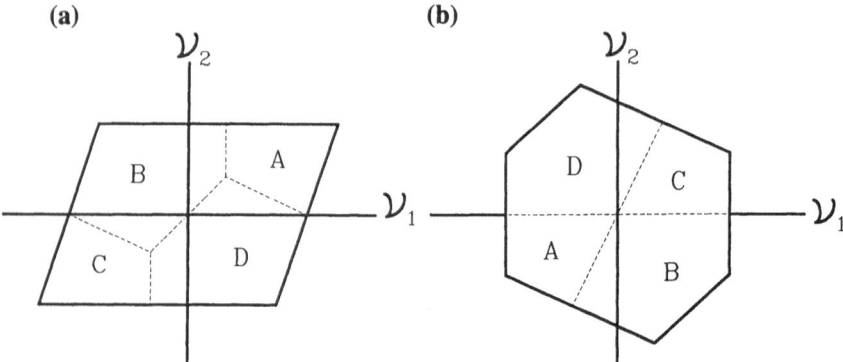

**FIGURE 3.4:** Two different cell shapes corresponding to the same periodicity matrix.

## 3.3.2 M–D GSE FORMULATION

Analogous to its 1-D counterpart. the M–D GSE starts with the partitioning of the baseband cell. denoted by $C_0$. The baseband cell for a 1-D $2\sigma$–band-limited function is the support of the function's spectrum, $(-\sigma, \sigma)$. For M–D band-limited functions, the baseband cell is a cell which encloses the zeroth-order replication of $\hat{F}(\vec{\nu})$. In the 1-D GSE, this baseband cell is partitioned into $m$ identical partitions. For the M–D GSE, each leg of the parallelopiped cell is partitioned into $k$ identical partitions. (This restriction will be relaxed later.) This partitioning of $C_0$ gives $L = k^N$ identical partitions. Each subcell is geometrically congruent to the cell $C$. And each dimension is scaled by a factor of $k$.

A subcell is a cell corresponding to the sampling matrix

$$\underline{V}_g = k\underline{V}.$$

The corresponding periodicity matrix is.

$$\underline{U}_g = \frac{1}{k}\underline{U}.$$

Following the procedure outlined in Section 3.2 , we need to locate a subcell as a reference subcell. In Section 3.2. the first partition in the region $(-\sigma, -\sigma + c)$ is taken as the reference subcell. For the M–D GSE, this reference subcell is located at one of the vertices of $C_0$. With reference to [211],

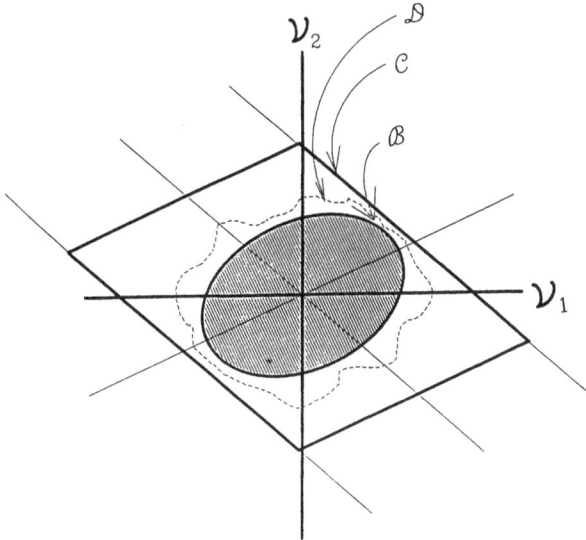

FIGURE 3.5: The parallelogram cell corresponding to the replication pattern in Fig. 3.3, $\mathcal{B}$ is the support of the spectrum, $\mathcal{C}$ the cell and $D$ the region enclosing the support.

this particular vertex is chosen to be at the point

$$\vec{V}_e = -\frac{1}{2}\sum_{i=1}^{N}\vec{u}_i. \qquad (3.3.3)$$

By denoting this reference subcell as $\mathcal{C}_{g0}$, the $i^{th}$ subcell is expressed as

$$\mathcal{C}_{gi} = \left\{ \vec{\nu} \,|\, \vec{\nu} \in (\mathcal{C}_{g0} \oplus \underline{U}_g \vec{q}_i) \right\}: \quad i = 0 \text{ to } L - 1, \qquad (3.3.4)$$

where the $\oplus$ denotes *offset by*. The vector $\vec{q}_i$, which denotes the position of $\mathcal{C}_{gi}$, is an $N$-dimensional vector of integers ranging from 0 to $k - 1$. In particular, the vector $\vec{q}_i$ is a $k$–ary representation of the integer $i$.

$$i = \sum_{p=0}^{N-1} q_{i,p} k^p.$$

The partitioning of the baseband cell also partitions the spectrum $F(\vec{\nu})$. For now, we require that the support of $F(\vec{\nu})$ be totally enclosed within $\mathcal{C}_0$. By positioning the partitioning mesh over $F(\vec{\nu})$. we obtain $L$ partitions

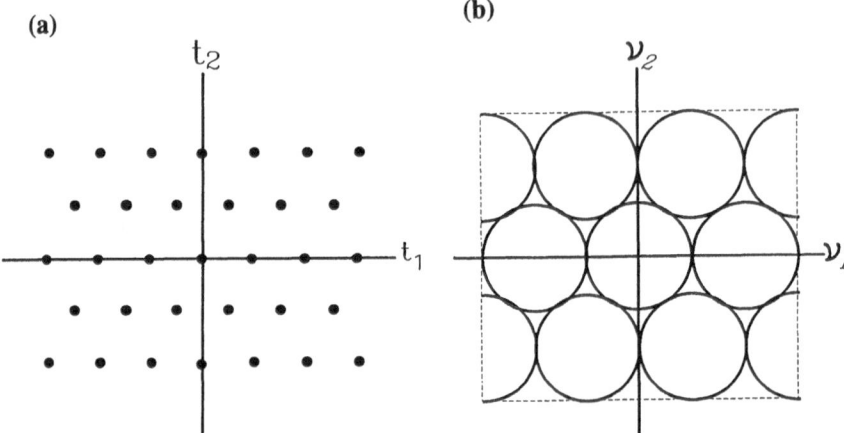

**FIGURE 3.6:** (a) The hexagonal sampling locations corresponds to the Nyquist density sampling of a hexagonal band-limited function. (b) The resulting spectral replications.

of $F(\vec{\nu})$. Let $F_i(\vec{\nu})$ be the partition of $F(\vec{\nu})$ enclosed within the $i^{th}$ subcell, $\mathcal{C}_{gi}$:

$$F_i(\vec{\nu}) = \begin{cases} F(\vec{\nu}). & \vec{\nu} \in \mathcal{C}_{gi} \\ 0 & \text{otherwise.} \end{cases}$$

Given $f(\vec{t})$. the common input, a total of $L$ sample sets are collected at the output of $L$ linear systems: $H_k(\vec{\nu})$, $k = 0$ to $L - 1$. The sample sets are obtained with the sampling geometry $[\underline{V}_g, \vec{0}]$. The sampled function from the output of the $k^{th}$ system is thus

$$\hat{g}_k(\vec{t}) = \sum_{\vec{n}} g_k(\underline{V}_g \vec{n}) \delta(\vec{t} - \underline{V}_g \vec{n}).$$

For $\vec{\nu} \in \mathcal{C}_{g0}$, the spectrum of this sample set is

$$\hat{G}_k(\vec{\nu}) = D_g \sum_{i=0}^{L-1} H_k(\vec{\nu} + \underline{U}_g \vec{q}_i) \hat{F}(\vec{\nu}), \qquad k = 0 \text{ to } L - 1. \tag{3.3.5}$$

where $\hat{F}_i(\vec{\nu})$ is the $i^{th}$ partition of $F(\vec{\nu})$ shifted to $\mathcal{C}_{g0}$,

$$\hat{F}_i(\vec{\nu}) = F_i(\vec{\nu} + \underline{U}_g \vec{q}_i)$$

and $D_g$ is the sampling density of each sample set:

$$D_g = \frac{1}{|\underline{V}_g|} = \frac{1}{L|\underline{V}|}.$$

The formulation of (3.3.5) parallels (3.2.2). Proceeding in a similar manner, we first write (3.3.5) as

$$\vec{G} = D_g \underline{H} \vec{F}, \tag{3.3.6}$$

where

$$\vec{G} = \left(\hat{G}_0(\vec{\nu}), \hat{G}_1(\vec{\nu}), \ldots, \hat{G}_{L-1}(\vec{\nu})\right)^T,$$

$$\vec{F} = \left(\hat{F}_0(\vec{\nu}), \hat{F}_1(\vec{\nu}), \ldots, \hat{F}_{L-1}(\vec{\nu})\right)^T.$$

Define the *carrier* vector $\vec{E}$, whose $(p+1)^{th}$ element is equal to

$$E_p = \exp(j2\pi \vec{q}_p^T \underline{U}_g^T \vec{t}), \qquad p = 0 \text{ to } L-1. \tag{3.3.7}$$

Secondly, parallel with (3.2.9), $\vec{Y}$ is solved from the following matrix equation:

$$\underline{H}^T \vec{Y} = \vec{E}. \tag{3.3.8}$$

where in this M–D extension

$$\vec{Y} = \left(Y_0(\vec{\nu}, \vec{t}), Y_1(\vec{\nu}, \vec{t}), \ldots, Y_{L-1}(\vec{\nu}, \vec{t})\right)^T.$$

Given $\vec{Y}$, the $L$ interpolation functions are obtained by

$$y_i(\vec{t}) = \frac{1}{D_g} \int_{C_{g0}} Y_i(\vec{\nu}, \vec{t}) e^{j2\pi \vec{\nu}^T \vec{t}} d\vec{\nu}, \qquad i = 0 \text{ to } L-1, \tag{3.3.9}$$

and the signal is restored by the interpolation formula

$$f(\vec{t}) = \sum_{i=0}^{L-1} \sum_{\vec{n}} g_i(\underline{V}_g \vec{n}) y_i(\vec{t} - \underline{V}_g \vec{n}). \tag{3.3.10}$$

Equations (3.3.8) to (3.3.10) form the core equations of the M–D GSE.

### 3.3.3  EXAMPLES

We now consider two fourth-order $(L = 4)$ sampling expansion examples of a 2–D function. The support is a circle of radius $\lambda$. Circularly band-limited functions can easily be generated by passing a coherent or incoherent image through an circular pupil [339]. In order to make the example simple. we use the rectangular sampling geometry where the sampling matrix is diagonal:

$$\underline{V} = \begin{bmatrix} 1/2\lambda & 0 \\ 0 & 1/2\lambda \end{bmatrix}. \tag{3.3.11}$$

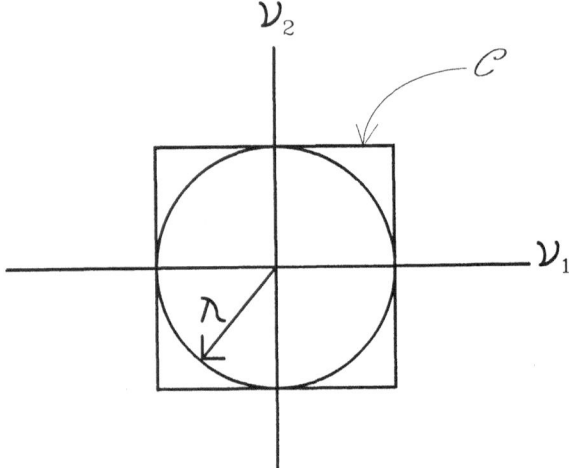

FIGURE 3.7: The rectangular cell corresponding to the rectangular sampling geometry used in Examples 1 and 2. The circular support is enclosed within the rectangular cell.

As shown in Fig. 3.7, the spectrum $F(\vec{\nu})$ is totally enclosed within a rectangular cell. The sampling matrix of the four sample sets is

$$\underline{V}_g = 2\underline{V}$$
$$= \begin{bmatrix} 1/\lambda & 0 \\ 0 & 1/\lambda \end{bmatrix}.$$

The corresponding periodicity matrix is

$$\underline{U}_g = \begin{bmatrix} \lambda & 0 \\ 0 & \lambda \end{bmatrix}.$$

The cell $\mathcal{C}$ is a square of dimension $2\lambda$ and the subcell $\mathcal{C}_g$ is a square of dimension $\lambda$.

**Example 1:** *Signal–Derivative Sampling*

The four linear systems used in this example to generate the four set of samples are respectively:

$$\begin{aligned}
H_0(\nu_1, \nu_2) &= 1, \\
H_1(\nu_1, \nu_2) &= j2\pi\nu_1, \\
H_2(\nu_1, \nu_2) &= j2\pi\nu_2, \\
H_3(\nu_1, \nu_2) &= -4\pi^2\nu_1\nu_2 = H_1(\nu_1, \nu_2)H_2(\nu_1, \nu_2).
\end{aligned} \qquad (3.3.12)$$

The second and the third systems are the first-order differentiators oriented in $t_1$ and $t_2$ directions, respectively. The fourth sample set is obtained by a direct array product of the second and the third sample sets. The sampling density $D_g$ of each sample set is $\lambda^2$ and the matrix $\underline{H}$ is

$$
\underline{H}^T = \begin{bmatrix}
1 & j2\pi\nu_1 & j2\pi\nu_2 & -4\pi^2\nu_1\nu_2 \\
1 & j2\pi(\nu_1+\lambda) & j2\pi\nu_2 & -4\pi^2(\nu_1+\lambda)\nu_2 \\
1 & j2\pi\nu_1 & j2\pi(\nu_2+\lambda) & -4\pi^2\nu_1(\nu_2+\lambda) \\
1 & j2\pi(\nu_1+\lambda) & j2\pi(\nu_2+\lambda) & -4\pi^2(\nu_1+\lambda)(\nu_2+\lambda)
\end{bmatrix}
$$

whose inverse is

$$
\left[\underline{H}^T\right]^{-1} = \frac{-1}{4\pi^2\lambda^2} \times
$$

$$
\begin{bmatrix}
-4\pi^2(\nu_1+\lambda)(\nu_2+\lambda) & 4\pi^2\nu_1(\nu_2+\lambda) & 4\pi^2\nu_1(\nu+\lambda) & 4\pi^2\nu_1\nu_2 \\
-j2\pi(\nu_2+\lambda) & -j2\pi(\nu_2+\lambda) & -j2\pi\nu_2 & -j2\pi\nu_2 \\
-j2\pi(\nu_1+\lambda) & j2\pi\nu_1 & j2\pi(\nu_1+\lambda) & -j2\pi\nu_1 \\
1 & -1 & -1 & 1
\end{bmatrix}
$$

The carrier vector $\vec{E}$ in this example is

$$
\vec{E} = \begin{pmatrix} 1 & e^{j2\pi\lambda t_1} & e^{j2\pi\lambda t_2} & e^{j2\pi\lambda(t_1+t_2)} \end{pmatrix}^T . \tag{3.3.13}
$$

According to (3.3.3), the region $C_{g0}$ lies at the lower left corner of $C$.

$$
C_{g0} = \{(\nu_1.\nu_2)| -\lambda \le \nu_1.\nu_2 < 0\}.
$$

The four interpolation functions are computed by using (3.3.11).

$$
\begin{aligned}
y_0(t_1.t_2) &= 16\frac{\sin^2(\lambda t_1/2)\sin^2(\lambda t_2/2)}{\lambda^4 t_1^2 t_2^2}. \\[4pt]
y_0(t_1.t_2) &= 16\frac{\sin^2(\lambda t_1/2)\sin^2(\lambda t_2/2)}{\lambda^4 t_1 t_2^2}. \\[4pt]
y_0(t_1.t_2) &= 16\frac{\sin^2(\lambda t_1/2)\sin^2(\lambda t_2/2)}{\lambda^4 t_1^2 t_2}. \\[4pt]
y_0(t_1.t_2) &= 16\frac{\sin^2(\lambda t_1/2)\sin^2(\lambda t_2/2)}{\lambda^4 t_1 t_2}.
\end{aligned}
\tag{3.3.14}
$$

Note that the expression for the four linear systems in this example can also be expressed as an *outer product* of four 1–D linear systems. Specifically, these four linear systems are the linear systems corresponding to two 1–D signal-derivative sampling problems.

Problem 1 :

$$
h_{10}(\nu_1) = 1 \quad \text{and} \quad h_{11}(\nu_1) = j2\pi\nu_1.
$$

Problem 2 :

$$h_{10}(\nu_2) = 1 \quad \text{and} \quad h_{11}(\nu_2) = j2\pi\nu_2.$$

An outer product of these four expressions produces the expression of the four linear systems used in Example 1.

$$
\begin{aligned}
h_0(t_1, t_2) &= h_{10}(t_1)h_{20}(t_2), \\
h_1(t_1, t_2) &= h_{11}(t_1)h_{20}(t_2), \\
h_2(t_1, t_2) &= h_{10}(t_1)h_{21}(t_2), \\
h_3(t_1, t_2) &= h_{11}(t_1)h_{22}(t_2).
\end{aligned}
$$

With reference to Papoulis [706], the interpolation functions for the two problems are:

Problem 1 :

$$y_{10}(t_1) = 4\frac{\sin^2(\lambda t_1/2)}{\lambda^2 t_1^2} \quad \text{and} \quad y_{20}(t_2) = 4\frac{\sin^2(\lambda t_2/2)}{\lambda^2 t_2^2}.$$

Problem 2 :

$$y_{11}(t_1) = 4\frac{\sin^2(\lambda t_1/2)}{\lambda^2 t_1} \quad \text{and} \quad y_{21}(t_2) = 4\frac{\sin^2(\lambda t_2/2)}{\lambda^2 t_2}.$$

Clearly, the four interpolation functions in (3.3.14) are also the outer product of the two sets of interpolation functions. This relationship is always true if the expression for the M–D systems are outer products of the expression of 1–D systems.

**Example 2:** *Interlaced Sampling*

The four linear systems in this example are

$$
\begin{aligned}
h_0(\nu_1.\nu_2) &= 1. \\
h_1(\nu_1.\nu_2) &= e^{j2\pi(\nu_1 d_{11} + \nu_2 d_{12})}. \\
h_2(\nu_1.\nu_2) &= e^{j2\pi(\nu_1 d_{21} + \nu_2 d_{22})}. \\
h_3(\nu_1.\nu_2) &= e^{j2\pi(\nu_1 d_{31} + \nu_2 d_{32})}.
\end{aligned}
\qquad (3.3.15)
$$

The $d_{ij}$ coefficients specify the offset of the sampling locations of each sample subset from the origin. It follows that

$$\underline{H}^T = \underline{H}_1\underline{H}_2.$$

where

$$
\underline{H}_1 = \begin{bmatrix}
1 & 1 & 1 & 1 \\
1 & e^{j2\pi\lambda d_{11}} & e^{j2\pi\lambda d_{21}} & e^{j2\pi\lambda d_{31}} \\
1 & e^{j2\pi\lambda d_{12}} & e^{j2\pi\lambda d_{22}} & e^{j2\pi\lambda d_{32}} \\
1 & e^{j2\pi\lambda(d_{11}+d_{12})} & e^{j2\pi\lambda(d_{21}+d_{22})} & e^{j2\pi\lambda(d_{31}+d_{32})}
\end{bmatrix}
$$

and

$$\underline{H}_2 = \operatorname{diag}\left[ 1 \; e^{j2\pi(\nu_1 d_{11}+\nu_2 d_{12})} \; e^{j2\pi(\nu_1 d_{21}+\nu_2 d_{22})} \; e^{j2\pi(\nu_1 d_{31}+\nu_2 d_{32})} \right]$$

Given the numerical values of $\lambda$ and the $d_{ij}$ coefficients, inversion of $\underline{H}_1$ can be done numerically. In particular, let $\beta_{ij}$, $i,j = 0$ to 3 be the $(i+1, j+1)^{th}$ element of $[\underline{H}^T]^{-1}$. Then, with reference to (3.3.8),

$$Y_i(\nu_1, \nu_2, t_1, t_2) = e^{-j2\pi(\nu_1 d_{i1}+\nu_2 d_{i2})} \sum_{j=0}^{3} \beta_{ij} e_j(t_1, t_2), \qquad (3.3.16)$$

where $e_j(t_1, t_2)$ is the $(j+1)^{th}$ element of the vector $\vec{E}$ in (3.3.13). Also, let

$$
\begin{aligned}
f_{dj}(t_1, t_2) &= \frac{\beta_{ij}}{\lambda^2} \int_{-\lambda}^{0} \int_{-\lambda}^{0} e_j(t_1, t_2) e^{j2\pi(\nu_1 t_1 + \nu_2 t_2)} d\nu_1 \nu_2 \\
&= \beta_{ij} \operatorname{sinc}(\lambda t_1) \operatorname{sinc}(\lambda t_2) e^{j2\pi\lambda(r_{j1}t_1 + r_{j2}t_2)}. \qquad (3.3.17)
\end{aligned}
$$

where the ordered pair $(r_{j1}, r_{j2})$ is the bipolar representation (binary representation except 0 is replaced by $-1$) of the integer $j$. The four interpolation functions follows as

$$y_i(t_1, t_2) = \sum_{j=0}^{3} f_{dj}(t_1 - d_{i1}, t_2 - d_{i2}), \qquad i = 0 \text{ to } 3. \qquad (3.3.18)$$

Clearly from this example, the evaluation of interpolation functions for 2–D interlaced sampling is not as trivial as its 1–D counterpart. For 1–D interlaced sampling, the matrix $\underline{H}_1$ is *Vandermonde* [513] and therefore the inversion can be evaluated straightforwardly in closed form. Whereas in M–D interlaced sampling, this is not generally the case. The inversion of $\underline{H}_1$ in most cases needs to be evaluated numerically.

An utilization of M–D GSE is for sampling density reduction [211, 215]. A first-order reduction is illustrated by the following example.

**Example 3:** *Sample Density Reduction*

If one of the partitions of $F(\vec{\nu})$, say $F_0(\vec{\nu})$, is identically zero, the first column of $\underline{H}$ in (3.3.6) can be eliminated and thus the system becomes an overdetermined system. Since, however, the system is also consistent, we can further eliminate a row of $\underline{H}$. Eliminating, for example, the first row, of $\underline{H}$ also eliminates an entry, $G_0(\vec{\nu})$, from $\vec{G}$. Thus, the sample group, $\{g_0(\underline{V}_g \vec{n})\}$ is dropped from being used to reconstruct $f(\vec{t})$. After all eliminations, Eq. (3.3.6) becomes a system of dimension $L-1$:

$$\vec{G}_0 = D_g \underline{H}_{0,0} \vec{F}_0.$$

The double subscripts denote respectively the indices of the eliminated
row and the eliminated column. Since one sample subset is eliminated, the
overall sampling density is reduced by a factor of $1/m$. Note that $F_0(\vec{\nu}) = 0$
can occur even at Nyquist densities (e.g., see Fig. 3.6 for circularly band-
limited functions). If this is the case, the removal of a sample set will result
a sampling density below that of Nyquist. The subject of such sampling
density reduction is treated more fully in Part II of this chapter.

## 3.4   Extension Generalization

In this section, we will consider relaxing the formulation of the matrix $\underline{V}_g$.
This relaxation can result in the creation of subcells which are no longer
congruent to the cell. In addition, more freedom in the sampling geometry
for the sample sets is allowed. We will also consider the cases where the
support of the spectrum does not match the cell shape and therefore cannot
totally be enclosed within a cell.

In a more general setting, we can express

$$\underline{V}_g = \underline{V}\,\underline{M}, \tag{3.4.1}$$

where $\underline{M}$ is a non-singular integer matrix of dimension $N$. So far we have
used $\underline{M} = k\underline{I}$. In this more general setting, $L$, the number of sample sets,
is equal to $|\underline{M}|$, which may not equal to $k^N$. The subcells, in general, need
not be congruent to the cell. Consider a circularly band-limited function
being sampled at a rectangular sampling geometry whose sampling matrix
is specified (3.3.11). A sampling geometry corresponding to the sampling
matrix, $\underline{V}_g$, where

$$\underline{M} = \begin{bmatrix} 1 & -1 \\ 1 & 1 \end{bmatrix}$$

is shown in Fig. 3.8. and the corresponding subcell is a rhombus, Fig. 3.8.
Clearly, this subcell is not congruent to the square cell. For this example.
$|\underline{M}| = 2$ and thus $L = 2$.

The support of the function may not be totally enclosed within a cell.
Consider our running example again. If a circularly band-limited function is
sampled rectangularly, the support of the function's spectrum is totally enclosed within a rectangular cell. If the function is sampled with a hexagonal
geometry

$$\underline{V} = \begin{bmatrix} 1/2\lambda & 1/4\lambda \\ 0 & 1/\sqrt{3}\lambda \end{bmatrix}. \tag{3.4.2}$$

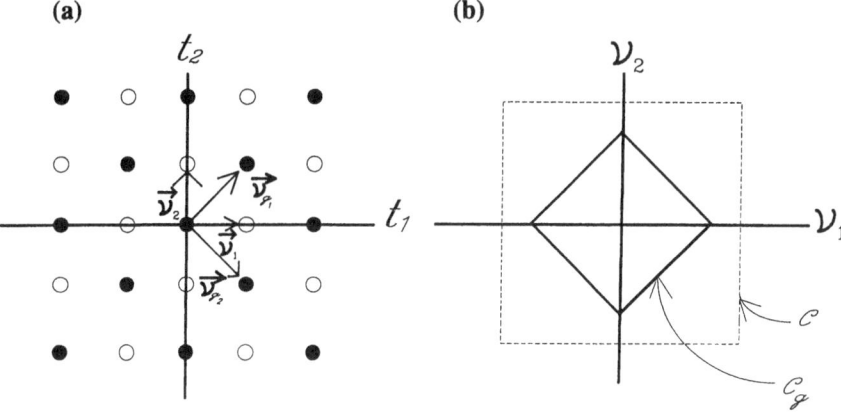

FIGURE 3.8: A sampling geometry in (a) which produces a rhombus, subcell which is not congruent to the rectangular cell as shown in (b).

which corresponds sampling at the Nyquist density, the corresponding cell shape is a parallelogram. As shown in Fig. 3.9, the circular support is not totally enclosed within this cell.

Restoring the function $f(\vec{t})$ under these two scenarios requires no extra formulation. As can be seen from (3.3.6) and (3.3.8), the M–D GSE always restores a period, though possibly not in the same shape of the original cell of the spectral replications, $\hat{F}(\vec{\nu})$. It is well-known that Fourier series coefficients are invariant to the cell shape [268]. The Fourier series coefficients here are the samples of $f(\vec{t})$ at locations $\left[\underline{V}.\vec{0}\right]$. We can therefore restore the function discretely by the M–D GSE:

$$
\begin{aligned}
f(\underline{V}\vec{n}) &= \sum_{m=-\infty}^{\infty} \sum_{k=0}^{L-1} g_k(\underline{V}_g\vec{m}) y_k(\underline{V}\vec{n} - \underline{V}_g\vec{m}) \\
&= \sum_{m=-\infty}^{\infty} \sum_{k=0}^{L-1} g_k(\underline{V}_g\vec{m}) y_k\big(\underline{V}(\vec{n} - \underline{M}\vec{m})\big). \quad (3.4.3)
\end{aligned}
$$

Restoring the function at those sampling locations is equivalent to restoring the samples' spectrum, $\hat{F}(\vec{\nu})$. Given that the samples are restored, the function is reconstructed by the traditional low-pass method,

$$
f(\vec{t}) = \sum_{n=-\infty}^{\infty} f(\underline{V}\vec{n}) h(\vec{t} - \underline{V}\vec{n}), \quad (3.4.4)
$$

where $h(\vec{t})$, an ideal low-pass filter impulse response of magnitude $|\underline{V}|$.

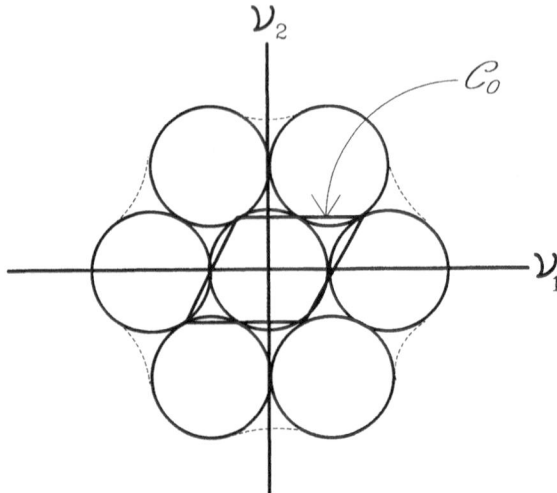

FIGURE 3.9: The hexagonal sampling geometry produces a parallelogram subcell which does not enclose the zeroth-order replication.

extracts only the zeroth-order replication. For our running example, $h(\vec{t})$ may have hexagonal support, as shown in Fig. 3.10.

## 3.5  Conclusion

In Part I of this chapter. we have extended Papoulis' generalized sampling expansion to multidimensional band-limited functions. For cases where the cells do not enclose the support of the function's spectrum, restoration of the function is achieved by discrete interpolation followed by a low-pass processing.

We have also demonstrated that the GSE formulation can be utilized to reduce sampling density if certain conditions are satisfied. We will expand on this point in Part II.

# Part II
# Sampling Multidimensional Band-Limited
# Functions at Minimum Densities

In contrast to the *Nyquist density sampling* which yields the most densely packed spectral replications, the *minimum density sampling* yields samples linearly independent of each other. In the following section, we will show

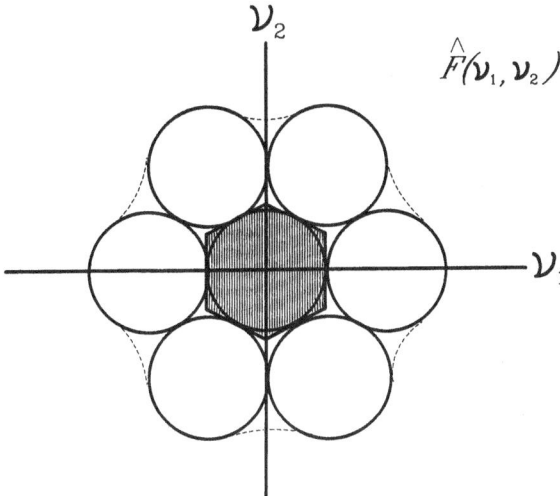

FIGURE 3.10: An ideal low-pass filter with the support of a hexagon is used to reconstruct the function from its samples.

that for 1–D band-limited functions, the two densities (or rates) are identical. For higher dimensional band-limited functions, Nyquist densities can be significantly higher than minimum densities.

## 3.6 Sample Interdependency

If the shape of the support of a function's spectrum is not a period of the spectral replications (or the support is not a cell $\mathcal{C}$), gaps exist among the replications. For 1–D band-limited functions, the support is a period of the replications and thus no gaps exist. For higher dimensional band-limited functions, gaps may exist even when the function is sampled at the Nyquist density. Consider sampling a 2–D circularly band-limited function at the Nyquist density. The corresponding sampling geometry has a hexagonal geometry. As shown in Fig. 3.6, gaps exist among the replications.

If gaps exist among the spectral replications, samples are linearly interdependent. To show this let $h(\vec{t}) \leftrightarrow H(\vec{\nu})$ be some ideal bandpass filter whose passband is only defined in gap regions. Thus,

$$H(\vec{\nu})\hat{F}(\vec{\nu}) = 0.$$

Correspondingly,

$$\sum_{\vec{n}} f(\underline{V}\vec{n})h(\vec{t} - \underline{V}\vec{n}) = 0, \qquad \forall \vec{t}.$$

Since $h(\vec{t})$ is nowhere identically zero, by definition the samples are linearly interdependent.

A sufficient condition for sampling a band-limited function at the minimum density is that no gaps are contained among the spectral replications. Clearly, the interdependency among samples is dictated by the support shape of $F(\vec{\nu})$. If the support shape is identical to the cell shape, gaps (e.g., rectangular or hexagonal) cease to exist and samples are linearly independent of each other.

The condition implies that the Nyquist rate is also the minimum sampling rate for 1–D band-limited functions. For higher dimensional cases, sampling at Nyquist densities may leave gaps among the spectral replications and hence the samples are linearly interdependent. Thus, the samples in our running example of sampling a 2–D circularly band-limited function are linearly interdependent.

Linearly interdependency among samples implies *oversampling* [588, 586, 596]. If there is no aliasing, we can straightforwardly define the oversampling index as

$$r = \frac{\text{area of } C}{\text{area of } F(\vec{\nu})} \geq 1. \qquad (3.6.1)$$

If $r = 1$, the function is sampled at the minimum density. Otherwise, the function is oversampled. The sampling of a 2–D circularly band-limited function at the Nyquist density corresponds to $r = 2/\sqrt{3} \approx 1.15$ [211, 721].

For 1–D cases. every support has a corresponding minimum sampling rate. Thus, if gaps exist in the replications. the sampling rate can be scaled and gaps closed. This, however, cannot be applied to higher dimensional cases. For our running example, the support shape is a circle. Clearly, this shape does not have a corresponding sampling geometry. As shown in Fig. 3.11, a direct down scaling of the sampling density on every sampling dimension will result an aliased spectrum. This example illustrates that sampling M–D band-limited functions at densities below that of Nyquist is not trivial.

## 3.7   Sampling Density Reduction Using M–D GSE

If gaps exist among the replications and if $L$ is large enough, the $L^{th}$-order M–D GSE allows the restoration of the function with less than $L$ sample

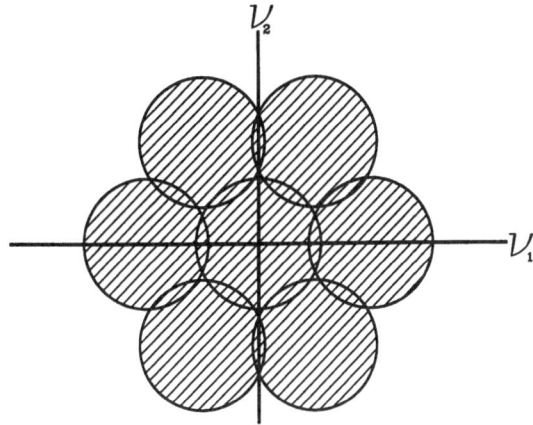

FIGURE 3.11: By decreasing the sampling period, we obtain an aliased replications. Note that gaps still exist among the aliased replications.

sets. Since fewer sample sets are used, the sampling density is reduced. If only $L-q$ sample sets are used, the sampling density is reduced by a fraction of $q/L$. As discussed in Example 3, given a function initially sampled at the Nyquist density, any reduction of sample sets will immediately result a density below that of Nyquist. We will show that the minimum sampling density is equal to the area of the support of the function's spectrum.

The formulation of the $L^{th}$-order M–D GSE starts with partitioning the baseband cell, $C_0$, into $L$ partitions. Each partition is referred to as a subcell. If gaps exist among the replications, gaps also exist in $C_0$. Let $A_D$ denotes the gap regions within $C_0$. If $L$ is large enough, then a number of the subcells, say $q$, will be totally subsumed within $A_D$. Consider the sampling of the 2–D circularly band-limited function. For illustration purposes, the function is sampled at a rectangular geometry as specified in (3.3.11). The corresponding baseband cell is a square of dimension $2\lambda$. As shown in Fig. 3.12, if we have $k \geq 7$, we have subcells totally enclosed within gap regions. In particular, for $k = 7$, we have $q = 4$ (Fig. 3.12a). For $k = 10$, we have $q = 12$ (Fig. 3.12b).

We will use these two cases as our running example in the discussion to follow. The $q = 4$ case will be referred to as the fourth–order case, and the $q = 12$ case the twelfth–order case.

Let $\mathcal{M}$ be the index set corresponding to the subcells subsumed within $A_D$.

$$\mathcal{M} = \{m \,|\, C_{gm} \in A_D\}. \tag{3.7.1}$$

The cardinality of $\mathcal{M}$ is $q$. Each subcell holds a partition of the baseband

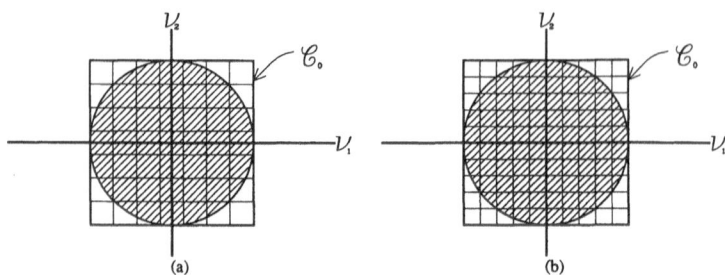

FIGURE 3.12: (a) The partitioning of $C_0$ corresponds to the case of $k = 7$. The total number of partitions is forty–nine, four of them are subsumed within gap regions. (b) By increasing the total number of partitions to one hundred, $k = 10$, we obtain twelve partitions subsumed within gap regions.

spectrum, $F(\vec{\nu})$. Hence. the partitions of $F(\vec{\nu})$ within those subcells specified in $\mathcal{M}$ are zero partitions:

$$\hat{F}_m(\vec{\nu}) = 0, \qquad m \in \mathcal{M}. \tag{3.7.2}$$

With reference to (3.3.6). this corresponds to $q$ zero entries within the vector $\vec{F}$. We can therefore remove these entries from $F$ and the corresponding columns from the matrix $\underline{H}$. The result is

$$\vec{G} = D_g \underline{H}_{\mathcal{M}} \vec{F}_{\mathcal{M}}. \tag{3.7.3}$$

The subscript denotes the set of the removed entries. The matrix $\underline{H}_{\mathcal{M}}$ has a dimension of $L \times (L - q)$ and thus (3.7.3) is overdetermined. Since the solution is consistent, a combination of $q$ entries in $\vec{G}$, and $q$ corresponding rows in $\underline{H}_{\mathcal{M}}$. can be deleted. Removal of $q$ entries in $\vec{G}$ corresponds to the removal of $q$ sample sets. Let $\mathcal{D}$ be the index set corresponding to the $q$ removed sample sets

$$\mathcal{D} = \{p \mid \{g_p(\underline{V}_g \vec{n})\} \text{ removed}\}. \tag{3.7.4}$$

Clearly, the removal of these sample sets also implies the sampling density is reduced by a fraction of $q/L$. We thus obtain a more compact system of dimension $L - q$ relating the non-zero partitions of $F(\vec{\nu})$ to the spectra of the $L - q$ remaining sample sets

$$\vec{G}_{\mathcal{D}} = D_g \underline{H}_{\mathcal{D},\mathcal{M}} \vec{F}_{\mathcal{M}}. \tag{3.7.5}$$

Note that the problem has transformed to a $(L - q)^{th}$-order M–D GSE problem. If $\underline{H}_{\mathcal{D},\mathcal{M}}$ is invertible for every $\vec{\nu}$ in the reference subcell $\mathcal{C}_{g0}$, we follow steps (3.3.6) and (3.3.10), and obtain the interpolation formula for restoring $f(\vec{t})$ from the remaining $L - q$ sample sets

$$f(\vec{t}) = \sum_{i \notin \mathcal{D}} \sum_{\vec{n}} g_i(\underline{V}_g\vec{n})y_i(\vec{t} - \underline{V}_g\vec{n}). \qquad (3.7.6)$$

where the spectra of the $L - q$ interpolation functions in $\mathcal{C}_{g0}$ are solved from the matrix equation

$$\underline{H}_{\mathcal{D},\mathcal{M}}^T \vec{Y}_{\mathcal{D}} = \vec{E}_{\mathcal{M}}. \qquad (3.7.7)$$

The methodology presented here is analogous to playing a jigsaw puzzle game. The puzzle is the spectrum $F(\vec{\nu})$ and the pieces are the $L$ partitions of $F(\vec{\nu})$. In assembling $F(\vec{\nu})$, the blank pieces (those partitions subsumed with gap regions) are discarded. Corresponding to every discarded partition, a sample set can be removed. The finished puzzle is the spectrum without blank pieces.

Thus, if gaps exist among the spectral replications, a subset of the sample sets can be removed. Given that $q$ subcells are totally subsumed within $\mathcal{A}_{\mathcal{D}}$, a combination of $q$ out of $L$ sample sets can be removed from reconstructing the function. The overall sampling density is reduced by $q/L$. The problem of removing $q$ out of $L$ sample sets in a M–D GSE setting is referred to as the $q^{th}$-*order reduction problem*.

## 3.7.1   SAMPLING DECIMATION

In this section, we will first demonstrate that regular sampling of $f(\vec{t})$ can always be formulated as an $L^{th}$-order M–D GSE problem. Thus, if gaps exist among the spectral replications, $\hat{F}(\vec{\nu})$, a number of sample sets can be removed without loss of information. Thus, by implementing the decimation at the sampling stage, the function can be sampled directly at reduced densities.

Consider the sampling of an $N$–D band-limited function, $f(\vec{t})$, at some regular geometry, $\left[\underline{V}.\vec{0}\right]$. The samples can be sectioned into $L = k^N$ subgroups in the following manner:

$$g_p(\underline{V}_g\vec{n}) = f(\underline{V}_g\vec{n} + \underline{V}\vec{k}_p), \qquad p = 0 \text{ to } L - 1, \qquad (3.7.8)$$

where $\underline{V}_g = k\underline{V}$. The vector $\vec{k}_p$ is an integer vector which is the $k$–ary representation of the integer $p$ (Subsection 3.3.2). Thus the $p^{th}$ sample subgroup can be obtained by sampling $f(\vec{t})$ with the geometry $\left[\underline{V}_g : \vec{k}_p\right]$. Here, we identify the $p^{th}$ sample subgroup as the $p^{th}$ sample set in the M–D GSE setting. The $L$ corresponding linear systems are therefore

$$H_p(\vec{\nu}) = e^{j2\pi\vec{\nu}^T\underline{V}\vec{k}_i}, \qquad p = 0 \text{ to } L-1. \tag{3.7.9}$$

Hence,

$$G_p(\vec{\nu}) = e^{j2\pi\vec{\nu}^T\underline{V}\vec{k}_p} \sum_{q=1}^{L-1} e^{j2\pi\vec{k}_q^T\vec{k}_p/k} F_q(\vec{\nu}), \qquad \begin{array}{l} \vec{\nu} \in \mathcal{C}_{d0}, \\ p = 0 \text{ to } L-1. \end{array} \tag{3.7.10}$$

Equation (3.7.10) can be written in matrix form as follows:

$$\vec{G} = D_g\underline{H}_1 \cdot \underline{H}_2\vec{F}. \tag{3.7.11}$$

where

$$\underline{H}_1 = \text{diag}\left[e^{j2\pi\vec{\nu}^T\underline{V}\vec{k}_p}|p = 0 \text{ to } L-1\right] \tag{3.7.12}$$

and

$$\underline{H}_2 = \left[pq^{th} \text{ element } = e^{j2\pi\vec{k}_q^T\vec{k}_p/k}\right]. \tag{3.7.13}$$

Note that $\underline{H}_2$ is an $N$–D DFT matrix and is therefore orthogonal.

$$H_2\left[H_2^*\right]^T = L\underline{I}.$$

The superscript * denotes complex conjugation. By letting

$$\begin{aligned} \hat{G} &= \underline{H}_1^{-1}\vec{G} \\ &= \left[e^{-2j\pi\vec{\nu}^T\underline{V}\vec{k}_p}G_p(\vec{\nu})|p = 0 \text{ to } L-1\right], \end{aligned} \tag{3.7.14}$$

which corresponds to offsetting the position of the $p^{th}$ sample subgroup by the vector $-\underline{V}\vec{k}_p$, we obtain

$$\hat{G} = D_g\underline{H}_2\vec{F}. \tag{3.7.15}$$

Following (3.7.2) to (3.7.3), given the index set $\mathcal{M}$, we delete the corresponding $q$ columns from $H_2$ in (3.7.15) and obtain

$$\hat{G} = D_g\underline{H}_{2,\mathcal{M}}\vec{F}_\mathcal{M}. \tag{3.7.16}$$

After the index set $\mathcal{D}$ is determined, $q$ columns in $H_{2,\mathcal{M}}$ are removed. We thus obtain a compact system of dimension $L - q$ relating the non-zero partitions of $F(\vec{\nu})$ to the spectra of the $L - q$ sample subgroups

$$\hat{G}_\mathcal{D} = D_g\underline{H}_{2,\mathcal{M},\mathcal{D}}\vec{F}_\mathcal{D}. \tag{3.7.17}$$

In the setting of sampling decimation, Eq. (3.7.7) becomes

$$\underline{H}^T_{2,\mathcal{M},\mathcal{D}}\vec{Y}_{\mathcal{D}} = \vec{E}_{\mathcal{M}}. \tag{3.7.18}$$

If $\underline{H}_{2,\mathcal{M},\mathcal{D}}$ is invertible, we can proceed to solve for the $L - q$ interpolation functions. Note that every element in $\underline{H}_{2,\mathcal{M},\mathcal{D}}$ is an Euler quantity and can be converted into a complex number. Thus, the inversion of $\underline{H}_{2,\mathcal{M},\mathcal{D}}$ can be also be computed numerically.

Let $\beta_{pq}, p \in \overline{\mathcal{D}}, q \in \overline{\mathcal{M}}$, to be the $pq^{th}$ element of $\left[\underline{H}^T_{\mathcal{D},\mathcal{M},2}\right]^{-1}$. Then

$$Y_p(\vec{\nu}.\vec{t}) = \sum_{q \in \overline{\mathcal{M}}} \beta_{pq}e^{j2\pi\vec{k}^T_q \underline{U}^T_g \vec{t}}, \qquad p \in \overline{\mathcal{D}}. \vec{\nu} \in \mathcal{C}_{d0}. \tag{3.7.19}$$

The overbar denotes the set complement. By substituting (3.7.19) into (3.3.9), and with reference to (3.3.4). we obtain the $p^{th}$ interpolation function for the $p^{th}$ sample subgroup:

$$y_p(\vec{t}) = \frac{1}{D_g} \sum_{q \in \overline{\mathcal{M}}} \int_{\vec{\nu} \in \mathcal{C}_{gq}} \beta_{pq}e^{j2\pi\vec{\nu}^T\vec{t}}d\vec{\nu}. \qquad p \in \overline{\mathcal{D}}. \tag{3.7.20}$$

The Fourier transform of $y_p(\vec{t})$ yields the spectrum

$$Y_p(\vec{\nu}) = \sum_{q \in \overline{\mathcal{M}}} \frac{\beta_{pq}}{D_g}\prod_q(\vec{\nu}). \tag{3.7.21}$$

where

$$\prod_q(\vec{\nu}) = \begin{cases} 1 & \vec{\nu} \in \mathcal{C}_{gq} \\ 0 & \text{otherwise.} \end{cases}$$

Thus, every $y_p(\vec{t})$ is a low-pass/bandpass function whose passband is defined within the subcells $\{\mathcal{C}_{gq} | q \in \overline{\mathcal{M}}\}$. The magnitude and phase of $Y_p(\vec{\nu})$ is staircase type over the passband, equal to $|\beta_{pq}|$ and $\angle\beta pq$, respectively, in $\mathcal{C}_{gq}, q \in \overline{\mathcal{M}}$. Such functions can be implemented modularly [397].

The interpolation formula follows as:

$$f(\vec{t}) = \sum_{p \notin \mathcal{D}} \sum_{\vec{n}} f(\underline{V}_g\vec{n} + \underline{V}\vec{k}_p)y_p(\vec{t} - \underline{V}\vec{k}_p - \underline{V}_g\vec{n}). \tag{3.7.22}$$

The existence of solutions relies on the invertibility of $\underline{H}_{2,\mathcal{M},\mathcal{D}}$. Since $\underline{H}_2$ is a orthogonal matrix, $\underline{H}_{2,\mathcal{M}}$ is an orthogonal column space. There always exists at least a set of $\mathcal{D}$ such that $\underline{H}_{2,\mathcal{M},\mathcal{D}}$ is invertible. In other words, there exists at least one combination of $q$ sample subgroups that can be removed without loss of information. The sampling density is thus reduced by a fraction of $q/L$. If the removal of sample subgroups are directly implemented at the sampling stage, the function is directly sampled at a reduced density. We refer this implementation as *sampling decimation*.

## 3.7.2   A Second Formulation for Sampling Decimation

An alternate formulation presented here for sampling decimation is to treat the removed samples as lost samples. We will formulate the interpolation formula to recover the lost samples only, instead of reconstructing the entirety of $f(\vec{t})$, from those remaining. When $q \ll L - q$, the computational complexity of this formulation can be much less than that the previous formulation.

Since $\underline{H}_2$ is an orthogonal matrix, with reference to (3.7.15),

$$\frac{1}{LD_g} [\underline{H}_2^*]^T \hat{G} = \vec{F} \tag{3.7.23}$$

or

$$\frac{1}{LD_g} \sum_{p=0}^{L-1} e^{-2j\vec{k}_p^T \vec{k}_q / k} e^{j2\pi \vec{\nu}^T \underline{V} \vec{k}_p} G_p(\vec{\nu}) = \hat{F}_q(\vec{\nu}), \qquad \begin{array}{l} \vec{\nu} \in \mathcal{C}_{d0}, \\ q = 0 \text{ to } L - 1. \end{array} \tag{3.7.24}$$

Given $\hat{F}_q(\vec{\nu}) = 0$, $q \in \mathcal{M}$, we have

$$\sum_{p=0}^{L-1} e^{-2j\vec{k}_p^T \vec{k}_q / k} e^{j2\pi \vec{\nu}^T \underline{V} \vec{k}_p} G_p(\vec{\nu}) = 0, \qquad \vec{\nu} \in \mathcal{C}_{d0}. \tag{3.7.25}$$

By separating the decimated sample subgroups from those remaining in the summation. we have

$$\sum_{p \in \mathcal{D}} e^{-2j\vec{k}_p^T \vec{k}_q / k} e^{j2\pi \vec{\nu}^T \underline{V} \vec{k}_p} G_p(\vec{\nu}) = - \sum_{p \notin \mathcal{D}} e^{-2j\vec{k}_p^T \vec{k}_q / k} e^{j2\pi \vec{\nu}^T \underline{V} \vec{k}_p} G_p(\vec{\nu}).$$
$$\vec{\nu} \in \mathcal{C}_{d0} \tag{3.7.26}$$

which in matrix form is

$$\hat{\underline{H}}_{2,\overline{\mathcal{M}},\overline{\mathcal{D}}} \hat{G}_{\overline{\mathcal{D}}} = \hat{\underline{H}}_{2,\overline{\mathcal{M}},\mathcal{D}} \hat{G}_{\mathcal{D}}. \tag{3.7.27}$$

Here, $\hat{\underline{H}}_{2,\overline{\mathcal{M}},\overline{\mathcal{D}}}$ is formed from $q$ columns of $\hat{\underline{H}}_{2,\overline{\mathcal{M}}}$ and therefore is a square matrix of dimension $q$. If it is invertible, the spectra of the removed sample subgroups are a linear combination of those remaining

$$\hat{G}_{\overline{\mathcal{D}}} = \left[\hat{\underline{H}}_{2,\overline{\mathcal{M}},\overline{\mathcal{D}}}\right]^{-1} \cdot \hat{\underline{H}}_{2,\overline{\mathcal{M}},\mathcal{D}} \hat{G}_{\mathcal{D}}. \tag{3.7.28}$$

Let $\gamma_{pq}$ be the $pq^{th}$ element in the matrix product $\left[\hat{\underline{H}}_{2,\overline{\mathcal{M}},\overline{\mathcal{D}}}\right]^{-1} \cdot \hat{\underline{H}}_{2,\overline{\mathcal{M}},\mathcal{D}}$. Then

$$G_p(\vec{\nu}) = - \sum_{q \notin \mathcal{D}} \gamma_{pq} e^{-j2\pi \vec{\nu}^T \underline{V}(\vec{k}_q - \vec{k}_p)}, \qquad \vec{\nu} \in \mathcal{C}_{g0}, p \in \mathcal{D}. \tag{3.7.29}$$

Since

$$g_p(\underline{V}_g\vec{n}) = \frac{1}{\mathcal{D}_g} \int_{\vec{\nu} \in C_{d0}} G_p(\vec{\nu}) e^{j2\pi\vec{\nu}^T \underline{V}_g\vec{n}} d\vec{\nu}. \tag{3.7.30}$$

Substituting (3.7.29) into (3.7.30), and with reference to (3.7.8), yields the interpolation formula to recover the $p^{th}$ sample set:

$$f(\underline{V}_g\vec{n} + \underline{V}\vec{k}_p) = \sum_{q \in \overline{\mathcal{D}}} \sum_{\vec{m}} f(\underline{V}_g\vec{m} + \underline{V}\vec{k}_q)$$

$$\times f_{pq}\left(\underline{V}_g(\vec{n} - \vec{m}) - \underline{V}(\vec{k}_q - \vec{k}_p)\right) \tag{3.7.31}$$

where

$$f_{pq}(\vec{t}) = \frac{1}{\mathcal{D}_g} \int_{\vec{\nu} \in C_{g0}} \gamma_{pq} e^{j2\pi\vec{\nu}^T \vec{t}} d\vec{\nu}. \tag{3.7.32}$$

We now apply this formulation to the fourth–order case example and the twelfth–order case example discussed in Example 5.

**Example 4:** *Fourth–Order Sampling Decimation*

Consider the case illustrated in Fig. 3.12 where a circularly band-limited is sampled at a rectangular geometry. When the samples are divided into $L = 49$ sample subgroups ($k = 7$), we obtain $q = 4$ subcells subsumed within gap region. Thus, four sample subgroups can be removed and the sampling density is reduced by a fraction 4/49 or around 8.16%.

Without loss of generality, we let $\lambda = 1/2$ such that the function is a unit circularly band-limited. Initially, the function is sampled at the rectangular sampling geometry $\left[\underline{V}, \vec{0}\right]$ where

$$\underline{V} = \begin{bmatrix} 1 & 0 \\ 0 & 1 \end{bmatrix}.$$

The sampling geometry of the $p^{th}$ sample subgroup is $\left[\underline{V}_g, \vec{k}_p\right]$ where $\underline{V}_g = 7\underline{V}$, $\vec{k}_p = [k_{p,1}, k_{p,2}]^T$, $k_{p,1}, k_{p,2} = 0$ to 6, and $p = 7k_{p,2} + k_{p,1}$.

The four subcells subsumed within the gap region reside at the four corners of $C_0$. The corresponding four position vectors for these four subcells are

$$\begin{aligned} \vec{k}_0 &= (0,0)^T, \\ \vec{k}_6 &= (6,0)^T, \\ \vec{k}_{42} &= (0,6)^T, \\ \vec{k}_{48} &= (6,6)^T. \end{aligned}$$

The indices of these four vectors constitutes the index set $\mathcal{M}$. The four sample subgroups chosen to be removed constitutes a rectangular decimation geometry. The four corresponding offset vectors are

$$
\begin{aligned}
\vec{k}_0 &= (0,0)^T, \\
\vec{k}_m &= (m,0)^T, \\
\vec{k}_{7n} &= (0,n)^T, \\
\vec{k}_{7n+m} &= (m,n)^T,
\end{aligned}
$$

where $1 \leq m, n \leq 6$. The indices of these four vectors constitutes the index set . The matrix $\hat{H}_{2,\overline{\mathcal{M}},\overline{\mathcal{D}}}$ follows as

$$
\hat{H}_{2,\overline{\mathcal{M}},\overline{\mathcal{D}}} =
\begin{bmatrix}
1 & 1 & 1 & 1 \\
1 & e^{-j2\pi 6m/7} & 1 & e^{-j2\pi 6m/7} \\
1 & 1 & e^{-j2\pi 6n/7} & e^{-j2\pi 6m/7} \\
1 & e^{-j2\pi 6m/7} & e^{-j2\pi 6m/7} & e^{-j2\pi 6(m+n)/7}
\end{bmatrix}.
$$

At this decimation geometry, the inverse of $\hat{H}_{2,\overline{\mathcal{M}},\overline{\mathcal{D}}}$ can be found in close form:

$$
\left[\hat{H}_{2,\overline{\mathcal{M}},\overline{\mathcal{D}}}\right]^{-1} = \frac{1}{(e^{-j2\pi 6m/7}-1)(e^{-j2\pi 6n/7}-1)}
$$

$$
\times
\begin{bmatrix}
e^{-j2\pi 6(m+n)/7} & -e^{-j2\pi 6m/7} & -e^{-j2\pi 6m/7} & 1 \\
-e^{-2\pi 6n/7} & -e^{-j2\pi 6n/7} & 1 & -1 \\
e^{-j2\pi 6m/7} & 1 & e^{-j2\pi 6m/7} & -1 \\
1 & -1 & -1 & 1
\end{bmatrix}.
$$

We obtain the four interpolation functions:

$$
\begin{aligned}
f_{0,q}(t_1.t_2) &= -\mathrm{sinc}\left(\frac{1}{7}[t_1 - k_{q,1}]\right)\mathrm{sinc}\left(\frac{1}{7}[t_2 - k_{q,2}]\right) \\
&\times \frac{\sin\left(\frac{6\pi}{7}[k_{q,1}-m]\right)\sin\left(\frac{6\pi}{7}[k_{q,2}-n]\right)}{\sin\left(\frac{6\pi}{7}m\right)\sin\left(\frac{6\pi}{7}n\right)}.
\end{aligned}
$$

$$
\begin{aligned}
f_{m,q}(t_1.t_2) &= \mathrm{sinc}\left(\frac{1}{7}[t_1 - k_{q,1}]\right)\mathrm{sinc}\left(\frac{1}{7}[t_2 - (k_{q,2}-n)]\right) \\
&\times \frac{\sin\left(\frac{6\pi}{7}[k_{q,1}-m]\right)\sin\left(\frac{6\pi}{7}k_{q,2}\right)}{\sin\left(\frac{6\pi}{7}m\right)\sin\left(\frac{6\pi}{7}n\right)}.
\end{aligned}
$$

$$
\begin{aligned}
f_{7n,q}(t_1.t_2) &= \mathrm{sinc}\left(\frac{1}{7}[t_1 - (k_{q,1}-m)]\right)\mathrm{sinc}\left(\frac{1}{7}[t_2 - k_{q,2}]\right) \\
&\times \frac{\sin\left(\frac{6\pi}{7}k_{q,1}\right)\sin\left(\frac{6\pi}{7}[k_{q,2}-n]\right)}{\sin\left(\frac{6\pi}{7}m\right)\sin\left(\frac{6\pi}{7}n\right)}.
\end{aligned}
$$

$$f_{7n+m,q}(t_1, t_2) = -\mathrm{sinc}\left(\frac{1}{7}[t_1 - (k_{q,1} - m)]\right)$$

$$\times \ \mathrm{sinc}\left(\frac{1}{7}[t_2 - (k_{q,2} - n)]\right) \frac{\sin\left(\frac{6\pi}{7}k_{q,1}\right)\sin\left(\frac{6\pi}{7}k_{q,2}\right)}{\sin\left(\frac{6\pi}{7}m\right)\sin\left(\frac{6\pi}{7}n\right)}.$$

Note that the four interpolation functions are real. This is because the four subcells subsumed within gap regions are symmetrical about the $\vec{\nu}_1-$ and $\vec{\nu}_2-$ axes.

**Example 5:** *Twelfth–Order Sampling Decimation*

As shown in Fig. 3.12, if $k$ is increased to twelve, we obtain twelve subcells subsumed with gap regions, thus $q = 12$. The corresponding twelve position vectors for these twelve subcells are

$$
\begin{aligned}
\vec{k}_0 &= (0,0)^T, \\
\vec{k}_1 &= (1,0)^T, \\
\vec{k}_8 &= (8,0)^T, \\
\vec{k}_9 &= (9,0)^T, \\
\vec{k}_{10} &= (0,1)^T, \\
\vec{k}_{19} &= (9,1)^T, \\
\vec{k}_{80} &= (0,8)^T, \\
\vec{k}_{89} &= (9,8)^T, \\
\vec{k}_{90} &= (0,9)^T, \\
\vec{k}_{91} &= (1,9)^T, \\
\vec{k}_{98} &= (8,9)^T, \\
\vec{k}_{99} &= (9,9)^T.
\end{aligned}
$$

The offset vectors for the twelve manually chosen sample subgroups are as follows:

$$
\begin{aligned}
\vec{k}_0 &= (0,0)^T, \\
\vec{k}_4 &= (4,0)^T, \\
\vec{k}_8 &= (8,0)^T, \\
\vec{k}_{22} &= (2,2)^T, \\
\vec{k}_{26} &= (6,2)^T, \\
\vec{k}_{40} &= (0,4)^T, \\
\vec{k}_{44} &= (4,4)^T, \\
\vec{k}_{48} &= (8,4)^T,
\end{aligned}
$$

$$\vec{k}_{62} = (2,6)^T,$$
$$\vec{k}_{66} = (6,6)^T,$$
$$\vec{k}_{80} = (0,8)^T,$$
$$\vec{k}_{89} = (9,8)^T.$$

The inversion of the matrix $\hat{\underline{H}}_{2,\overline{\mathcal{M}},\overline{\mathcal{D}}}$ is carried out numerically. The numerical values of $\left[\hat{\underline{H}}_{2,\overline{\mathcal{M}},\overline{\mathcal{D}}}\right]^{-1}$ are available elsewhere [211].

By truncating the extent of the $L-q$ interpolation functions, the interpolation formula in (3.7.31) becomes a FIR (Finite Impulse Response) estimation of the sample $f(\underline{V}_g\vec{n} + \underline{V}\vec{k}_p)$:

$$f(\underline{V}_g\vec{n} + \underline{V}\vec{k}_p) = \sum_{q\in\overline{\mathcal{D}}}\sum_{\vec{m}\in\mathcal{W}_{p,n}} f(\underline{V}_g\vec{m} + \underline{V}\vec{k}_q)$$
$$\times f_{pq}\left(\underline{V}_g(\vec{n}-\vec{m}) - \underline{V}(\vec{k}_q - \vec{k}_p)\right) \quad (3.7.33)$$

where $\mathcal{W}_{p,n}$ is the extent of the FIR filter window centered at the position of the sample $f(\underline{V}_g\vec{n} + \underline{V}\vec{k}_p)$. In particular, $\mathcal{W}_{p,n} = \mathcal{W}_{0,0} \oplus (\underline{V}_g\vec{n} + \underline{V}\vec{k}_p)$ and $\mathcal{W}_{0,0}$ is the window centered at the origin:

$$\mathcal{W}_{0,0} = \left\{\vec{m} = (m_1, m_2, \ldots, m_N)^T \mid m_i = -M \text{ to } M, i = 1 \text{ to } N\right\}.$$

The output of the FIR filter at each deleted sample's location is the least square estimate of the decimated sample at that location [268]. An example of this FIR estimation is given in Chapter 4 in [211]. The quality of the recovery of the decimated samples depends on the condition of $\hat{\underline{H}}_{2,\overline{\mathcal{M}},\overline{\mathcal{D}}}$, which in turn depends on three factors: 1. the interdistance between the deleted samples [248], 2. the uniformity of the decimation geometry [596], and 3. the total number of decimated samples enclosed within the FIR's window. Ching [219] performed an extensive empirical examination on the effect of these three factors on the quality of restoring the decimated samples.

# 3.8   Computational Complexity of the Two Formulations

The bulk of computational overhead in the two formulations of sampling decimation lies in the inversion of matrices. As mentioned in Section 3.7, the inversion of both matrices can be carried out numerically. For the first formulation, the matrix requires inversion of $\underline{H}_{2,\mathcal{M},\mathcal{D}}$, whose dimension is

$L - q$. For the second formulation, the matrix is $\hat{H}_{2,\overline{\mathcal{M}},\overline{\mathcal{D}}}$, whose dimension is $q$.

If $q < L - q$, the second formulation will have less computational overhead and may be more preferable over the first formulation. For the fourth–order case, if the first formulation were used, we would have to invert a matrix of dimension forty–five, instead of four with the second formulation. On the other hand, for the $q = 12$ case, the dimension will increase to twelve and eighty–eight, respectively, for the second formulation and the first formulation. Clearly, the second formulation is more favored for both cases.

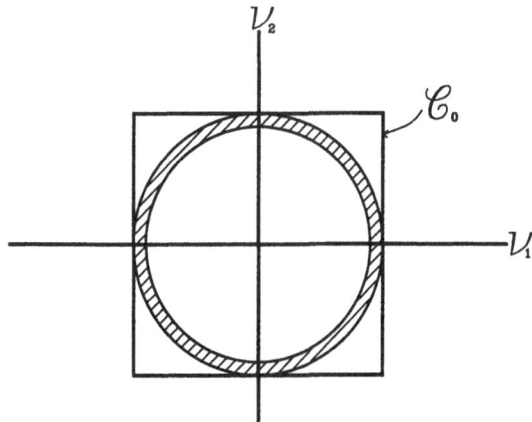

FIGURE 3.13: An example of a spectrum with sparse spectral support. The shape of the support is a donut shape.

For similar reasons, the first formulation is favored when $q > L - q$. This corresponds to most of $\mathcal{C}_0$ being occupied by $\mathcal{A}_D$. As shown in Fig. 3.13, a donut (annulus) shape spectrum is an example of such a case.

## 3.8.1 GRAM–SCHMIDT SEARCHING ALGORITHM

In exercising either approach to the decimation problem, we encounter a task of deciding which $q$ sample subgroups are to be removed. As already stated, if $\mathcal{M}$ is known, there exists at least one combination of $\mathcal{D}$ such that either the $\underline{H}_{2,\mathcal{M},\mathcal{D}}$ in the first formulation or $\underline{\hat{H}}_{2,\overline{\mathcal{M}},\overline{\mathcal{D}}}$ in the second formulation is invertible. If the inversion of both matrices is to be done numerically, the condition of the two matrices becomes important. We desire a combination wherein the condition of either of the two matrices is the best overall combination.

In a $q^{th}$-order decimation problem, the total number of combinations is equal to $L!/[(L-q)!q!]$. Normally, this number can be quite large even if $q$ is small. Consider the fourth-order decimation problem in Example 4: the total number of combinations is equal to 211,876. Clearly, an exhaustive search for the best combination is impractical. In this section, we will discuss a search algorithm which will locate the optimal combination. The algorithm is iterative. For a $q^{th}$-order decimation problem, the number of iterations is $q$.

Since the objective is the same regardless of which formulation is used, we will use the second formulation in our discussion. With reference to (3.7.25) and (3.7.26), $\underline{\hat{H}}_{2,\overline{\mathcal{M}},\overline{\mathcal{D}}}$ is formed by the $q$ columns of $\underline{\hat{H}}_{2,\overline{\mathcal{M}}}$. The desired $\underline{\hat{H}}_{2,\overline{\mathcal{M}},\overline{\mathcal{D}}}$ is an orthogonal matrix which has the best condition. One translation of this desired property is that the $q$ column vectors chosen out of $\underline{\hat{H}}_{2,\overline{\mathcal{M}}}$ be orthogonal or almost orthogonal to each other. Using this as the criterion, one can apply the Gram–Schmidt procedure to single out $q$ such column vectors. This procedure has been successfully employed in adaptive least mean square signal estimation [7] and neural network training [213].

The algorithm is as follows.

Let $\{\hat{h}_i | i = 1 \text{ to } L\}$ be the $L$ column vectors of $\underline{\hat{H}}_{2,\overline{\mathcal{M}}}$, and $\{\vec{z}_j | j = 1 \text{ to } q\}$ be the $q$ chosen vectors.

1. We arbitrarily pick $\hat{h}_k$ to be $\vec{z}_1$.

2. The $m, 1 \le m < q$ chosen vectors, $\{\vec{z}_j | j = 1 \text{ to } m\}$, span a linear subspace $\mathcal{L}_m$. The projection matrix of projecting onto $\mathcal{L}_m$ is equal to

$$\underline{P}_m = \underline{F}_m (\underline{F}_m^T \underline{F}_m)^{-1} \underline{F}_m^T. \tag{3.8.1}$$

where

$$\underline{F}_m = [\vec{z}_1 | \vec{z}_2 | \cdots | \vec{z}_m]. $$

3. The projection of every column vectors in $\underline{\hat{H}}_{2,\overline{\mathcal{M}}}$ onto $\mathcal{L}_m$ corresponds to a matrix product:

$$\underline{R}_m = \underline{P}_m \underline{\hat{H}}_{2,\overline{\mathcal{M}}}. \tag{3.8.2}$$

The $k^{th}$ column vector of $\underline{R}_m$, $\vec{r}_{m,k}$, is the projection of the $k^{th}$ column vector of $\underline{\hat{H}}_{2,\overline{\mathcal{M}}}$ onto $\mathcal{L}_m$. We then form the error vector:

$$\vec{e}_k = \vec{r}_{m,k} - \hat{h}_k, \qquad k = 1 \text{ to } L,$$

and the error norm is

$$n_k = \vec{e}_k^T \vec{e}_k.$$

The column vector, say $\hat{h}_n$ which yields the largest error norm is chosen as $\vec{z}_{m+1}$. The projection matrix is then updated via the Gram–Schmidt procedure:

$$\underline{P}_{m+1} = \underline{P}_m + \frac{\vec{e}_n \vec{e}_n^T}{\vec{e}_n^T \vec{e}_n}. \tag{3.8.3}$$

4. Go to step 3 and repeat until $m = q$.

For the fourth-order decimation problem in Example 4, the algorithm found that one of the best combinations is when $m = n = 4$, which corresponds to

$$\mathcal{D} = \{0, 4, 28, 32\}.$$

For the twelfth-order decimation problem in Example 5, the algorithm found the following twelve sample subgroups as one of the best combinations:

$$\mathcal{D} = \{0, 15, 23, 28, 41, 45, 58, 63, 70, 75, 93, 97\}.$$

Clearly, this set is very different from the one in Example 5, where the set is arbitrarily chosen.

## 3.9   Sampling at the Minimum Density

In this section, we consider expanding the decimation order to infinity. As $k \to \infty$, the area of each subcell becomes infinitesimally small. The ratio of $q$ to $L$ approaches asymptotically to the ratio of the area of $\mathcal{A}_D$ to the area of $\mathcal{C}_0$:

$$\lim_{k \to \infty} \frac{q}{L} = \frac{\text{area of } \mathcal{A}_D}{\text{area of } \mathcal{C}_0}. \tag{3.9.1}$$

Let $\tilde{D}$ be the sampling density after $q$ sample subgroups are removed,

$$\tilde{D} = D - q D_g. \tag{3.9.2}$$

Since $D =$ area of $\mathcal{C}_0$, and also, as $k \to \infty$, $q D_g \to$ area of $\mathcal{A}_D$, we have

$$\lim_{k \to \infty} \tilde{D} = \text{area of } \mathcal{C}_0 - \text{area of } \mathcal{A}_D$$

$$= \text{area of the support of } F(\vec{\nu}). \tag{3.9.3}$$

Thus, the minimum sampling density of a M–D band-limited function is equal to the area of the support of the spectrum of the function.

For 2–D unit–circularly (radius $= 1/2$) band-limited functions, the minimum density is $\pi/4 \approx 0.785$. This minimum density can be achieved arbitrarily close by expanding the decimation order. The following table lists the ratio of $1 - q/L$ and $r$, the oversampling index, as $k$ increases.

| $k$ | $L = k^2$ | $q$ | $1 - q/L$ | $r$ |
|-----|-----------|-----|-----------|-----|
| 7 | 49 | 4 | 0.918 | 1.170 |
| 10 | 100 | 12 | 0.880 | 1.120 |
| 16 | 225 | 32 | 0.858 | 1.092 |
| 20 | 400 | 56 | 0.860 | 1.095 |
| 30 | 900 | 144 | 0.840 | 1.070 |
| 50 | 2,500 | 460 | 0.816 | 1.039 |
| 100 | 10,000 | 1976 | 0.804 | 1.022 |
| $\infty$ | $\infty$ | $\infty$ | 0.785 | 1.000 |

Clearly, the sampling density approaches the minimum density as the decimation order increases.


## 3.10  Discussion

Note that the minimum density can be achieved arbitrarily closely by the sampling decimation technique regardless of any initial regular sampling geometry. For our running example, this technique is applicable to both the rectangular sampling geometry as well as the hexagonal sampling geometry, which yields the sampling density equal to that of Nyquist. For this case, Cheung [211, 215] demonstrated that, if $k$ is increased to eight, one sample subgroup can be removed. The sampling density is reduced by 1/64. After the decimation, the function is sampled at a density below that of Nyquist.

The decimation technique can also be applied to sample multiband functions [131] at the minimum rate or density. Like band-limited functions, the support of the spectra of multiband functions is finite but fragmented. Bandpass functions are examples of multiband functions. There are multiband functions of higher dimensions. In particular, the TV chrominance signal is a 3–D multiband function: 2–D spatial and 1–D temporal. Dubois [267] posed such a challenge to sampled TV chrominance signal at the minimum density. Dubois showed that the 3–D spectrum of the TV chrominance signal has a diamond-shaped band centered at the origin and eight smaller diamond-shaped bands symmetrically stationed at the eight corners of a cube. The decimation technique can be applied to close the gaps among the diamond-shaped bands and thus avoid the unnecessary waste of the band space.

Our decimation and the interpolation formulation is also a M–D extension of the digital signal processing methodology discussed by Crochiere and Rabiner [236, 235]. In this sense, Part II of this chapter has laid a foundation of extending the multirate digital signal processing to process functions of higher dimensions, such as images.

## 3.11 Conclusion

In the second half of this chapter, we have developed a sampling decimation technique based on the formulation of the M–D GSE. A band-limited function is first sampled at some sampling geometry $[\underline{V}, \vec{p}]$. If gaps exist among the spectral replications, then, by employing the decimation technique, samples can be deleted without loss of information. The sampling density is thus reduced. The function can be restored with the remaining samples. The decimation can be implemented at the sampling stage such that the function can be directly sampled at the reduced density. This reduction can even be applicable when sampling is performed at Nyquist densities. The minimum density is equal to the area of the support of the function's spectrum.

# 4

# Nonuniform Sampling

Farokh Marvasti

## 4.1 Preliminary Discussions

In this chapter we present an overview of the theory of nonuniform sampling. Examples of nonuniform sampling include the loss of one or more samples from a sampled signal, uniform sampling with jitter, and periodic nonuniform sampling. [1]

One might argue that nonuniform sampling is the natural way for the discrete representation of a continuous-time signal. For example, consider a non-stationary signal with high instantaneous frequency components in certain time intervals and low instantaneous frequency components in others. It is more efficient to sample the low frequency regions at a lower rate than the high frequency regions. This implies that with fewer samples per interval, one might approximate a signal with appropriate nonuniform samples. In general, fewer samples mean data compression: i.e., it signifies less memory and processing time for a computer and faster transmission time and/or lower bandwidth for digital transmission. This observation is an underlying reason for some vocoders in speech data compression application, [2] sample-data control, antenna design, [3] nonuniform tap delay lines and filter design from nonuniform samples in the frequency domain. Another potential application is in the area of error correction codes, where oversampling and discarding the erased or erroneous samples is a potential alternative to error correcting codes [632] . Nonuniform sampling recon-

---

[1] In each period there is a finite number of nonuniform samples which are repeated in other intervals. Therefore, periodic nonuniform samples can be regarded as a combination of a finite number of uniform samples.

[2] A voiced sound can be represented by three or four formants which are nonuniformly spaced in the frequency domain.

[3] In fact, the classical paper by Yen in 1956 [980] was motivated by antenna design considerations reported by the same author in another paper [981].

struction techniques are also important in the demodulation of frequency modulated [954], phase modulated, pulse position modulated [500], and delta modulated signals [320].

In addition, there are some cases where there is no choice but to process nonuniform data. Some of the examples are:

1. Data measured in a moving vehicle with fluctuations in speed for applications in seismology and oceanography, where random or uniform samples with jitter are inevitable.

2. Data tracked in digital flight control.

3. Data read from and recorded on a tape or disk with speed fluctuations.

4. Data loss due to channel erasures and additive noise.

In this chapter, we address a number of interesting aspects of nonuniform sampling including the conditions under which nonuniform samples represent a signal uniquely, evaluation of spectra from nonuniform samples, and interpolation techniques from nonuniform samples. For a more comprehensive treatment of nonuniform sampling, see the monograph written by the author [628].

## 4.2   General Nonuniform Sampling Theorems

Unlike uniform sampling, there is no guarantee of the uniqueness of a band-limited signal reconstructed from arbitrary nonuniform samples. This is true even when the average sampling rate is equal to the Nyquist rate. For example, suppose there is one solution to a set of nonuniform samples at instances $\{t_n\}$, and assume that it is possible to interpolate a band-limited function of the same bandwidth at the zero-crossings $\{t_n\}$. Now, if we add this interpolated function to the first solution, we get another band-limited function (of the same bandwidth) having the same nonuniform samples. Thus, nonuniform samples do not specify a unique band-limited signal. For a given bandwidth, this ambiguity is related to the set $\{t_n\}$. Therefore, we must choose the $\{t_n\}$ instances such that the existence of a unique solution is guaranteed. A set of these sampling instances that assures unique reconstruction is called *a sampling set*. We thus derive the following lemma [758]:

**Lemma 1** If the nonuniform sample locations $\{t_n\}$ satisfy the Nyquist rate on the average, they uniquely represent a band-limited signal if the sample locations are not the zero-crossings of a band-limited signal of the same bandwidth.

Various authors have proposed sufficient conditions for $\{t_n\}$ [541, 84]. Lemma 1 implies the following important corollary:

**Corollary 1** If the average sampling rate of a set of sample location $\{t_n\}$ is higher than the Nyquist rate, the samples uniquely specify the signal and $\{t_n\}$ is a sampling set.

**Proof** The average density of zero-crossings (real zeros) of a signal band-limited to $W$ is always less than or at most equal to the Nyquist rate $(2W)$ for deterministic [903] and random signals [764]. The sampling positions, therefore, cannot be the zero-crossings of a signal band-limited to $W$. From Lemma 1, we conclude that the samples are a sampling set. This has been proved by Beutler [85] in a different way for both deterministic and random signals.

Another interesting observation is that even if the average sampling rate is less than the Nyquist rate. if $\{t_n\}$ is a sampling set, the reconstruction is unique. [4] This fact has no parallel in the uniform sampling theory [85].

### 4.2.1 LAGRANGE INTERPOLATION

To find an interpolation function corresponding to nonuniform samples, we look at $\{e^{j\omega t_n}\}$ as a basis function in the frequency domain [85, 391]. The basis function $\{e^{j\omega t_n}\}$ is complete if any signal band-limited to $W$ can be represented in the frequency domain as

$$X(f) \;=\; \sum_{n=-\infty}^{\infty} c_n e^{j\omega t_n}, \qquad |f| \le W. \tag{4.2.1}$$

The inverse Fourier transform of (4.2.1) is

$$x(t) \;=\; \sum_{n=-\infty}^{\infty} c_n \operatorname{sinc}[2W(t - t_n)], \tag{4.2.2}$$

where $c_n$ is the inner product of $x(t)$ with another function $\Psi_n(t)$ – which is called the biorthogonal of $\operatorname{sinc}[2W(t - t_n)]$, i.e.. [391]

$$c_n \;=\; \int_{-\infty}^{\infty} x(t)\Psi_n(t)dt, \tag{4.2.3}$$

where

$$\int_{-\infty}^{\infty} \Psi_k(t) \operatorname{sinc}[2W(t - t_n)]dt \;=\; \begin{cases} 1, & k = n \\ 0, & k \ne 0. \end{cases}$$

---

[4]Notice that for the case of undersampling, no band-limited signal with a bandwidth smaller than or equal to W can be found where $\{t_n\}$ is a subset of its zeros.

The problem with (4.2.2) is that $c_n$ is not in sampling form, i.e., $c_n \neq x(t_n)$ except when $t_n = nT$. Equation (4.2.2) can be written in the sampling representation as

$$x(t) \quad = \quad \sum_{n=-\infty}^{\infty} x(t_n)\Psi_n(t). \tag{4.2.4}$$

The above equation can be proved by multiplying both sides by $\mathrm{sinc}[2W(t - t_k)]$ and then integrating both sides. By invoking the Parseval relationship and the definition of the bi-orthogonal function in (4.2.3), Eq. (4.2.4) is verified. The above equation cannot be of any practical use unless some criteria can be found on $\{t_n\}$ guaranteeing that $\{e^{j\omega t_n}\}$ is indeed a complete basis for the band-limited signal in the frequency domain. Our previous discussion (Lemma 1) that the sampling set $\{t_n\}$ cannot be the zero-crossings of another band-limited signal (with the proper bandwidth) also guarantees that $\{e^{j\omega t_n}\}$ is a complete basis function. Another hurdle is the explicit evaluation of $\Psi_n(t)$ from $\mathrm{sinc}[2W(t - t_n)]$. An explicit $\Psi_n(t)$ is not known for a general sampling set $t_n$. But if we limit $t_n$ around $nT$ (where $T = \frac{1}{2W}$ at the Nyquist rate) such that [465]

$$|t_n - nT| \leq \quad D \quad < \frac{T}{4}, \quad n = 0, \pm 1, \pm 2, \ldots . \tag{4.2.5}$$

then $\{e^{j\omega t_n}\}$ is a basis for band-limited signals in the frequency domain, [5] and $\Psi_n(t)$ can be shown to be the Lagrange interpolation function [698]

$$\Psi_n(t) = \frac{H(t)}{H'(t_n)(t - t_n)}, \tag{4.2.6}$$

$$\Psi_n(t_n) = 1, \quad \text{and} \quad \Psi_n(t_k) \neq 0 \text{ for } k \neq n,$$

where

$$H(t) \quad = \quad (t - t_0) \prod_{k=-\infty,\, k\neq 0}^{\infty} \left(1 - \frac{t}{t_k}\right). \tag{4.2.7}$$

Condition (4.2.5) guarantees the convergence of (4.2.6). [6] $H(t)$ cannot be a band-limited function; however, $\Psi_n(t)$ is a band-limited interpolation function as explained by Requicha [758]. The $\Psi_n(t)$ functions, unlike sinc functions, have their maximum at $t \neq t_n$. This might create a dynamic

---

[5]Equation (4.2.5) is a sufficient condition. There are other sufficient conditions discussed in [85] and [392].

[6]The Lagrange interpolation in this case is uniformly and pointwise convergent.

range problem when $t_n - t_{n-1}$ is large [480]. In general, $H(t)$ may not have a closed form; however, under certain conditions, $H(t)$ can be written explicitly in a closed form. For some examples, see Higgins [391]. If one replaces the sampling set $\{t_n\}$ by a finite set, say $\{t_n; 0 \le n = N\}$, one derives the classical Lagrange interpolation of polynomials of degree less than or equal to $N$. Hence, (4.2.4) and (4.2.6) can be considered as the generalized Lagrange interpolation.

If condition (4.2.5) is not satisfied but $t_n$ is higher than the Nyquist rate on the average and satisfies

$$|t_n - nT| < L < \infty, \quad |t_n - t_m| > \delta > 0, \quad n \ne m, \qquad (4.2.8)$$

then the Lagrange interpolation – (4.2.4) and (4.2.6)– is still valid but the product in (4.2.7) does not converge. However, the following product converges pointwise and uniformly:

$$H(t) \;=\; e^{at}(t - t_0)\prod_{k \ne 0}\left(1 - \frac{t}{t_k}\right)e^{t/t_k}. \qquad (4.2.9)$$

where

$$a \;=\; \sum_{k \ne 0}\frac{1}{t_k}.$$

The omission of a finite number of samples does not change the uniqueness. and (4.2.9) is still valid.

The above Lagrange interpolation provides a general interpolation for any sampling scheme. Indeed, many interpolating functions can be shown to be special cases of (4.2.4), (4.2.6) and (4.2.9). The following examples illustrate this point.

• Uniform Sampling Interpolation

The well-known Shannon sampling theorem can be derived from the Lagrange interpolation by taking the sampling set $\{t_n\}$ to be $\{nT\}$. In this case, the product in (4.2.7) converges to $\frac{T}{\pi}\sin(\frac{\pi t}{T})$. Equation (4.2.6) becomes [590]

$$\Psi_n(t) \;=\; \frac{(-1)^n \sin(\frac{\pi t}{T})}{\frac{\pi}{T}(t - nT)} = \operatorname{sinc}\left(\frac{t}{T} - n\right).$$

• Kramer's Generalized Sampling Theorem

Let $I = [a, b]$ be some finite closed interval and $L_2$ the class of square integrable functions over $I$. Suppose that for a real $t$, we have [503]

$$x(t) \;=\; \int_I K(s, t)g(s)ds. \qquad (4.2.10)$$

where $g(s)$ and $K(s,t) \in L_2(I)$. If the sampling set $\{t_n\}$ is such that $K\{s,t_n\}$ forms a complete orthogonal set of functions in $L_2(I)$, we have

$$x(t) \;=\; \sum_{n=-\infty}^{\infty} x(t_n)S_n(t). \qquad (4.2.11)$$

where

$$S_n(t) \;=\; \frac{\int_I K(s,t)K^*(s,t_n)ds}{\int_I |K(s,t_n)|^2 ds}, \qquad (4.2.12)$$

where $K^*(s,t)$ is the complex conjugate of $K(s,t)$. A proof of the above interpolation is to write $g(s)$ in terms of the orthogonal functions $K^*(s,t_n)$, i.e.,

$$g(s) \;=\; \sum_{n=-\infty}^{\infty} c_n K^*(s,t_n),$$

where

$$c_n \;=\; \frac{\int_I g(s)K(s,t_n)ds}{\int_I |K(s,t_n)|^2 ds} = \frac{x(t_n)}{\int_I |K(s,t_n)|^2 ds}.$$

By multiplying the above equation by $K(s,t)$ and formally integrating term by term, we obtain Eqs. (4.2.11) and (4.2.12) [440].

Zayad, Ilinsen and Butzer have recently shown that Kramer's interpolation can be represented by Lagrange interpolation when the kernel $K(s,t)$ arises from a Sturm-Liouville boundary-value problem [999].

• Bessel Interpolation

Candidates for orthogonal basis functions in Kramer sampling interpolation are Bessel functions. If the function $x(t)$ is the Hankel (Bessel) transform

$$x(t) \;=\; \int_I sJ_m(st)X(s)ds.$$

where $I$ implies that $x(t)$ is $J_m$–Bessel band-limited to the interval $I = [0, W]$, the interpolation function is

$$x(t) \;=\; \sum_{n=-\infty, n\neq 0} x(t_n)\frac{J_m(t)}{(t_n - t)J_{m+1}(t_n)}, \qquad (4.2.13)$$

where $J_m(t)$ is the $m^{th}$-order Bessel function of the first kind and $\{t_n\}$ are the zero-crossings of $J_m(t)$. For extensions to sampling at $t_n$ with derivatives, see Jerri [440].

To derive the above equation from Lagrange interpolation, we know that the zero-crossings of $J_m(t)$ satisfy the sufficiency condition [940] given in (4.2.5). Therefore, $H(t)$ in (4.2.7) can be modified as

$$H(t) = t^m \prod_{k=-\infty,k\neq0}^{\infty} \left(1 - \frac{t}{t_k}\right),$$

where $t^m$ is used because the Bessel function $J_m(t)$ has an $m^{th}$-order zero at $t = 0$. From Watson [940],

$$H(t) = J_m(t)2^m\Gamma(m+1). \qquad (4.2.14)$$

where $\Gamma$ is the Gamma function. Using $\dot{J}_m(t_n) = -J_{m+1}(t_n)$, one can derive (4.2.13) from (4.2.6) and (4.2.14).

Rawn [754] has shown that a $J_{2k}$ – Bessel band-limited signal can be expanded similarly to the Lagrange interpolation given in (4.2.4) and (4.2.6) provided that $|t_n - (n-1/4)| < 1/4$ and $t_n = -t_{-n}, n = 1, 2, \ldots$. This verifies the statement by Zayed [999] that the Kramer sampling interpolation can be represented by Lagrange interpolation under certain conditions.

• Migration of a Finite Number of Uniform Points

Suppose in a uniform sampling scheme (sampled at the Nyquist interval $T$), $N$ uniform samples are migrated to new positions $(t_1, \cdots, t_n)$ [628]. Yen has given an explicit reconstruction (interpolation) formula which can be derived from the Lagrange interpolation, i.e., $H(t)$ from (4.2.7) becomes [980]

$$H(t) = \prod_{k=1}^{N} \left(1 - \frac{t}{t_k}\right) \prod_{k<0,\,k\geq N} \left(1 - \frac{t}{kT}\right) = \frac{\prod_{k=1}^{N}\left(1 - \frac{t}{t_k}\right)\sin(\frac{2\pi t}{T})}{t\prod_{k=1}^{N-1}\left(1 - \frac{t}{kT}\right)}.$$

The following relations are derived if the above equation is substituted in (4.2.6), viz.,

$$x(t) = \sum_{m=-\infty}^{\infty} x(t_m)\psi_m(t), \qquad (4.2.15)$$

where $t_m$ is equal to the nonuniform instances when $1 \leq m \leq N$ and is equal to $mT$ otherwise; $\psi_m(t)$ is the interpolating function and is given by

$$\psi_m(t) = \begin{cases} \dfrac{\prod_{q=1}^{n}(t-t_q)\prod_{i=1}^{N}mT-iT}{\prod_{i=1}^{N}(t-iT)\prod_{q=1}^{N}(mT-t_q)} \cdot \text{sinc}(\frac{t}{T}-m) & \begin{array}{l} m \leq 0 \\ m > N \end{array} \\[4mm] \dfrac{\prod_{q\neq m}(t-t_q)\prod_{i=1}^{N}(mT-iT)}{\prod_{i=1}^{N}(t-iT)\prod_{q\neq m}(m_m-t_q)} \cdot \dfrac{\sin(2\pi\frac{t}{T})}{\sin(2\pi\frac{t_m}{T})} & 1 \leq m \leq N. \end{cases}$$

$$(4.2.16)$$

Yen proved the above interpolation from the uniform sampling theorem [980].

When all the uniform samples $(x(t_m); \ m \le 0, m > N)$ are zero, the interpolation function (4.2.16) represents a band-limited function interpolated over $N$ nonuniform samples in an interval of $NT$ having uniform zeros outside the interval, Eq. (4.2.15) becomes

$$x(t) \ = \ \sum_{m=1}^{N} x(t_m)\psi_m(t). \tag{4.2.17}$$

The interpolation over $N$ nonuniform samples is not unique. Yen used a "minimum energy" signal criterion for unique interpolation. An interpolating function on $N$ nonuniform samples with minimum energy $(\int_{-\infty}^{\infty} x^2(t)dt)$ is

$$\psi_m(t) = \sum_{q=1}^{N} \alpha_{qm} \operatorname{sinc}[2W(t - t_q)]. \tag{4.2.18}$$

where $\alpha_{qm}$'s are the elements of the inverse of a matrix whose elements are

$$\operatorname{sinc}[2W(t_m - \tau_q)]. \quad m, q = 1, 2, \ldots, N.$$

Chen and Allebach [207] showed that this interpolation is a minimum mean-squared estimate. Applying the projection theorem, Yeh and Stark [979] and Calvagio and Munson [186] have shown the optimality of Yen's interpolation in the sense of Mean Square Error (MSE).

• Sampling with a Single Gap in an Otherwise Uniform Distribution

Let us take the special case as shown in Fig. 4.1, i.e.,

$$t_n = \begin{cases} +nT. & n \le 0 \\ \Delta + nT. & n > 0 . \end{cases}$$

If $\Delta < T$, then $x(t)$ is uniquely specified. The interpolation function is

$$x(t) = \sum_{n=-\infty}^{\infty} x(t_n)\psi_n(t).$$

where

$$\psi_n(t) \ = \ \frac{(-1)^n \Gamma(2W\Delta + n)}{\Gamma(2Wt)\Gamma[2W(\Delta - t)]n!} \times \begin{cases} (n + 2Wt)^{-1}. & n \le 0 \\ (n + 2W(\Delta - t))^{-1}. & n > 0. \end{cases} \tag{4.2.19}$$

where $\Gamma$ denotes the Gamma function.

Yen [980] actually proved the above theorem using uniform sampling theory and solving a set of infinite linear equations. However, Eq. (4.2.19) can

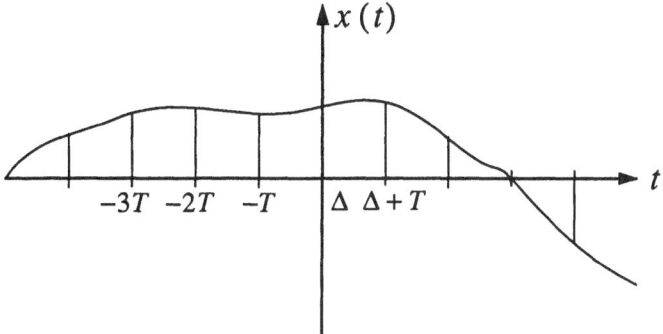

FIGURE 4.1: Shifted positive uniform samples.

be also derived from Lagrange interpolation (4.2.9) although the derivation is not as obvious.

• Periodic Nonuniform Sampling

For a given set of $N$ delays, $\tau_k$, let the sampling times be as illustrated in Fig. 4.2, i.e.,

$$t_{nk} = nNT + \tau_k, \qquad \begin{aligned} k &= 1, 2, \ldots, N. \\ n &= 0, \pm 1, \pm 2, \ldots \\ T &= \tfrac{1}{2W}. \end{aligned} \qquad (4.2.20)$$

The interpolation formula is

$$x(t) = \sum_{n=-\infty}^{\infty} \sum_{k=1}^{N} f(t_{nk}) v_{nk}(t), \qquad (4.2.21)$$

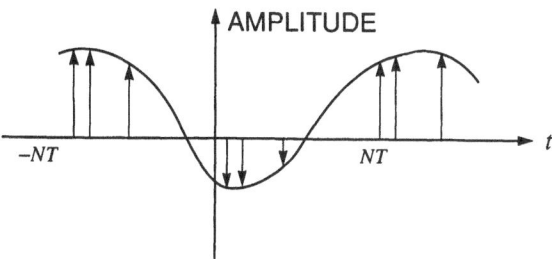

FIGURE 4.2: Periodic nonuniform sampling.

where

$$\psi_{nk} = \frac{N \prod_{i=1}^{N} \sin \frac{2\pi W}{N}(t - t_{ni})}{2\pi W(t - t_{nk}) \prod_{i=1, i \neq k}^{N} \sin \frac{2\pi W}{N}(\tau_k - \tau_i)}. \qquad (4.2.22)$$

Unlike the sinc function. $\psi_{nk}$ does not take its maximum at the sampling times; but rather it attains its maximum between nonuniform samples.

The above equation can be derived from the Lagrange interpolation, i.e., we can write $H(t)$ in (4.2.7) as (see Fig. 4.2)

$$H(t) = \prod_{k=-\infty}^{\infty} \left(1 - \frac{t}{t_k}\right), \qquad t_0 \neq 0. \qquad (4.2.23)$$

or

$$H(t) = \prod_{|k|=0,N,2N,\cdots} \left(1 - \frac{t}{t_k}\right) \prod_{|k|=1,1+N,1+2N,\ldots} \left(1 - \frac{t}{t_k}\right) \prod \cdots \qquad (4.2.24)$$

Each product in (4.2.24) converges to

$$K_k \sin \frac{2\pi W}{N}(t - \tau_k), \qquad (4.2.25)$$

where $K_k$ is a scale factor equal to $K_k = -[\sin(\frac{2\pi W}{N} t_k)]^{-1}$. which can be determined from $H(0) = 1$. Hence. Eq. (4.2.24) can be written as

$$H(t) = K \prod_{k=0}^{N-1} \sin \frac{2\pi W}{N}(t - \tau_k). \qquad (4.2.26)$$

where $K = \prod_{k=0}^{N-1} K_k$. From (4.2.26) we can find $\dot{H}(t_n)$, i.e.,

$$\dot{H}(t_n) = \frac{K 2\pi W}{N} \prod_{k \neq n} \sin 2\pi W(t_n - \tau_k). \qquad (4.2.27)$$

Equation (4.2.6) then becomes

$$\Psi_n(t) = \frac{N \prod_{k=0}^{N-1} \sin \frac{2\pi W}{N}(t - \tau_k)}{2\pi W(t - t_n) \prod_{k \neq n} \sin 2\pi W(t_n - \tau_k)}. \qquad (4.2.28)$$

A comparison of (4.2.28) with (4.2.22) shows that these two interpolating functions are equivalent.

Papoulis [707] has derived (4.2.22) from the generalized sampling function (see Marks [590] for an extensive discussion on Papoulis' generalized sampling). The periodic nonuniform samples are regarded as $N$ different uniform samples of $x(t)$ interlaced in time (bunched samples). For the special case when $N = 2$ and $\tau_1 = -\tau_2 = \tau$, the following relationship is derived:

$$x(t) = \frac{\cos 2\pi W\tau - \cos 2\pi Wt}{2\pi W \sin 2\pi W\tau}$$
$$\times \sum_{n=-\infty}^{\infty} \left[ \frac{x(2nT + \tau)}{t - 2nT - \tau} - \frac{x(2nT - \tau)}{t - 2nT + \tau} \right]. \qquad (4.2.29)$$

Using Papoulis' powerful method, one can generalize (4.2.21) and (4.2.22) to a combination of periodic nonuniform samples of a signal at instances $t_k$ and samples of $N - 1$ derivatives at the same instances taken at $\frac{1}{N}$ times the Nyquist rate. Another application of Papoulis' generalized sampling theorem is a nonuniform scheme interlaced among the signal samples and a combination of $n^{th}$-order derivative samples. Cheung et al. [215, 590] have shown that such kinds of sampling under the umbrella of Papoulis generalized sampling theorem might be severely sensitive to noise (i.e., ill-posed) at the Nyquist rate. They have found out that, in certain cases, by oversampling, the sensitivity to noise goes away.

### 4.2.2  INTERPOLATION FROM NONUNIFORM SAMPLES OF A SIGNAL AND ITS DERIVATIVES

The interpolation of a band-limited signal from the samples of the signal and its derivatives was known back in 1960s [545]. Recently, Rawn [755] has shown an interpolation based on Lagrange interpolators: the theorem is as follows.

Let $x(t)$ be band-limited with bandwidth $W$ and let the samples of $x(t)$ and its $R - 1$ derivatives be known at $t_n$. Let $t_n$ be such that $|t_n - nRT| < \frac{T}{4R}$. Then

$$x(t) = \sum_{n=-\infty}^{\infty} [x(t_n) + \left(1 - \frac{t}{t_n}\right) x_1(t_n)$$
$$+ \ldots + \left(1 - \frac{t}{t_n}\right)^{R-1} x_{R-1}(t_n)][\Psi_n(t)]^R, \qquad (4.2.30)$$

where $\Psi_n(t)$ is the Lagrange interpolator as defined in (4.2.6) and

$$x_k(t_n) = \frac{(-1)^k t_n^k}{k!} \frac{d^k}{dt^k} \left[ \frac{x(t)}{[\Psi_n(t)]^R} \right]\Bigg|_{t=t_n}, \quad 0 \le k \le R - 1.$$

Convergence of the series in (4.2.30) is uniform on $(-\infty, \infty)$. For the case $t_n = nTR$, (4.2.30) reduces to the uniform derivative sampling interpolation given by Linden and Abramson [545] .

## 4.2.3   Nonuniform Sampling for Nonband-Limited Signals

We can show that certain classes of non-band-limited signals can be represented by a set of uniform samples violating the Nyquist rate (see, for example, Marvasti and Jain [629]). By the same token, there is a class of non-band-limited signals that can be represented uniquely by a set of nonuniform samples. We will discuss some special cases first and subsequently give a general theorem. Assume that a band-limited signal goes through a monotonic non-linear distortion, $y(t) = f[x(t)]$. Although $y(t)$ is a non-band-limited signal, it can be represented by nonuniform samples if $t_n$ is a sampling set for the band-limited signal $x(t)$. The reconstruction is shown in Fig. 4.3. Necessary and sufficient conditions for such non-band-limited signals have been established [629]. Another example is a set of

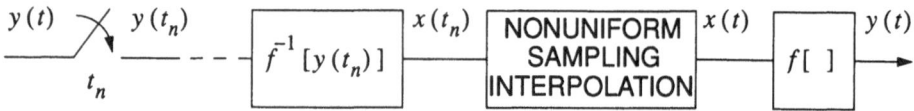

FIGURE 4.3: Reconstruction of a non-band-limited signal from nonuniform samples.

non-band-limited signals generated by a time varying filter when the input is a band-limited signal. If the system has an inverse, then the samples of the output(the non-band-limited signal) are sufficient to reconstruct the signal. This non-band-limited signal is essentially a time-warped version of the band-limited signal. A non-linear time varying example is a frequency modulated signal that can be reconstructed from a set of samples that satisfy the Nyquist rate for the modulating band-limited signal. e.g., the zero-crossings of the FM signal [628].

Uniqueness of a non-band-limited finite energy signal (or in general an entire function) by a set of nonuniform samples has been discussed [758]. Basically, if the sampling set $t_n$ cannot be the zero-crossings of a non-band-limited signal of finite energy (or an entire function of the same class), then the sampling set $t_n$ uniquely represents the non-band-limited signal. This uniqueness theorem, however, does not give a constructive method of interpolation of the non-band-limited signal from its nonuniform samples.

An exact interpolation for a certain class of non-band-limited signals is possible and is the topic of the next section.

## 4.2.4   JITTERED SAMPLING

By jittered samples, we mean nonuniform samples that are clustered around uniform samples either deterministically or randomly with a given probability distribution. This random jitter is due to uncertainty of sampling at the transmitter end. We will not consider the jitter of uniform samples at the receiver end due to channel delay distortion.

For deterministic jitter, Papoulis has proposed an interesting method for the recovery [701]. The problem is recovery of $x(t)$ from the jittered samples, $x(nT - \mu_n)$, where $\mu_n$ is a known deviation from $nT$. The main idea is to transform $x(t)$ into another function $g(\tau)$ such that the nonuniform samples at $t_n = nT - \mu_n$ are mapped into uniform samples $\tau = nT$. Consequently, $g(\tau)$ can be reconstructed from $g(nT)$ if $g(\tau)$ is band-limited ($W \le \frac{1}{2T}$). Now. $x(t)$ can be found from $g(\tau)$ if the transformation is one-to-one (Fig. 4.4). Let us take the one-to-one transformation as $t = \tau - \theta(\tau)$.

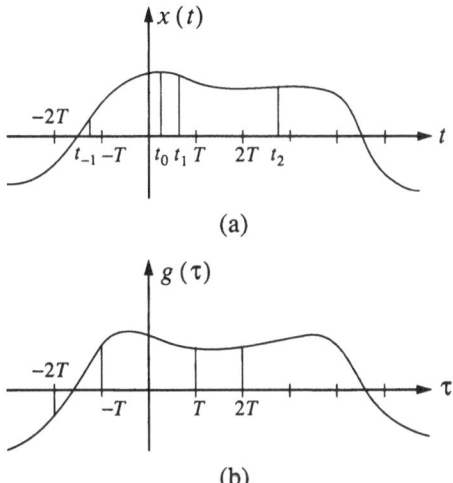

(a)

(b)

FIGURE 4.4: The deterministic jittered samples.

where $\theta(\tau)$ is a band-limited function defined as

$$\theta(\tau) = \sum_{n=-\infty}^{\infty} \mu_n \operatorname{sinc}[2\pi W_{2i}(\tau - nT)],$$

where $W_2 \le \frac{1}{2T}$ is the bandwidth of $\theta(\tau)$. Since we assume the transformation is one-to-one. the inverse exists and is defined by $\tau = \gamma(t)$. Note that $\theta(nT) = \mu_n$ and $t_n = nT - \theta(nT)$. Let us assume

$$
\begin{aligned}
g(\tau) \quad &= \quad x[\tau - \theta(\tau)] \\
&\longrightarrow \quad g(nT) = x[nT - \theta(nT)] = x(t_n).
\end{aligned}
\qquad (4.2.31)
$$

However, $g(\tau)$ is not band-limited in general. But if $\mu_n$ and, as a result, $\theta_n(\tau)$ are assumed to be small. $g(\tau)$ is approximately a band-limited function [701] . Using the uniform sampling interpolation for $g(\tau)$, we get

$$g(\tau) = \sum_{-\infty}^{\infty} g(nT) \operatorname{sinc}\left[\frac{\pi}{T}(\tau - nT)\right].$$

Using the substitution $\tau = \gamma(t)$, we derive

$$x(t) = g[\gamma(t)] \approx \sum_{n=-\infty}^{\infty} x(t_n) \operatorname{sinc}\left[\frac{\pi}{T}(\gamma(t) - nT)\right]. \qquad (4.2.32)$$

Comparing Eq. (4.2.32) to the Lagrange interpolation (4.2.4), we conclude that $\Psi_n(t) \approx \operatorname{sinc}[\frac{\pi}{T}(\gamma(t) - nT)]$. The Lagrange interpolation and the sinc function are both equal to 1 when $t = t_n$. and they are both equal to 0 when $t = t_k$. $k \neq n$.

Clark et al. [224] have suggested that the Papoulis transformation for jitter (4.2.31) and (4.2.32) can also be extended to a certain class of non-band-limited signals. If $x(t)$ is band-limited, $g(\tau)$ – in Eq. (4.2.31) – cannot be band-limited in general and, therefore, (4.2.32) is only an approximation. But if $g(\tau)$ is band-limited, then $x(t)$ cannot be band-limited; hence (4.2.32) is an exact representation for this class of non-band-limited signals, i.e.,

$$x(t) = \sum_{n=-\infty}^{\infty} x(t_n) \operatorname{sinc}\left[\frac{\pi}{T}(\gamma(t) - nT)\right]. \qquad (4.2.33)$$

## 4.2.5  PAST SAMPLING

From Lemma 1. we conclude that when the average sampling rate is higher than that of Nyquist, the nonuniform samples represent uniquely the band-limited signal. As a special case, these nonuniform samples could be the past samples. The past samples at a rate higher than the Nyquist is a sampling set because no signal of bandwidth $W$ can be found that has zero-crossings of density greater than Nyquist in an infinite interval. In fact, these past samples could be uniform. That is, past uniform samples of a band-limited signal that are slightly higher than the Nyquist rate uniquely represent the signal [627, 628]. The topic of prediction by samples from the past is discussed in depth in the chapter by Butzer and Stens in this volume.

An infinite number of samples in a finite interval also form a sampling set for a band-limited signal. For example, the nonuniform samples at $t_n = \frac{1}{n}$. $n = 1, 2, \ldots$ form a sampling set but have no practical value since they have the same problem as the extrapolation of a signal from the knowledge of a portion of the signal in an interval. We show in the next section that these kinds of sampling sets do not provide for stable restoration.

## 4.2.6 STABILITY OF NONUNIFORM SAMPLING INTERPOLATION

By stability, we mean that a slight perturbation of nonuniform sample amplitudes due to noise leads to a bounded interpolation error [516]. The necessary and sufficient condition for nonuniform samples of a finite energy signal to be stable is [976].

$$E_x = \int |x(t)|^2 dt \le C \sum_{n=-\infty}^{\infty} |x(t_n)|^2, \qquad (4.2.34)$$

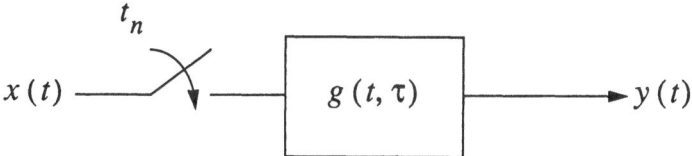

FIGURE 4.5: System model for an interpolator.

where $x(t)$ is any signal band-limited to $W$, and $E$ and $C$ are, respectively, the energy of $x(t)$ and a finite positive constant. In order to see why Eq. (4.2.34) conforms to our definition of stability, let us take the error in the samples as $e(t_n)$. The energy of the error signal, $e(t)$, is small if the errors in the samples are small. This is because $e(t)$ is a band-limited function derived from a linear interpolation such as Lagrange (4.2.4) (see Fig. 4.5). Therefore, $e(t)$ satisfies (4.2.34), viz.,

$$E_e = \int |e(t)|^2 dt \le C \sum_{n=-\infty}^{\infty} |e(t_n)|^2.$$

Equation (4.2.34) may be derived from inequalities discussed by Duffin and Schaeffer [270]. Specifically, we can show that for a condition similar to (4.2.8), i.e.,

$$\begin{cases} |t_n - nT| \le L < \infty, & n = 0, \pm 1, \pm 2, \dots, \\ |t_n - t_m| > \delta > 0. & n \ne m , \end{cases} \qquad (4.2.35)$$

the following inequalities are valid:

$$A \le \frac{\sum_{n=-\infty}^{\infty} |x(t_n)|^2}{\int_{-\infty}^{\infty} |x(t)|^2 dt} \le B. \qquad (4.2.36)$$

where $A$ and $B$ are positive constants which depend exclusively on $t_n$ and the bandwidth of $x(t)$. Equation (4.2.36) implies that under the condition (4.2.35), $\sum_{n=-\infty}^{\infty} |x(t_n)|^2$ cannot be zero or infinite. The exact value can be calculated from Parseval relationship as will be shown later when we get to spectral analysis of nonuniform samples.

Yao and Thomas [976] have shown that whenever **Lagrange** interpolation is possible, the nonuniform samples are stable; but for cases such as past sampling, or infinite number of nonuniform samples in a finite interval, the sampling set $\{t_n\}$ is not stable.

## 4.2.7  INTERPOLATION VIEWED AS A TIME VARYING SYSTEM

It is instructive to compare interpolation for uniform and nonuniform samples. Consider the system in Fig. 4.5, where $g(t, \tau)$ is a time varying interpolating function. If the sampling process is uniform and $g(t, \tau)$ is a sinc function representing an ideal low-pass filter, the output is equal to the input and the overall system is a linear time-invariant (LTI) system. If the sampling process is nonuniform and the same low-pass filter is used for $g(t, \tau)$, the system is not LTI anymore but rather is a linear time varying system. The reason is that if the input is delayed, the output is not equal to its delayed version. In general, the output is given by

$$y(t) \;=\; \int x_s(\tau)g(t,\tau)d\tau = \sum_{n=-\infty}^{\infty} x(t_n)g(t,t_n), \qquad (4.2.37)$$

where $x_s(\tau)$ is the sum of nonuniform impulsive samples. The overall system (sampler and the time varying interpolator in Fig. 4.5) is modeled as

$$y(t) \;=\; \int_{-\infty}^{\infty} h(t,\tau)x(\tau)d\tau. \qquad (4.2.38)$$

If the nonuniform samples satisfy the conditions in (4.2.5) or (4.2.8), we can use Lagrange interpolation. Comparing Lagrange interpolation (4.2.4) to (4.2.37), we conclude that

$$g(t, t_n) = \Psi_n(t), \qquad (4.2.39)$$

where $\Psi_n$ is defined as in (4.2.6) or (4.2.9). In this case, the output is equal to the input ($y(t) = x(t)$), and the system $h(t, \tau)$, as a whole, is LTI – although $g(t, \tau) = \Psi(t, \tau)$ describes a time varying system. In order to find a relationship between $g(t, \tau)$ and $h(t, \tau)$, we proceed as follows: From (4.2.37), we have

$$y(t) \;=\; \int_{-\infty}^{\infty} \sum_{n=-\infty}^{\infty} \delta(\tau - t_n)g(t, t_n)x(\tau)d\tau. \qquad (4.2.40)$$

Comparing (4.2.40) to (4.2.38), we conclude that

$$h(t, \tau) \;=\; \sum_{n=-\infty}^{\infty} g(t, t_n)\delta(\tau - t_n). \qquad (4.2.41)$$

The bi-frequency transfer function of the time varying system, $h(t, \tau)$, is defined as a bivariate Fourier transform

$$H(f, \lambda) = \int_{-\infty}^{\infty} \int_{-\infty}^{\infty} h(t, \tau) e^{-j2\pi ft} e^{j2\pi\lambda\tau} dt \, d\tau. \qquad (4.2.42)$$

From (4.2.41), the bi-frequency transfer function for the interpolator becomes

$$H(f, \lambda) = \int \int \sum_{n} g(t, t_n) \delta(\tau - t_n) e^{-j2\pi ft} e^{j2\pi\lambda\tau} dt \, d\tau. \qquad (4.2.43)$$

The above equation can be simplified as

$$H(f, \lambda) = \sum_{n=-\infty}^{\infty} \Phi_n(f) \exp(j2\pi\lambda t_n), \qquad (4.2.44)$$

where $\Phi_n(f)$ is the Fourier transform of $g(t, t_n)$. Equations (4.2.41) and (4.2.44) are valid for any interpolating function including Lagrange interpolation. However, since the Lagrange interpolation is exact, i.e., $y(t) = x(t)$ in the system shown in Fig. 4.5, we expect to get an impulsive bi-frequency transfer function $(H(f, \lambda) = \delta(f - \lambda))$ because the output is given by

$$Y(f) = \int X(\lambda) H(f, \lambda) d\lambda. \qquad (4.2.45)$$

In order to see this, from (4.2.39), we write (4.2.43) as

$$H(f, \lambda) = \int_{-\infty}^{\infty} e^{-j2\pi ft} \sum_{n} \Psi_n(t) e^{j2\pi\lambda t_n} dt. \qquad (4.2.46)$$

Since $\Psi_n$ is an interpolating function for any signal band-limited to $W$, we have

$$\sum_{n} \Psi_n(t) \exp(j2\pi\lambda t_n) = \exp(j2\pi\lambda t). \quad \lambda < W. \qquad (4.2.47)$$

From Eq. (4.2.46) and (4.2.47), we conclude that

$$H(f, \lambda) = \delta(f - \lambda). \qquad (4.2.48)$$

As an example, consider an ideal low-pass filter as an interpolator for a set of nonuniform samples. Then, $g(t, t_n)$ in (4.2.41) becomes

$$g(t, t_n) = \text{sinc}[2W(t - t_n)]. \qquad (4.2.49)$$

The bi-frequency transfer function can be derived from (4.2.43). The result is

$$H(f, \lambda) = \sum_{n} \exp[j2\pi t_n(\lambda - f)]. \quad |f|, \ |\lambda| < W. \qquad (4.2.50)$$

If we substitute (4.2.50) in (4.2.45), we obtain the following expected result

$$Y(f) = \sum_n x(t_n) \exp(-j2\pi t_n f), \quad |f| < W. \qquad (4.2.51)$$

If the set of samples are uniform, (4.2.50) reduces to (4.2.48) for $|\lambda| < W$, *i.e.*, the output becomes equal to the input.

Calvagio and Munson [186] have plotted the bi-frequency transfer function of different interpolating functions for a finite set of nonuniform samples. The plots give a better insight on properties of interpolating functions on a comparative basis.

## 4.2.8   RANDOM SAMPLING

All the theorems we have thus far presented can also be applied to random sampling of random signals. This has been shown by a number of researchers [85, 90. 89].

Beutler and Leneman [90] have proven that random samples with Poisson distribution. uniform samples with jitter. and uniform samples with a few skips are all valid sampling sets. We shall explain these processes in the next section on spectral analysis of random samples.

Removal of a finite number of samples $[t_n, n \neq 1, 2, \dots N]$ does not affect the uniqueness property of the set when the sampling rate is higher than the Nyquist rate. This observation also agrees with Lemma 1 and Corollary 1 since removal of a finite number of samples does not affect the average rate and hence the set $[t_n, n \neq 1. 2, \dots, N]$ is still a sampling set.

# 4.3   Spectral Analysis of Nonuniform Samples and Signal Recovery

The spectrum of a periodic impulse train is well-known and is given below:

$$\sum_{n=-\infty}^{\infty} \delta(t - nT) \leftrightarrow \sum_{n=-\infty}^{\infty} e^{j\omega nT} = \frac{1}{T} \sum_{n=-\infty}^{\infty} \delta\left(f - \frac{n}{T}\right),$$

where $f = \frac{\omega}{2\pi}$. The above equation implies that the Fourier transform of a periodic impulse train is another periodic impulse train in the frequency domain. However. the spectrum of a set of nonuniform impulses, in general, is not another set of (uniform or nonuniform) impulses [7], i.e.,

$$\sum_{n=-\infty}^{\infty} \delta(t - t_n) \leftrightarrow \sum_{n=-\infty}^{\infty} e^{j\omega t_n} = G(f).$$

---

[7]The exception is periodic nonuniform samples. The spectrum consists of uniform but complex weighted impulses.

The value of $G(f)$ is not known in general. It may resemble white noise or it may not be convergent at all. The above relationship implies that the spectrum of nonuniform samples might be totally smeared (aliased) in the frequency domain. This implies that we cannot use any linear time-invariant filter for signal recovery. Although the spectrum of the form $\sum_{n=-\infty}^{\infty} x(t_n)e^{j\omega t_n}$ seems to be totally smeared in the frequency domain, under some conditions we can determine the approximate shape of the spectrum. These conditions are related to the observation that jittered uniform samples, or nonuniform samples at a much higher rate than that of the Nyquist, can be low-pass filtered and the original signal can be recovered with negligible error. Before we look into the shape of the spectra of the samples, we develop a general theoretical background for future use.

**Lemma 2** The spectrum of a finite number of nonuniform samples is bounded and continuous and belongs to the class of uniformly almost periodic (a.p.) functions [8].

**Proof** The spectrum of a finite number of nonuniform samples can be represented by

$$\sum_{k=1}^{N} x(t_k)e^{j\omega t_k}.$$

Since the above equation is a finite trigonometric polynomial, it is an almost periodic function [98].

**Lemma 3** The spectrum of a set of infinite number of nonuniform samples of an $L^2$ band-limited signal that satisfy (4.2.8) and/or condition (4.2.5) is a bounded almost periodic function.

**Proof** If condition (4.2.8) is satisfied, Lagrange interpolation is possible.

---

[8]A function $f(t)$ is defined uniformly a.p. when, for every $\epsilon > 0$ , there exists a relatively dense set of translation numbers $\tau$ of $f(x)$ corresponding to $\epsilon$ [98]. A translation number $\tau$ of $f(x)$ for each $\epsilon$ is defined as a real number such that $|f(t + \tau) - f(t)| \leq \epsilon$ for $-\infty < t < \infty$. The set $E$ of all real numbers $\tau$ that satisfy the above condition is called relatively dense if there are no arbitrary large gaps among $\tau$'s or, to be exact, if some length $L$ exists such that every interval $\alpha < t < \alpha + L$ of this length contains at least one number $\tau$ of the set $E$.

The fundamental theorem of a.p. functions state that the Fourier series of a.p. functions is given by

$$f(t) \quad = \quad \sum_{n=-\infty}^{\infty} a_n e^{j\omega_n t}, \qquad (4.3.1)$$

where $a_n = M[f(t)e^{j\omega_n t}] = \lim_{T\to\infty} \frac{1}{T}\int_0^T f(t)e^{-j\omega_n t}dt$ , $M[\ ]$ is the mean operator and $\omega_n$'s are discrete real numbers.

Therefore, the following relationship holds:

$$x_s(t)\Psi_n(t) = x(t_n)\delta(t - t_n), \qquad (4.3.2)$$

where $x_s(t)$ is the set of impulsive nonuniform samples. The Fourier transform of (4.3.2) is

$$X_s(f) * \Phi_n(f) = x(t_n)e^{-j2\pi f t_n}, \qquad (4.3.3)$$

where $X_s(f)$ and $\Phi_n(f)$ are the corresponding Fourier transforms of $x_s(t)$ and $\Psi_n(t)$. From (4.3.3), we deduce that $X_s(f)$ is either a distribution function or a bounded function since $\Phi_n(f)$ is a band-limited function and is zero when $|f| > W$. However, the set of samples of an $L^2$ band-limited signal that satisfies (4.2.8) is a square summable sequence as shown in (4.2.36). This fact implies that $X_s(f)$ belongs to the space of signals of finite energy over any finite interval [286].

**Theorem 1 (Reisz-Fischer)**  To any series $\sum_n x(t_n)e^{j2\pi f t_n}$ for which $\sum_n |x(t_n)|^2$ converges, there corresponds an almost periodic function which converges in the mean ($B^2$ a.p.).

**Proof**  See the reference book by Besicovitch [83].

Now, we can discuss the extension of the Parseval theorem to nonuniform samples.

## 4.3.1   EXTENSION OF THE PARSEVAL RELATIONSHIP TO NONUNIFORM SAMPLES

The Poisson Sum Formula for uniform samples is given by

$$\sum_{n=-\infty}^{\infty} |x(nT)|^2 = \frac{1}{T}\sum_{i=-\infty}^{\infty} \left|X(\frac{i}{T})\right|^2,$$

where $\{x(nT)\}$ are samples of a real signal and $\{X(\frac{i}{T})\}$ are samples of the Fourier transform $X(f)$. The above equation reduces to the following when $\frac{1}{T} \geq 2W$, where $W$ is the bandwidth of $x(t)$,

$$\sum_{n=-\infty}^{\infty} |x(nT)|^2 = \frac{1}{T}\int_{-\infty}^{\infty} |x(t)|^2 dt = \frac{1}{T}\int_{-W}^{W} |X(f)|^2 df.$$

The above equation is the same as the Parseval relation for discrete signals when $e^{j\omega}$ is used instead of $(f)$. For the extension of the Poisson Sum Formula to nonuniform samples, see Marvasti [607].

This reference discusses the nonuniform sampling set that follows the zero-crossings of an FM signal or the nonuniform positions of a pulse position modulated signal.

It is our objective to show the Parseval relationship for a general stable sampling set $t_n$ that satisfies condition (4.2.8) and hence (4.2.5) . We would like to show that

$$\sum_{n=-\infty}^{\infty} |x(t_n)|^2 = \frac{1}{T} \int_{-\infty}^{\infty} x(t)x_{lp}(t)dt \qquad (4.3.4)$$

$$= \frac{1}{T} \int_{-\infty}^{\infty} X(f)X_{lp}^*(f)df,$$

where $x_{lp}(t)$ is the low-pass filtered version of the nonuniform samples, $T = \frac{1}{2W}$, and $X_{lp}(f)$ is the corresponding Fourier transform.

**Proof** $x_{lp}(t)$ is obtained by the following interpolation:

$$x_{lp}(t) = \sum_{n} x(t_n) \operatorname{sinc}[2W(t - t_n)]. \qquad (4.3.5)$$

The Fourier transform of (4.3.5) is

$$X_{lp}(f) = TX_s(f) \prod \left( \frac{f}{2W} \right), \qquad (4.3.6)$$

where $X_s(f)$ is the Fourier transform of the ideal nonuniform samples (the Fourier transform exists because of Lemma 3) and $\prod(\frac{f}{2W})$ is an ideal low-pass filter with a bandwidth of $W$. Substituting (4.3.6) into (4.3.4), we get

$$\frac{1}{T} \int_{-\infty}^{\infty} X(f)X_{lp}^*(f)df = \int_{-\infty}^{\infty} X(f)X_s^*(f) \prod \left( \frac{f}{2W} \right) df$$

$$= \int_{-\infty}^{\infty} X(f)X_s^*(f)df. \qquad (4.3.7)$$

After invoking the Parseval theorem for $L^2$ signals, Eq. (4.3.7) becomes

$$\int_{-\infty}^{\infty} X(f)X_s^*(f)df = \int_{-\infty}^{\infty} x(t) \sum_{n=-\infty}^{\infty} x(t_n)\delta(t - t_n)dt$$

$$= \int_{-\infty}^{\infty} \sum_{n=-\infty}^{n=\infty} |x(t_n)|^2 \delta(t - t_n)dt$$

$$= \sum_{n=-\infty}^{n=\infty} |x(t_n)|^2. \qquad (4.3.8)$$

and we have proved Eq. (4.3.4). The Parseval relationship for uniform samples is also a special case of Eq. (4.3.7) and (4.3.8) since $X_{lp}(f) =$

$\prod(\frac{f}{2W})X_s(f) = X(f)$, for $\frac{1}{T} > 2W$ and

$$\frac{1}{T}\int_{-\infty}^{\infty} X(f)X_{lp}(f)df \quad = \quad \frac{1}{T}\int_{-\infty}^{\infty} X(f)X^*(f)df$$

$$= \quad \frac{1}{T}\int_{-\infty}^{\infty} |X(f)|^2 df. \qquad (4.3.9)$$

That is, for any set of uniform or nonuniform samples which satisfy condition (4.2.8) and/or (4.2.5), we have

$$\sum_{n=-\infty}^{\infty} |x(t_n)|^2 \quad = \quad \int_{-W}^{W} X(f)X_s(f)df$$

$$= \quad \frac{1}{T}\int_{-W}^{W} X(f)X_{lp}^*(f)df$$

$$= \quad \frac{1}{T}\int_{-\infty}^{\infty} x(t)x_{lp}(t)dt.$$

## 4.3.2  ESTIMATING THE SPECTRUM OF NONUNIFORM SAMPLES

The Fourier series expansion of nonuniform samples can be derived as follows:

$$x_s(t) = x(t)x_p(t). \qquad (4.3.10)$$

where

$$x_p(t) = \sum_k \delta(t - t_k). \qquad (4.3.11)$$

From the theory of generalized functions, we can write

$$\delta(t - t_k) = |\dot{g}(t)|\delta[g(t)]. \qquad (4.3.12)$$

provided that $g(t_k) = 0$, $\dot{g}(t_k) \neq 0$ and $g(t)$ has no other zeros than $t_k$. One possible $g(t)$ can be written as

$$g(t) = t - kT - \theta(t). \qquad (4.3.13)$$

where $\theta(t)$ is any function such that

$$g(t_k) = t_k - kT - \theta(t_k) = 0,$$

i.e., $\theta(t_k) = t_k - kT$ is the deviation of the samples from the uniform positions. From (4.3.12) and (4.3.13), we get

$$\delta(t - t_k) = |1 - \dot{\theta}(t)|\delta[t - kT - \theta(t)]. \qquad (4.3.14)$$

provided that $1 \neq \dot{\theta}(t)$ (see Marvasti and Lee [633] for the sufficient condition that assures $\dot{\theta}(t)$ is not equal to 1). Substituting (4.3.14) in (4.3.11), we get

$$x_p(t) = |1 - \dot{\theta}(t)| \sum_{k=-\infty}^{\infty} \delta[\Phi - kT], \qquad (4.3.15)$$

where $\Phi = t - \theta(t)$. The Fourier expansion of (4.3.15) in terms of $\Phi$ is

$$
\begin{aligned}
x_p(t) &= |1 - \dot{\theta}(t)| \frac{1}{T} \sum_{k=-\infty}^{\infty} e^{jk\frac{2\pi}{T}\Phi} \\
&= \frac{|1 - \dot{\theta}(t)|}{T} \left[ 1 + 2 \sum_{k=1}^{\infty} \cos\left( \frac{2\pi kt}{T} - \frac{2\pi k\theta(t)}{T} \right) \right]. \quad (4.3.16)
\end{aligned}
$$

Equation (4.3.16) reveals that $x_p(t)$ has a DC component $|\frac{1-\dot{\theta}}{T}|$ plus har-

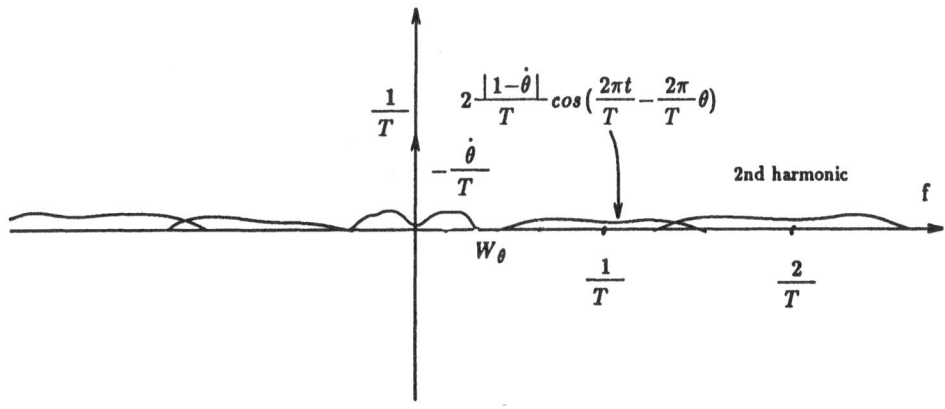

FIGURE 4.6: Spectrum of nonuniform positions with unit area.

monics that resemble phase modulated (PM) signals. The index of modulation is $\frac{2\pi k}{T}$. The bandwidth is proportional to the index of modulation ($\frac{2\pi k}{T}$), the bandwidth of $\theta(t)$ ($\leq \frac{1}{2T}$) and the maximum amplitude of $\theta(t)$, which in our case is related to $\theta(t_k) = t_k - kT$. The spectrum of $x_p(t)$ is sketched in Fig. 4.6 for the case $1 > \dot{\theta}$. The spectrum of nonuniform samples can be determined from (4.3.10) and (4.3.16) i.e..

$$
\begin{aligned}
x_s(t) &= x(t)x_p(t) \qquad\qquad\qquad\qquad\qquad\qquad\qquad (4.3.17) \\
&= x(t)\frac{|1 - \dot{\theta}(t)|}{T} \left[ 1 + 2 \sum_{k=1}^{\infty} \cos\left( \frac{2\pi k}{T} - \frac{2\pi k}{T}\theta(t) \right) \right].
\end{aligned}
$$

The spectrum of (4.3.17) is sketched in Fig. 4.7 for the case $1 > \dot{\theta}$. Now if we

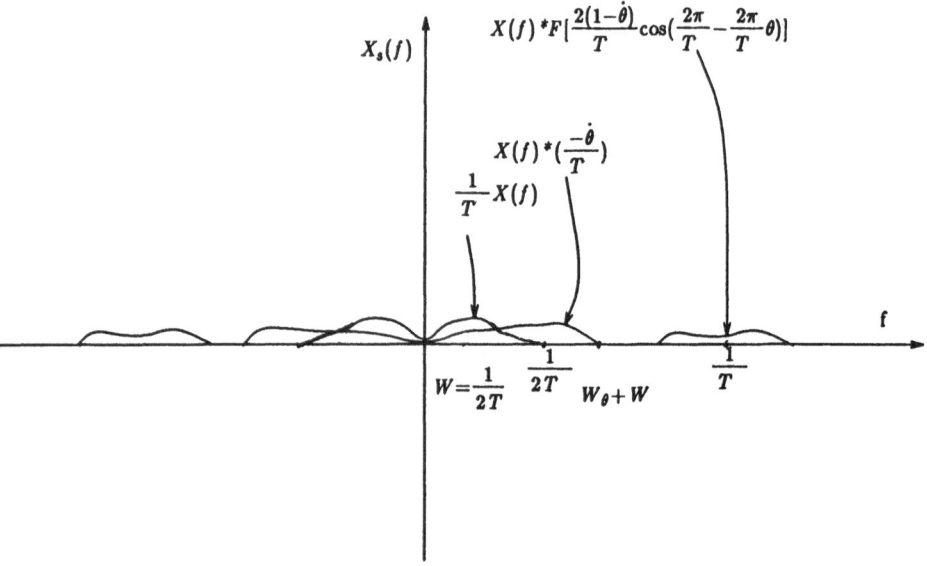

FIGURE 4.7: Spectrum of nonuniform samples $x_s(t)$.

assume the bandwidth of $\theta(t)$ is less than $\frac{1}{2T}$, the phase modulated signal at the carrier frequency $\frac{1}{T}$ is a narrow band PM and has a bandwidth of approximately twice the bandwidth of $\theta(t)$. $i.e., \leq \frac{1}{T}$. Therefore, low-pass filtering $x_p(t)$ (see Fig. 4.6 ), yields

$$x_{p_{lp}}(t) = \frac{1 - \dot{\theta}}{T}, \tag{4.3.18}$$

where we have assumed that $1 > \dot{\theta}$.

If the bandwidth of $\theta(t)$ is taken to be $W_\theta$, as long as $\frac{1}{T} - W_\theta - W > W + W_\theta$, there is no overlap between the narrow band PM signal and $X(f)^* (\frac{-\dot{\theta}}{T})$ (see Fig. 4.7). Thus, low-pass filtering the nonuniform samples, $x_s(t)$ (with a bandwidth of $W + W_\theta$), gives

$$x_{s_{lp}}(t) = x(t)\frac{(1 - \dot{\theta})}{T}. \tag{4.3.19}$$

The above spectral analysis reveals a reconstruction method that will be discussed in a separate section.

### 4.3.3  SPECTRAL ANALYSIS OF RANDOM SAMPLING

In the case of random sampling, similar results can be derived [610, 611]. We can show that the power spectrum of the random samples consists of the power spectrum of the original signal plus additive uncorrelated noise. It is instructive – for the purpose of spectral analysis and signal recovery – to model a general random sampling process as depicted in Fig. 4.8. This figure consists of the input signal $x(t)$ (deterministic or a wide sense

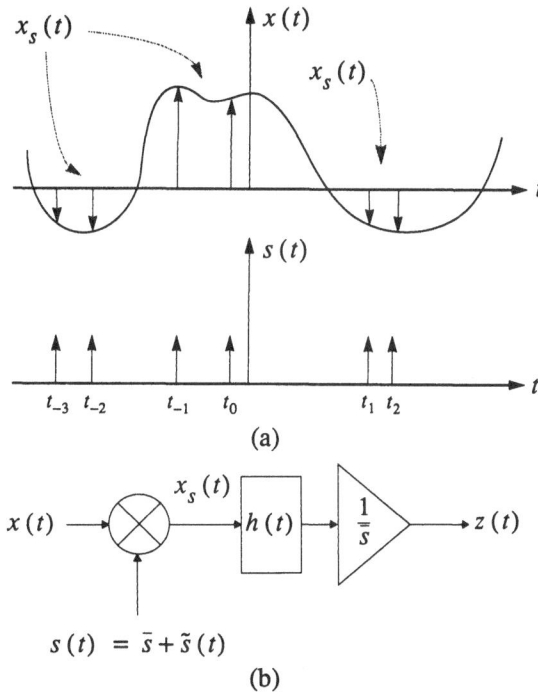

FIGURE 4.8: (a) A general random sampling process ($s(t) = \hat{s} + \tilde{s}x_s(t)$ is the random sampling process). (b) The model of (a).

stationary continuous-time process band-limited to $W$) and the random point process [9] (a non-zero mean wide sense stationary process) with $\tilde{s}(t)$ equal to the AC component and $\bar{s}$ equal to the DC component. In Fig. 4.8, $x_s(t)$ is the random sampled process. The impulse response $h(t)$ in this figure is an ideal low-pass filter with a gain of 1 and a bandwidth of $W$. This filter recovers the signal from the random samples $x_s(t)$. The gain after the low-pass filter is a scale factor. Now, $x_s(t)$ can be written as

$$x_s(t) = \bar{s}x(t) + \tilde{s}(t)x(t) = \bar{s}x(t) + n(t). \tag{4.3.20}$$

The above equation implies that the set of random samples is proportional to the original signal plus "additive noise." It can be shown that if the point process has a Poisson or uniform probability distribution, the noise

---

[9]By a point process we mean unit pulses or impulses at random instances $t_k$.

term $n(t)$ is uncorrelated to the signal $x(t)$. Therefore, the power spectrum of (4.3.20) is

$$
\begin{aligned}
P_{x_s}(f) &= \bar{s}^2 P_x(f) + P_n(f) \\
&= \bar{s}^2 P_x(f) + P_{\tilde{s}}(f) * P_x(f),
\end{aligned} \tag{4.3.21}
$$

where $P(f)$ is the power spectrum [10] and $*$ is the convolution operator.

Equation (4.3.21) is illustrated in Fig. 4.9. The shape of $P_n(f)$ depends

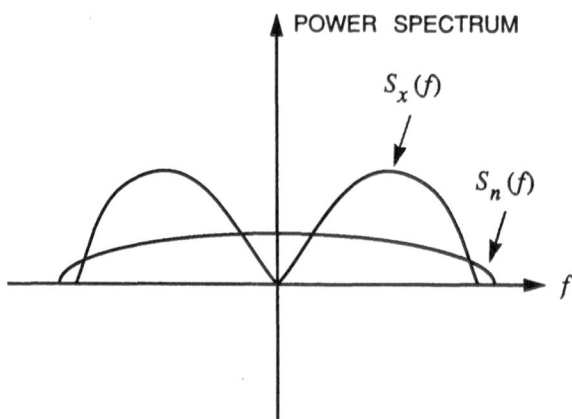

FIGURE 4.9: The power spectrum of random samples.

on the type of pulse used to represent the point process $s(t)$. For the case of Poisson or uniformly distributed impulsive sampling, the power spectrum of the point process, $s(t)$, is

$$
P_s(f) = \lambda + \lambda^2 \delta(f). \tag{4.3.22}
$$

where $\bar{s} = \lambda$ is the average number of impulses per unit time. Equation (4.3.21) becomes

$$
P_{x_s}(f) = \lambda^2 P_x(f) + \lambda R_x(0). \tag{4.3.23}
$$

where $R_x(0)$ is the total signal power. The above equation shows that $P_n(f)$ is the spectrum of white noise for an impulsive Poisson or uniformly distributed random samples. Likewise, we can derive the same relationship as (4.3.21) for rectangular pulses, random samples of variable pulse widths, uniform sampling with some missing samples, and finally, uniform samples with jitter [610, 611]. We have shown [611] that the mean square error of the recovered signal decreases with increasing $\lambda$. The results are summarized in Table 4.I. The reader should note that we have assumed that $s(t)$ in

---

[10]If $x(t)$ is deterministic, then $P_x(f) = |X(f)|^2$.

| Type of sampling | $P_{x_s}(f)$ | Comments |
|---|---|---|
| Impulsive | $p_x(f) + \dfrac{1}{\lambda}r_x(0)$ | see 1 |
| Rectangular | $P_x(f) + \dfrac{1}{\lambda}\text{sinc}^2 f\tau * P_x(f)$ | see 1 and 2 |
| Variable pulse width | $P_x(f) + \left[\dfrac{2\lambda(1-p)}{p\lambda^2 + p(2\pi f)^2}\right] * P_x(f)$ | see 3 |
| Uniform sampling : with skips | $P_x(f) + \displaystyle\sum_{k\neq 0} P_x(f-\dfrac{k}{T}) + \dfrac{T(1-p)}{p}R_x(0)$ | see 3 |
| Uniform sampling : with jitter | $P_x(f) + \displaystyle\sum_{k\neq 0} P_x(f-\dfrac{k}{T})G(\dfrac{k\tau}{T}) + H(f)$ | see 4 |

TABLE 4.I: Power spectrum of random samples of $x(t)$.

Comments :

1. Poisson or uniform distribution: $\lambda$ is the average number of pulse/unit time, $x$ is a band-limited signal, $P(f)$ is the power spectrum and $R(0)$ is the total signal power.

2. $\tau$ is the pulse width.

3. $p$ is the probability of having a pulse.

4. $G(f)$ is the characteristic function (Fourier transform of the probability density) of jitter. $H(f)$ is equal to $T[G(f\tau) - |G(f\tau)|^2 * P_y(f)]$.

the model of Fig. 4.8 has a DC component. Therefore, any point process such as alternately reversed pulses would lose the signal component $P_x(f)$.

In summary, we have shown that nonuniform or random samples of a signal consist of the signal component and a noise term. Thus in the frequency domain, we generally expect the original signal to be corrupted by noise. This observation leads to practical reconstruction methods to be discussed in the next section.

## 4.4 Discussion on Reconstruction Methods

There are several methods for signal recovery. We shall only discuss the most promising ones, namely, the non-linear technique and the iterative technique discussed in [628] .

### 4.4.1 SIGNAL RECOVERY THROUGH NON-LINEAR METHODS

Equation (4.3.19) reveals that, by low-pass filtering the nonuniformly spaced samples, we get the original signal plus an additive distortion. Comparing (4.3.19) to (4.3.18), we conclude that simple division yields a good estimate of the signal. The reconstruction procedure is depicted in Fig. 4.10. As a special case of nonuniform sampling, we have considered a set of uniform samples where some of the samples are lost. Simulating this special set of nonuniform samples for speech signals, we have obtained impressive results [624]. The signal to noise ratio for a recovered speech signal from lost samples is shown in Fig. 4.11.

The dotted line is the recovered signal using low-pass filtering as an interpolator; the solid line is the improved signal to noise (S/N) ratio using the non-linear technique illustrated in Fig 4.10.

### 4.4.2 ITERATIVE METHODS FOR SIGNAL RECOVERY

An iterative procedure has been proposed by Wiley [956] which can recover a signal from a set of nonuniform pulses without any distortion after infinite iterations. This method is an extension of demodulation of wideband FM signals [954]. Each iteration improves the signal to distortion ratio. The iteration is based on the theorems of Duffin and Schaeffer [270] and Sandberg [789]. The extension of this method to ideal [11] nonuniform sampling [622, 623], and random sampling of a random process [611] has been done by the author. Here, we try to present a unified treatment for both nonuniform

---

[11]Other forms of sampling such as sample-and-hold, flat top sampling, and natural sampling are also considered.

$$x_s(t) = \sum_{k=[i]} x(kT)\,\delta(t-kT)$$

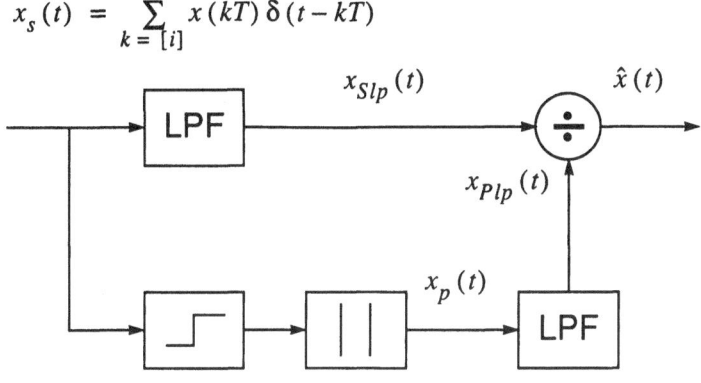

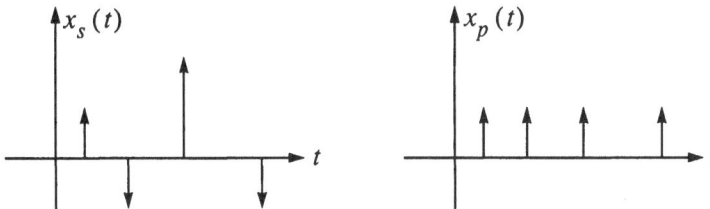

FIGURE 4.10: A non-linear method for signal recovery from nonuniform samples.

and random samples using the model illustrated in Fig. 4.8. The result
of low-pass filtering the nonuniform (or random) samples (Eqs. (4.3.19)
or (4.3.20)) is [12]

$$z(t) = x(t) + e(t).$$

where $e(t)$ is the error term. Now, if the error norm is less than the signal norm, [13] the error norm can be reduced with additional iterations
[610, 611]. The iterations are depicted in Fig. 4.12. For the case of power
deterministic signals and random signals, the signal to noise ratio after $i$
iterations is

$$\frac{S_i}{N_i} = i \cdot \frac{S_1}{N_1} \text{ dB.} \qquad (4.4.1)$$

Equation (4.4.1) shows that improvement is possible if $\frac{S_1}{N_1} > 0\,\text{dB}$. For
instance, for Poisson random sampling, at the output of the low-pass filter,

---

[12]After proper scaling by $T$ for the deterministic case and $1/\bar{s}$ for the random
case.
[13]For deterministic energy signals, $E_e = ||e||^2 = \int_{-\infty}^{\infty} e^2(t)dt$ and for random
power signals $S_e = ||e||^2 = E[e^2(t)]$.

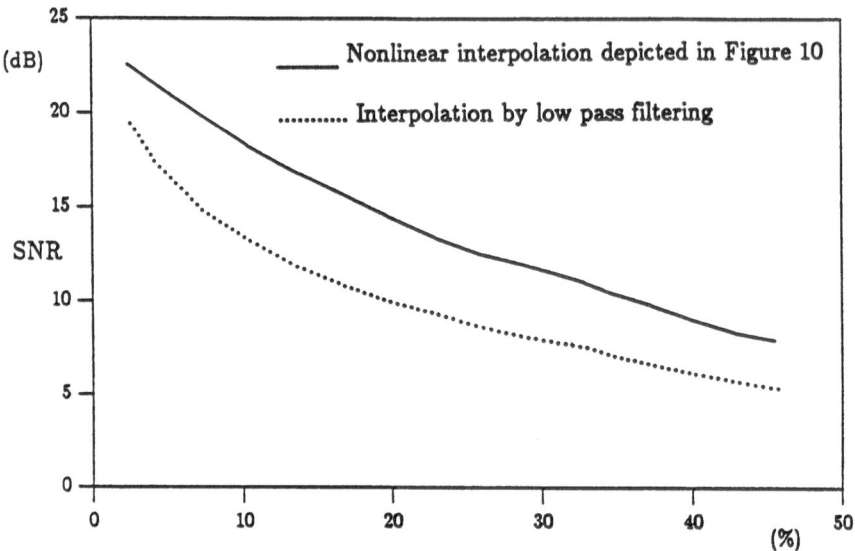

FIGURE 4.11: S/N ratio versus the percentage of lost samples.

shown in Fig. 4.8, we have (from (4.3.23))

$$\frac{S}{N} = \frac{\lambda}{2W}. \tag{4.4.2}$$

We deduce from (4.4.2) that the average sampling rate should be greater than the Nyquist rate. The above analysis gives a unified treatment to both deterministic nonuniform and random sampling. It also accommodates impulsive and other types of sampling ($s(t) = \tilde{s}(t) + \bar{s}$ in the sampling model of Fig. 4.8 is valid for any pulse shape sampling).

For deterministic nonuniform sampling, the iterative method converges if set $t_n$ is relatively separable, i.e.,

$$(t_{n+1} - t_n) \geq d > 0$$

and

$$|(t_n - nT)| < L.$$

where $T$ is the average Nyquist interval and $d$ and $L$ are real positive numbers. The above conditions are sufficient for $\{t_n\}$ to satisfy (4.2.36) and be a stable sampling set (Eq. 4.2.34). This stability satisfies Sandberg's inequalities, i.e., for any pair of band-limited square-integrable signals $x(t)$ and $y(t)$ [789], we have

$$\int_{-\infty}^{\infty} (Sx - Sy)(x - y)dt \geq k_1 \int_{-\infty}^{\infty} (x - y)^2 dt \tag{4.4.3}$$

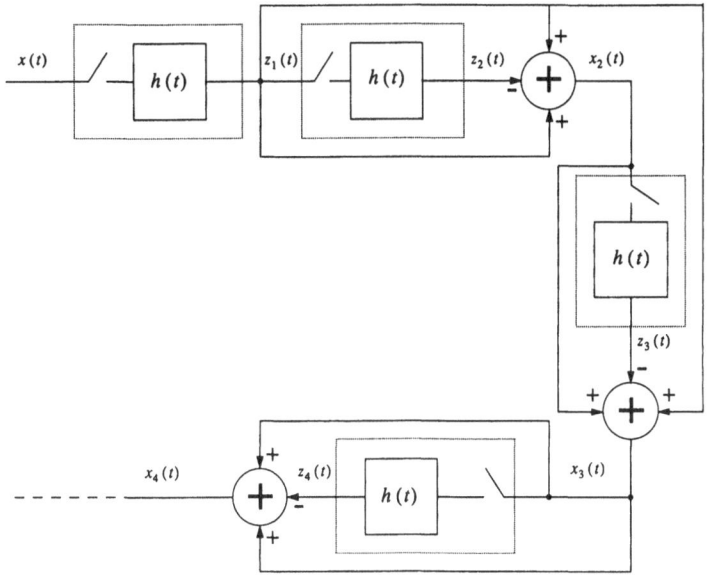

FIGURE 4.12: The iterative method.

and

$$\int_{-\infty}^{\infty} (PSx - PSy)^2 \le k_2 \int_{-\infty}^{\infty} (x - y)^2 dt. \qquad (4.4.4)$$

where $P$ is the band-limiting operator and $S$ is the sampling process. If (4.4.3) and (4.4.4) are satisfied, according to Sandberg's theorem, we have

$$x(t) = \lim_{n \to \infty} x_n(t),$$

where

$$x_{n+1}(t) = \frac{k_1}{k_2}(PSx - PSx_n) + x_n, \qquad (4.4.5)$$

and

$$x_0 = 0.$$

For simulation results, the reader is referred elsewhere [622, 623]. Some of the results are shown in Figs. 4.13-4.16 and Tables 4.II-4.IV.

At rates lower than the Nyquist rate, the iterative method diverges.

The simulation is based on a low-pass signal band-limited to $W = 100$ Hz. The original signal and the ideal nonuniform samples, at the Nyquist rate, are shown in Fig. 4.13. The nonuniform samples are initially taken at the

| Mean Square Error | | | | |
|---|---|---|---|---|
| Iteration | Variable S&H | Constant S&H | Ideal | Natural |
| 0 | 1.864412e-1 | 4.365404e-1 | 1.580624e-1 | 5.428200e-1 |
| 1 | 2.684877e-2 | 9.017097e-2 | 2.535609e-3 | 1.208514e-1 |
| 2 | 4.488294e-3 | 2.132896e-2 | 1.746579e-3 | 2.655716e-2 |
| 3 | 1.975845e-3 | 6.128620e-3 | 1.607121e-3 | 6.746200e-3 |
| 4 | 2.217043e-3 | 2.903171e-3 | 1.575803e-3 | 3.343687e-3 |
| 5 | 2.644699e-3 | 2.641249e-3 | 1.522073e-3 | 3.220765e-3 |
| 6 | 2.940791e-3 | 3.070048e-3 | 1.475997e-3 | 3.551968e-3 |
| 7 | 3.122244e-3 | 3.567942e-3 | 1.439057e-3 | 3.817421e-3 |
| 8 | 3.233513e-3 | 3.977919e-3 | 1.411968e-3 | 3.975198e-3 |
| 9 | 3.305488e-3 | 4.280155e-3 | 1.392150e-3 | 4.059322e-3 |
| 10 | 3.356229e-3 | 4.491736e-3 | 1.378551e-3 | 4.101327e-3 |

TABLE 4.II: Mean square error at the Nyquist rate; the nonuniform samples are to within $\left(\frac{T}{4}\right)$ intervals from uniform positions.

| Mean Square Error | | | |
|---|---|---|---|
| Iteration No. | Nyquist rate | 2x Nyquist | 3x Nyquist |
| 0 | 1.580624e-1 | 1.905335e-1 | 4.7484000e-1 |
| 1 | 2.535609e-3 | 2.485873e-2 | 4.2737980e-2 |
| 2 | 1.746579e-3 | 5.125920e-3 | 1.2164360e-2 |
| 3 | 1.607121e-3 | 1.089340e-3 | 4.5758700e-3 |
| 4 | 1.575803e-3 | 7.573805e-4 | 2.1398690e-3 |
| 5 | 1.522073e-3 | 5.363611e-4 | 1.1480990e-3 |
| 6 | 1.475997e-3 | 5.348017e-4 | 5.3484040e-4 |
| 7 | 1.439057e-3 | 4.930253e-4 | 4.4935560e-4 |
| 8 | 1.411968e-3 | 4.781214e-4 | 2.7248240e-4 |
| 9 | 1.392150e-3 | 4.545103e-4 | 3.1401990e-4 |
| 10 | 1.378551e-3 | 4.353163e-4 | 2.5662360e-4 |

TABLE 4.III: Mean square error at the Nyquist rate.

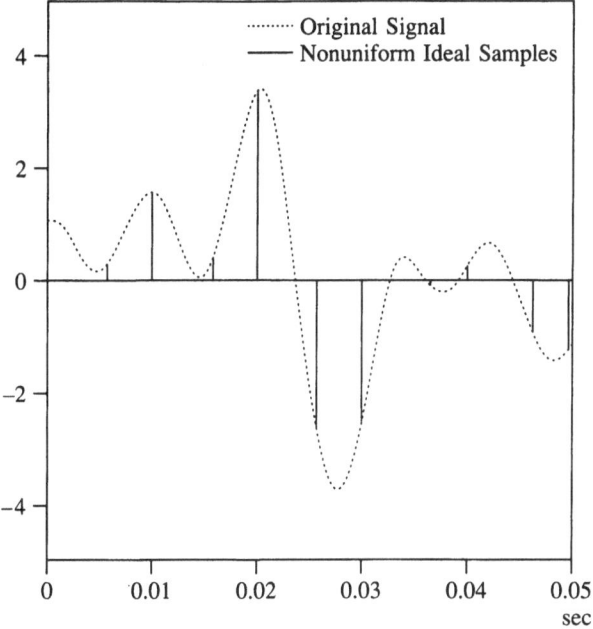

FIGURE 4.13: The original signal and its nonuniform samples (to within a $\frac{T}{4}$).

Nyquist rate. The instances are chosen randomly such that $|t_k - kT| < \frac{T}{4}$. This is a sufficient condition to ensure a stable sampling set. The reconstructed signal from the ideal nonuniform samples at the Nyquist rate is shown in Fig. 4.14 after 10 iterations. The mean square error (MSE) for the first 10 iterations is shown in Table 4.II under the column "Ideal." If we relax the sufficient condition $|t_k - kT| < \frac{T}{4}$, there is no guarantee that the sampling set converges at the Nyquist rate. For a specific sampling set that is to be within $|t_k - kT| < \frac{T}{2}$, the iterative technique slowly converges; Fig. 4.15 shows the result after 10 iterations. [14] When the samples are totally random, we observe an even slower convergence as shown in Table 4.IV. [15]

Obviously, if the average sampling rate is higher than the Nyquist rate, the convergence is guaranteed [16] and is faster. For instance, at twice or

---

[14]Figures 4.13-4.15 are optimized for $\frac{k_1}{k_2}$ experimentally; the optimum values are somewhere between 0.5 and 1.

[15]Clearly, the results for random sampling are different for each realization of the random samples. At the Nyquist rate, it may or may not converge. Even if it converges, the convergence rate depends on the configuration of the random samples. If the samples are clustered in a small interval, the convergence is slow and the segmental MSE is high at other regions of the waveform where there are no samples.

[16]Random samples at the higher than the Nyquist rate is a sampling set and uniquely determine the signal.

*Farokh Marvasti*

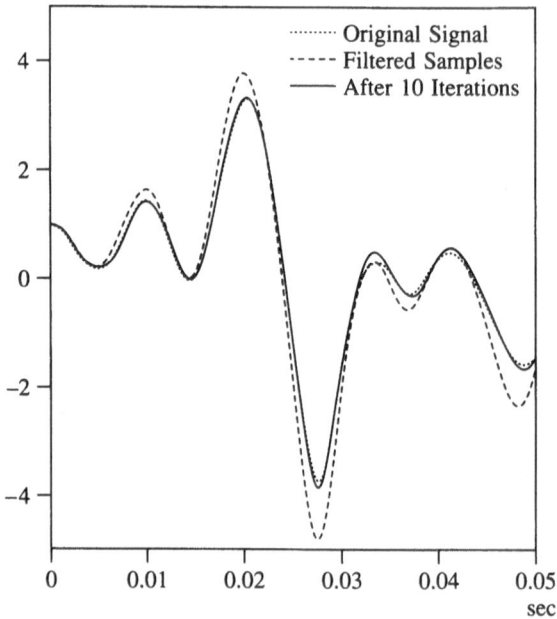

FIGURE 4.14: Reconstruction from ideal nonuniform samples at the Nyquist rate ($\frac{T}{4}$ interval).

three times the Nyquist rate, from Table 4.III we can deduce that if the nonuniform samples are restricted to within $\frac{T}{4}$, where $T$ is the Nyquist interval, we have a faster convergence after 10 iterations compared to the Nyquist sampling. This conclusion is also true for Table 4.IV where we sample randomly at twice or three times the Nyquist rate.

The reconstructed signal from the nonuniform samples with sample-and-hold (S & H- constant and variable width ) and the reconstructed signal from natural samples are listed in Table 4.II. [17] This table shows that for the cases of ideal samples, the iteration converges slightly faster than other sampling schemes. For good signal recovery, 5 to 30 iterations are sufficient in most cases, depending on the sampling rate.

Another iterative method is shown by the method of projection onto convex sets [979]. This topic is treated separately in this volume in the chapter by Stark.

The iterative method can be used for interpolation of uniform samples when some of them are lost. The signal to noise ratio for this type of

---

[17]The oscillations shown in this table for non-ideal sampling cases are due to quantization error and approximation of a convolution integral with a discrete one.

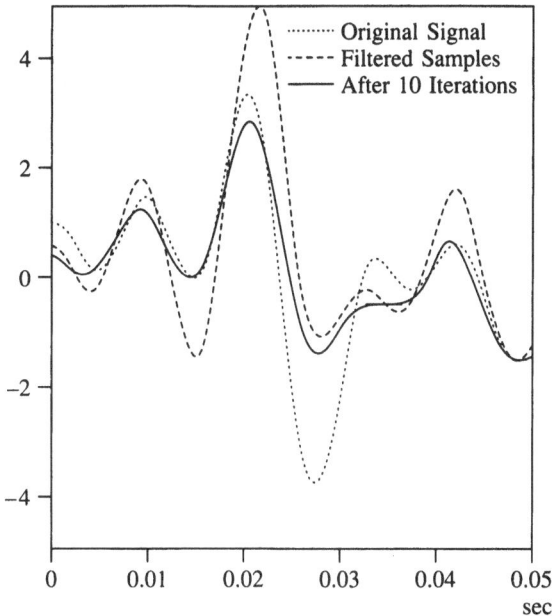

FIGURE 4.15: Reconstruction from ideal nonuniform samples at the Nyquist rate ($\frac{T}{2}$ interval).

sampling (after low-pass filtering) can be evaluated from Table 4.I, i.e.,

$$\frac{S_1}{N_1} = \frac{p}{2T(1-p)W},\qquad (4.4.6)$$

where $p$, $T$ and $W$ are, respectively, the probability of having a sample, the sampling interval and the signal bandwidth. At the Nyquist rate, $T = \frac{1}{2W}$, we obtain

$$\frac{S_1}{N_1} = \frac{p}{1-p}.$$

From (4.4.1) we know that the iterative procedure can improve upon this process if $\frac{S_1}{N_1} > 1$, which implies $p > \frac{1}{2}$. This analysis signifies that less than half the uniform samples (at the Nyquist rate) can be lost without any loss of information. The same iterative method can be used for uniform samples with time jitter.

## Acknowledgments

I sincerely appreciate the review and comments of my students. Specifically, I would like to thank Gilbert Adams and Mostafa Analoui for their helpful comments.

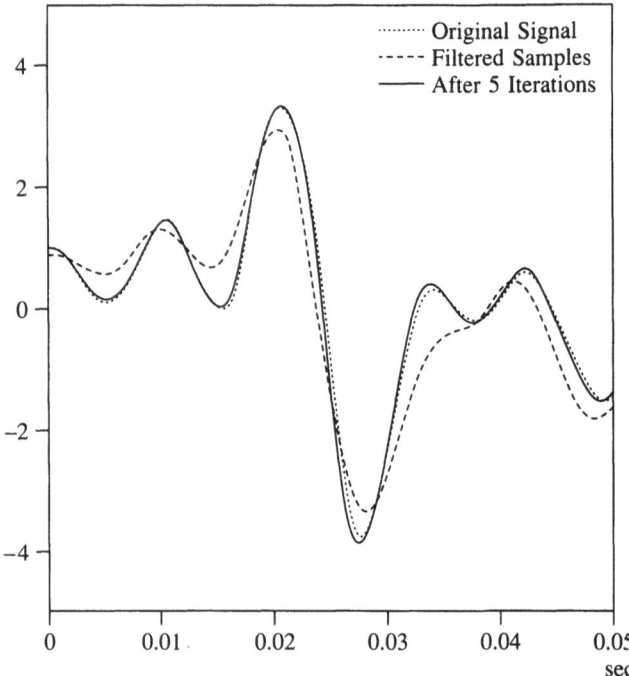

FIGURE 4.16: Reconstruction from ideal nonuniform samples at the Nyquist rate ($\frac{T}{4}$ interval).

| Mean Square Error | | | |
|---|---|---|---|
| Iteration No. | Nyquist rate | 2x Nyquist | 3x Nyquist |
| 0 | 1.672375e+0 | 4.450645e-1 | 2.719940e-1 |
| 1 | 1.143412e+0 | 2.491319e-1 | 9.645810e-2 |
| 2 | 6.669778e-1 | 1.789243e-1 | 4.881012e-2 |
| 3 | 9.370443e-1 | 1.405293e-1 | 2.707859e-2 |
| 4 | 8.414629e-1 | 1.131710e-1 | 1.639921e-2 |
| 5 | 8.303990e-1 | 9.394870e-2 | 1.104171e-2 |
| 6 | 7.694325e-1 | 7.923566e-2 | 8.194261e-3 |
| 7 | 7.596434e-1 | 6.719316e-2 | 6.596941e-3 |
| 8 | 7.190652e-1 | 5.758550e-2 | 5.629838e-3 |
| 9 | 7.089961e-1 | 4.926686e-2 | 4.992424e-3 |
| 10 | 6.810131e-1 | 4.243471e-2 | 4.530937e-3 |

TABLE 4.IV: Mean square error for random sampling at different sampling rates.

# 5

# Linear Prediction by Samples from the Past

P.L. Butzer and R.L. Stens

## 5.1  Preliminaries

If a signal $f$ is band-limited to $[-\pi W, \pi W]$ for some $W > 0$, then $f$ can be completely reconstructed for all values of $t \in \mathbf{R}$ from its sampled values $f(k/W)$, $k \in \mathbf{Z}$, taken just at the nodes $k/W$, equally spaced apart on the whole $\mathbf{R}$, in terms of

$$f(t) = \sum_{k=-\infty}^{\infty} f\left(\frac{k}{W}\right) \frac{\sin \pi(Wt - k)}{\pi(Wt - k)} \qquad (t \in \mathbf{R}). \qquad (5.1.1)$$

This is the famous sampling theorem, often associated with the names of E. T. Whittaker (1915), V. A. Kotel'nikov (1933) and C. E. Shannon (1940/49). However, K. Ogura (1920), H. Raabe (1939) and I. Someya (1949) could just as well be associated with it. (Concerning the history of the sampling theory see. e.g. [177, 393, 179, 558, 853, 590].)

The samples in (5.1.1) are taken not only from the whole past but also from the future, relative to some time $t = t_0$. But in practice a function or signal $f$ is only known in the past, i.e., if $t_0$ is the present instance, then only the values $f(t)$ for $t < t_0$ are at one's disposal. So the question arises whether it is possible to reconstruct a signal $f$, at least in the band-limited case, from samples taken exclusively from the past of $t_0$. Obviously this is a problem of prediction or forecasting of a time-variant process. Although this problem is often treated in a statistical (or stochastical) frame, let us begin by considering it in a deterministic setting and later carry it over into a stochastic one.

One answer to this question is the following: If the signal $f$ is band-limited to $[-W'\pi, W'\pi]$, then for each $0 < T < 1$ one can find so-called predictor coefficients $a_{kn} \in \mathbf{R}$ such that $f$ can be uniquely determined from its samples taken at the instants $t_0 - T/W$, $t_0 - 2T/W$, $t_0 - 3T/W$, $\ldots$, in terms of

$$f(t_0) = \lim_{n \to \infty} \sum_{k=1}^{n} a_{kn} f\left(t_0 - \frac{kT}{W}\right) \qquad (t_0 \in \mathbf{R}). \qquad (5.1.2)$$

This would yield the value of $f$ at the present instance $t = t_0$.

Here a new parameter comes into play, namely, $T$, which can be chosen arbitrarily from the interval $(0, 1)$. Obviously, the closer $T$ is to 1 the wider apart are the sampling points $kT/W$, $k \in \mathbf{N}$. But in any case the sampling rate $T/W$ in (5.1.2) is strictly greater than the Nyquist rate $1/W$ in (5.1.1). Note that the limiting case $T = 1$ has to be excluded here; just apply (5.1.2) with $T = 1$ and $t_0 = 0$ to the function $(\sin \pi W t)/\pi W t$.

Concerning the proof of (5.1.2), one turns to the frequency space and obtains by Schwarz's inequality that (see Section 5.2.1 for the definition of the Fourier transform $f^\wedge$)

$$\left| f(t) - \sum_{k=1}^{n} a_{kn} f\left(t - \frac{kT}{W}\right) \right|$$

$$= \left| \frac{1}{2\pi} \int_{-\pi W}^{\pi W} f^\wedge(v) e^{ivt} \left(1 - \sum_{k=1}^{n} a_{kn} e^{-ivkT/W}\right) dv \right|$$

$$\leq \left\{ \int_{-\pi W}^{\pi W} |f^\wedge(v)|^2 dv \right\}^{1/2} \left\{ \frac{W}{2\pi T} \int_{-\pi T}^{\pi T} \left|1 - \sum_{k=1}^{n} a_{kn} e^{ikv}\right|^2 dv \right\}^{1/2}.$$

$$(5.1.3)$$

The problem of predicting band-limited signals from past samples amounts to that of approximating the constant function 1 by (one-sided) trigonometric polynomials $\sum_{k=1}^{n} a_{kn} \exp(ikt)$ in the $L^2[-\pi T, \pi T]$-norm. By applying a result due to G. Szegö (1920) or a more general one due to N. Levinson (1940) one can show (see, e.g., [855, pp. 27, 28] or [275, p. 189ff.].) that for each $T$ with $0 < T < 1$ there *exist* coefficients $a_{kn}$ such that the latter integral in (5.1.3) tends to zero for $n \to \infty$ and hence (5.1.2) holds uniformly in $t_0 \in \mathbf{R}$.

Regarding the problem of the actual *construction* of the $a_{kn}$, the optimal solution would consist in trying to minimize the right-hand side of (5.1.3) for each fixed $n$. Using the orthogonality of the trigonometric system the optimal coefficients can be found to be the solutions of the linear system

$$\sum_{k=1}^{n} a_{kn} \frac{\sin \pi(k-j)T}{(k-j)} = \frac{\sin \pi j T}{j} \qquad (1 \leq j \leq n). \qquad (5.1.4)$$

It is of Toeplitz structure [369] and there exist algorithms for computing the solution: see, e.g., [254]. However. since these have many handicaps, several authors sought to determine *non-optimal* coefficients $a_{kn}$.

Thus L. A. Wainstein and V. D. Zubakov [932] (1962) showed that (5.1.2) is valid with $a_{kn} \approx (-1)^{k+1} \binom{n}{k}$ provided $0 < T < 1/3$. This means that for these $a_{kn}$ the sampling rate has to be three times as large as for the $a_{kn}$ of (5.1.4). J. L. Brown Jr. [122] (1972) extended the range of $T$ to $0 < T < 1/2$ for the coefficient choice $a_{kn} \equiv (-1)^{k+1} \binom{n}{k} (\cos \pi T)^k$: the coefficients now depend on $T$. Then W. Splettstößer [855, 851] (1981/82) extended this result even further, showing that (5.1.2) holds uniformly in $t_0 \in \mathbf{R}$ for $a_{kn} \equiv (-1)^{k+1} \binom{n+k-1}{k} 4^{-k}$ with $0 < T < \frac{1}{\pi} \arccos\left(-\frac{1}{8}\right) = 0.5399\ldots$ . This result improves the possible sampling rate in (5.1.2) to approximately twice the Nyquist rate, and the coefficients $a_{kn}$ are even independent of $T$. He also constructed predictor sums for larger $T$, namely, $1/2 < T < 2/3$, but the corresponding $a_{kn}$ are much more complicated and dependent on $T$. For a continuation of the foregoing approach of Splettstößer in the matter the reader is referred especially to Mugler and Splettstösser [664, 666, 665, 667]. The $T$-independent coefficients in the range $0 < T \leq 1/2$ have been derived [134]. A survey of the field has also been penned [170, pp. 35–43].

Now it is known that a function being band-limited is a rather restrictive condition. In fact, beginning with band-limited functions $f \in L^2(\mathbf{R})$, then $f$ can be extended to the complex plane as an entire function of exponential type $\pi W$, so it is extremely smooth. Further, such a function cannot be simultaneously duration limited, and it is the latter class of functions which actually occurs in practice. The next question therefore is whether prediction can be carried out for functions that are *not* necessarily band-limited. In this respect Splettstößer [854] showed that if the $(r+1)$th derivative $f^{(r+1)}$ is uniformly continuous and bounded on $\mathbf{R}$, then

$$
\sup_{t \in \mathbf{R}} \left| f(t) - \sum_{k=1}^{n} (-1)^{k+1} \binom{n}{k} (\cos \pi T)^k f\left(t - \frac{kT}{W}\right) \right|
$$

$$
= \mathcal{O}\left( (1 + \cos \pi T)^n W^{-r-1} + (\sin \pi T)^n \sqrt{W} \right) \qquad (n, W \to \infty)
$$

(5.1.5)

for each $0 < T < 1/2$. Since both terms on the right of (5.1.5) contain a factor tending to zero and one to infinity for $n, W \to \infty$, one has to choose $n$ independent of $W$ (or vice versa) such that both terms still tend to zero.

The disadvantages in the prediction procedures described so far are

(i)   The sampling rates are just $T/W$ with $0 < T < 1$ instead of the Nyquist rate $1/W$.

(ii)  The sample points depend on $t_0$ (or $t$), thus all the sample values have to be computed or measured anew when the series is evaluated for another $t_0$.

(iii) In the case of prediction of not necessarily band-limited functions the number of samples *plus* the distance between the sample points generally has to be appropriately regulated (recall (5.1.5)).

(iv)  To improve the approximation of $f$ by the series in (5.1.2) or (5.1.5) the number $n$ of samples has to be increased.

(v) Neither the sampling series in (5.1.2) nor that in (5.1.5) have the classical convolution structure for sums as given by the Shannon series (5.1.1).

To avoid these disadvantages, let us try to reconstruct functions from past samples by the convolution series

$$(S_W^{\varphi} f)(t) \equiv \sum_{k=-\infty}^{\infty} f\left(\frac{k}{W}\right) \varphi(Wt - k) \qquad (5.1.6)$$

for $W \to \infty$, where the kernel $\varphi$ will be assumed to be continuous and have compact support contained in $[T_0, T_1]$ for some $0 < T_0 < T_1$. This means that $\varphi(Wt - k) \neq 0$ at most for those $k \in \mathbf{Z}$ for which $k/W \in (t - T_1/W, t - T_0/W)$, so that only a *finite* number of samples taken from the past will be needed to evaluate (5.1.6), and this number will be fixed for all $f$, $W$ and $t$. Increasing $W$ in the series (5.1.6) will only mean that the distance between the sample points will decrease. Further, $f$ need not necessarily be band-limited. Of course, the coefficients $\varphi(Wt - k)$ depend on $t$, but the evaluation of $\varphi$ should be simpler than that of the signal $f$ to be sought.

Note that our results enable one to predict or extrapolate the value of a signal even arbitrarily far ahead of the sample values, at least theoretically.

Connections of the present study with the seminal work of A. N. Kolmogorov [497] (1941), N. Wiener [951] (1949) as well as of M. G. Krein [506] (1954) in the subject sketched in [176]. For text book coverage of this rather difficult approach, see, e.g., [276, pp. 2–9, 82–96, 146–278, 279–291] and [530, pp. 354–439]. For further literature on prediction theory see, e.g., the extensive reference list in the commentaries on the work of N. Wiener. edited by P. Masani [952].

Concerning possible applications, one of the main ones is to speech processing (see Markel and Gray, e.g., [578]). including differential pulse-code modulation [383]. Further applications are to economic prediction and forecasting (see Box and Jenkins [107]. to geophysics and medicine. see, e.g., [566]).

## 5.2   Prediction of Deterministic Signals

### 5.2.1   GENERAL RESULTS

Concerning notations, let $\mathbf{R}$ be the real axis, $\mathbf{N_0}$, $\mathbf{N}$, $\mathbf{Z}$ be the sets of all nonnegative integers, all naturals, and all integers, respectively. For an arbitrary finite or infinite subinterval $I \subset \mathbf{R}$ let $C(I)$ denote the space of all uniformly continuous and bounded functions $f \colon I \mapsto \mathbf{R}$ endowed with the supremum norm $\|f\|_{C(I)} \equiv \sup_{t \in I} |f(t)|$, and $C^{(r)}(I) \equiv \{f \in C(I) \colon f^{(r)} \in C(I)\}$ for $r \in \mathbf{N_0}$. The space $C_{00}(\mathbf{R})$ consists of those $f \in C(\mathbf{R})$ which have compact support; $C_{00}^{(r)}(\mathbf{R})$ is defined analogously. If $g: \mathbf{R} \mapsto \mathbf{R}$ is integrable over $\mathbf{R}$ with respect to Lebesgue measure (i.e., $g \in L^1(\mathbf{R})$), then its Fourier transform is defined by $g\hat{\,}(v) \equiv \int_{-\infty}^{\infty} g(u)e^{-ivu}\,du$, $v \in \mathbf{R}$. Some elementary properties of the respective operational calculus are collected in the following lemma (cf. [159, Chapter 5]).

**Lemma 1** *a) For $g \in L^1(\mathbf{R})$ and $h \in \mathbf{R}$ there holds*

$$[g(u+h)]\hat{\,}(v) = e^{ivh}g\hat{\,}(v) \qquad (v \in \mathbf{R}).$$

*b) If $g \in L^1(\mathbf{R}) \cap C^{(r)}(\mathbf{R})$ is such that $g^{(r)} \in L^1(\mathbf{R})$ for some $r \in \mathbf{N}$, then*

$$[g^{(r)}]\hat{\,}(v) = (iv)^r g\hat{\,}(v) \qquad (v \in \mathbf{R}).$$

*c) If $g \in L^1(\mathbf{R})$ is such that $u^r g(u) \in L^1(\mathbf{R})$ for some $r \in \mathbf{N}$, then $g\hat{\,} \in C^{(r)}(\mathbf{R})$ and*

$$(g\hat{\,})^{(r)}(v) = [(-iu)^r g(u)]\hat{\,}(v) \qquad (v \in \mathbf{R}).$$

For $\varphi \in C_{00}(\mathbf{R})$ and $f: \mathbf{R} \mapsto \mathbf{R}$ consider the sampling series

$$(S_W^{\varphi} f)(t) \equiv \sum_{k=-\infty}^{\infty} f\left(\frac{k}{W}\right)\varphi(Wt - k) \qquad (t \in \mathbf{R}: W > 0). \tag{5.2.1}$$

Since $\varphi$ has compact support, the series consists of only a finite number of non-zero terms, namely, for those $k \in \mathbf{Z}$ for which $Wt - k$ belongs to the support of $\varphi$. It defines a continuous function of $t \in \mathbf{R}$ for each fixed $W > 0$. Moreover, one easily verifies that $\{S_W^{\varphi}\}_{W>0}$ is a family of bounded linear operators from $C(\mathbf{R})$ into itself with operator norm

$$\|S_W^{\varphi}\|_{[C(\mathbf{R})]} = m_0(\varphi) \qquad (W > 0).$$

$m_r(\varphi)$ denoting the absolute sum moment of order $r \in \mathbf{N}_0$, namely,

$$m_r(\varphi) \equiv \sup_{u \in \mathbf{R}} \sum_{k=-\infty}^{\infty} |u - k|^r \, |\varphi(u - k)|.$$

We are interested in conditions on the kernel function $\varphi$ such that

$$\lim_{W \to \infty} (S_W^{\varphi} f)(t) = f(t) \qquad (t \in \mathbf{R}). \tag{5.2.2}$$

In this respect one has:

**Theorem 1** *If $\varphi \in C_{00}(\mathbf{R})$ is such that*

$$\sum_{k=-\infty}^{\infty} \varphi(u-k) = 1 \qquad (u \in \mathbf{R}), \qquad (5.2.3)$$

*then there holds (5.2.2) for each function $f: \mathbf{R} \mapsto \mathbf{R}$ at each point $t \in \mathbf{R}$ where $f$ is continuous. Furthermore, the operator family $\{S_W^\varphi\}_{W>0}$ defines a strong approximation process on $C(\mathbf{R})$, i.e.,*

$$\lim_{W \to \infty} \|f - S_W^\varphi f\|_{C(\mathbf{R})} = 0 \qquad (f \in C(\mathbf{R})). \qquad (5.2.4)$$

**Proof** In view of (5.2.3) one can estimate for any $\delta > 0$ and $W > 0$

$$|f(t) - (S_W^\varphi f)(t)|$$

$$\leq \left( \sum_{|Wt-k|<\delta W} + \sum_{|Wt-k|\geq\delta W} \right) |f(t) - f\left(\frac{k}{W}\right)| \, |\varphi(Wt-k)|$$

$$\equiv S_1 + S_2,$$

say. Since $f$ is continuous at the point $t$, to each $\varepsilon > 0$ there exists a $\delta > 0$ such that $|f(t) - f(k/W)| < \varepsilon$ for all $|t - k/W| < \delta$. This implies $S_1 < \varepsilon m_0(\varphi)$. Now take $\delta$ fixed. If $W$ is so large that the support of $\varphi$ is contained in $[-\delta W, \delta W]$, then $S_2 = 0$, and (5.2.2) follows. The proof that (5.2.3) implies (5.2.4) is quite similar, since the uniform continuity of $f$ in this case implies that $\delta$ can be chosen independently of $t$. ∎

It should be noted that condition (5.2.3) is not only sufficient for (5.2.2) or (5.2.4) to hold but is also necessary. This can be seen by taking $f(t) = 1$.

In practice it may be difficult to decide whether a function $\varphi \in C_{00}(\mathbf{R})$ satisfies (5.2.3) or not. The following lemma is useful in this respect.

**Lemma 2** *For $\varphi \in C_{00}(\mathbf{R})$ the condition*

$$\varphi^\wedge(2\pi k) = \begin{cases} 1, & k = 0 \\ 0, & k \in \mathbf{Z} \setminus \{0\} \end{cases} \qquad (5.2.5)$$

*is equivalent to (5.2.3).*

**Proof** By Poisson's summation formula (cf. [159, pp. 201, 202]) there holds

$$\sum_{k=-\infty}^{\infty} \varphi(u-k) \sim \sum_{k=-\infty}^{\infty} \varphi^\wedge(2\pi k) e^{i2\pi ku}. \qquad (5.2.6)$$

where the sign $\sim$ means that the right-hand side is the Fourier series of the 1–periodic function on the left. Hence, if (5.2.3) holds, then the Fourier series of the left-hand side in (5.2.6) reduces to the term for $k = 0$, which must be equal to 1. Conversely, if (5.2.5) holds, then the Fourier series on the right of (5.2.6) is finite and hence represents the continuous function on the left, i.e., there results (5.2.3).

At this point one should mention that the sampling series (5.2.1) can be regarded as a discrete analog of the singular convolution integral

$$(I_W^\varphi f)(t) \equiv W \int_{-\infty}^{\infty} f(u)\varphi(W(t-u))\, du \qquad (t \in \mathbf{R}; W > 0). \qquad (5.2.7)$$

In fact, (5.2.1) is a Riemann sum of the integral (5.2.7). For the latter it is well-known (cf. [159, Lemma 3.1.5]) that the condition

$$\varphi^\wedge(0) \equiv \int_{-\infty}^{\infty} \varphi(u)\, du = 1,$$

i.e., (5.2.5) for $k = 0$ only, is a necessary and sufficient one for

$$\lim_{W \to \infty} \|f - I_W^\varphi f\|_{C(\mathbf{R})} = 0 \qquad (f \in C(\mathbf{R})). \qquad (5.2.8)$$

This means that (5.2.8) holds provided (5.2.5) is satisfied for $k = 0$ only, whereas for (5.2.4) to be satisfied the whole of (5.2.5) is needed. In other words, (5.2.4) for sums implies (5.2.8) for integrals, but not conversely.

With the aid of Lemma 2 it is easy to give examples of kernels $\varphi$ satisfying (5.2.3), and hence (5.2.2) and (5.2.4). The most convenient examples are the so-called central $B$-splines. Recall that a function $q: I \mapsto \mathbf{R}$ is called a (polynomial) spline of order $n \in \mathbf{N}$ (degree $n - 1$) with knots $a_1 < a_2 < \cdots < a_m$ in $I$, if it coincides with a polynomial of degree $n - 1$ on each of the intervals $(a_\mu, a_{\mu+1})$.

The central $B$-splines of order $n \in \mathbf{N}$ are defined by

$$M_n(t) \equiv \frac{1}{(n-1)!} \sum_{j=0}^{n} (-1)^j \binom{n}{j} \left(\frac{n}{2} + t - j\right)_+^{n-1} \qquad (t \in \mathbf{R}) \qquad (5.2.9)$$

where $x_+^r \equiv \max\{x^r, 0\}$. They have knots at the points $0, \pm 1, \pm 2, \ldots, \pm n/2$ in case $n$ is even and at $\pm 1/2, \pm 3/2, \ldots, \pm n/2$ in case $n$ is odd, and their support is the compact interval $[-n/2, n/2]$. The Fourier transform of the $M_n$ has the simple form

$$M_n^\wedge(v) = \left(\frac{\sin v/2}{v/2}\right)^n \qquad (v \in \mathbf{R}; n \in \mathbf{N}), \qquad (5.2.10)$$

(see [800, pp. 67–71], [801, pp. 175, 176], [803, p. 12] or [805, §4.4]). If $n = 2$, then $M_n$ is the familiar roof-function, namely, $M_2(t) = (1 - |t|)_+$, and for $n = 3, 4$ one has the piecewise polynomial representations

$$M_3(t) = \begin{cases} \frac{1}{2}(|t| + \frac{3}{2})^2 - \frac{3}{2}(|t| + \frac{1}{2})^2, & |t| \leq \frac{1}{2} \\ \frac{1}{2}(-|t| + \frac{3}{2})^2, & \frac{1}{2} < |t| \leq \frac{3}{2} \\ 0, & |t| > \frac{3}{2}, \end{cases}$$

$$M_4(t) = \begin{cases} \frac{1}{6}(-|t| + 2)^3 - \frac{4}{6}(-|t| + 1)^2, & |t| \leq 1 \\ \frac{1}{6}(-|t| + 2)^3, & 1 < |t| \leq 2 \\ 0, & |t| > 2. \end{cases}$$

For higher $n$ such representations can be deduced likewise from (5.2.9). But it should be noted that in order to evaluate $M_n$ one usually uses the recurrence formula (see, e.g., de Boor [102, Chapter X] or Schumaker [805, Chapter 5])

$$M_n(t) = \frac{(\frac{n}{2} + t)M_{n-1}(t + \frac{1}{2}) + (\frac{n}{2} - t)M_{n-1}(t - \frac{1}{2})}{n - 1}. \tag{5.2.11}$$

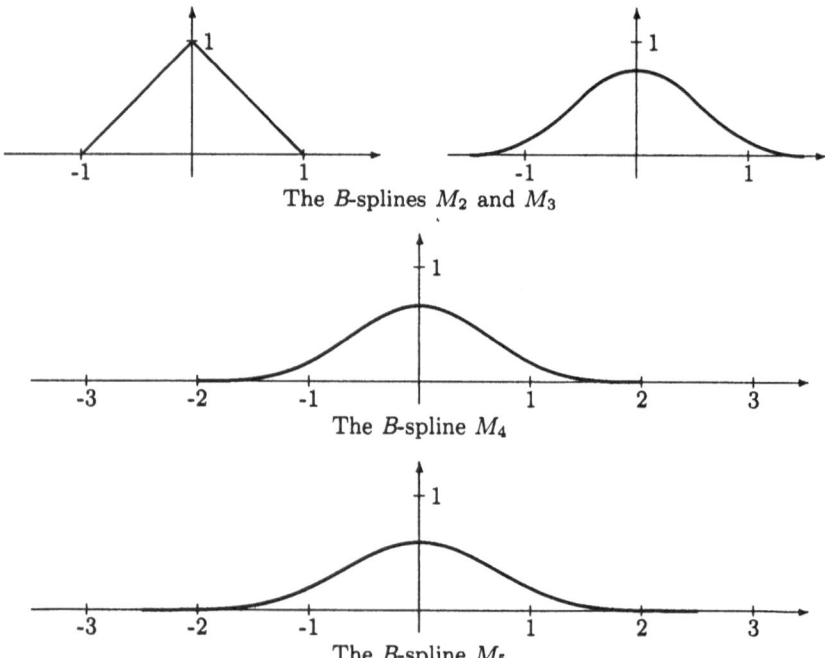

The B-splines $M_2$ and $M_3$

The B-spline $M_4$

The B-spline $M_5$

    In order to study the basic question as to the rate of approximation in (5.2.4), one has to assume that in addition to (5.2.3) certain higher order sum moments of $\varphi$ vanish.

**Theorem 2** *Let $\varphi \in C_{00}(\mathbf{R})$. If for some $r \in \mathbf{N}$*

$$\sum_{k=-\infty}^{\infty} (u-k)^j \varphi(u-k) = \begin{cases} 1, & j = 0 \\ 0, & j = 1, 2, \ldots, r-1, \end{cases} \qquad (5.2.12)$$

*for all $u \in \mathbf{R}$, then*

$$\|f - S_W^{\varphi} f\|_{C(\mathbf{R})} \leq \frac{m_r(\varphi)}{r!} \|f^{(r)}\|_{C(\mathbf{R})} W^{-r} \qquad (5.2.13)$$

$$(f \in C^{(r)}(\mathbf{R}); W > 0).$$

**Proof** An application of the operator $S_W^{\varphi}$ to the expansion

$$f(u) = \sum_{\nu=0}^{r-1} \frac{f^{(\nu)}(t)}{\nu!} (u-t)^{\nu} + \frac{1}{(r-1)!} \int_t^u f^{(r)}(y)(u-y)^{r-1} dy \qquad (5.2.14)$$

considered as a function of $u$ yields, by (5.2.12),

$$(S_W^{\varphi} f)(t) - f(t)$$

$$= \sum_{k=-\infty}^{\infty} \frac{1}{(r-1)!} \int_t^{k/W} f^{(r)}(y) \left(\frac{k}{W} - y\right)^{r-1} dy \, \varphi(Wt - k).$$

Now the integrals can be estimated by

$$\left| \int_t^{k/W} f^{(r)}(y) \left(\frac{k}{W} - y\right)^{r-1} dy \right| \leq \frac{W^{-r}}{r} |k - Wt|^r \|f^{(r)}\|_{C(\mathbf{R})}.$$

This immediately yields (5.2.13). ∎

For $r = 1$, condition (5.2.12) reduces to (5.2.3). This means that (5.2.13) for $r = 1$ holds already under the assumptions of Theorem 1.

**Corollary 1** *If $\varphi \in C_{00}(\mathbf{R})$ satisfies (5.2.12) for some $r \in \mathbf{N}$, then*

$$(S_W^{\varphi} p_{r-1})(t) = p_{r-1}(t) \qquad (t \in \mathbf{R})$$

*for each algebraic polynomial $p_{r-1}$ of degree $r - 1$.*

The proof follows by the same arguments as in the proof of (5.2.13), noting that the remainder in the Taylor expansion vanishes for polynomials of degree $r - 1$.

As in the case of condition (5.2.3) there exists an equivalent characterization of (5.2.12) in terms of the Fourier transform of $\varphi$.

**Lemma 3** *For $\varphi \in C_{00}(\mathbf{R})$ the condition*

$$(\varphi^{\widehat{\phantom{x}}})^{(j)}(2\pi k) = \begin{cases} 1, & k = j = 0 \\ 0, & k \in \mathbf{Z} \setminus \{0\}, \quad j = 0 \\ 0, & k \in \mathbf{Z}, \quad j = 1, 2, \ldots, r-1 \end{cases} \tag{5.2.15}$$

*is equivalent to (5.2.12).*

**Proof** One uses again Poisson's summation formula, this time applied to the function $(-iu)^j \varphi(u)$. By Lemma 1(c) one has

$$(-i)^j \sum_{k=-\infty}^{\infty} (u-k)^j \varphi(u-k) \sim \sum_{k=-\infty}^{\infty} (\varphi^{\widehat{\phantom{x}}})^{(j)}(2\pi k) e^{i2\pi ku}.$$

and then one proceeds as in the proof of Lemma 2. ∎

Conditions of type (5.2.5) or (5.2.15) are to be found in connection with finite element approximation; see, e.g., Fix and Strang [297].

The foregoing approach can be generalized to kernels having unbounded support; [767], [170, Section 4.1, 4.2]. Reconstruction of signals in terms of splines using a finite number of samples from the past as well as from the future has also been considered [150]. It is also possible to approximate functions having jump discontinuities by sampling sums [163].

### 5.2.2 SPECIFIC PREDICTION SUMS

In this subsection we use series of type (5.2.1) in order to predict a signal at time $t \in \mathbf{R}$. provided that only sample values $f(k/W)$ for $k/W < t$ are known. In order to apply the results of Subsection 5.2.1 one has to assume that the support of the kernel function $\varphi$ is contained in $(0, \infty)$, since in this instance $\varphi(Wt - k)$ vanishes for all $k$ with $k/W \geq t$, and so the series (5.2.1) can be rewritten as

$$(S_W^{\varphi} f)(t) \equiv \sum_{k/W < t} f\left(\frac{k}{W}\right) \varphi(Wt - k) \qquad (t \in \mathbf{R}; W > 0). \tag{5.2.16}$$

The problem now is to construct kernels $\varphi$ satisfying (5.2.12) for different values of $r$ having support contained in $(0, \infty)$. The following theorem gives a general approach to this problem, starting off with the $B$-splines $M_n$ of (5.2.9).

**Theorem 3** *For $r \in \mathbf{N}$, $r \geq 2$, let $\varepsilon_0 < \varepsilon_1 < \cdots < \varepsilon_{r-1}$ be any given real numbers, and let $a_{\mu r}$, $\mu = 0, 1, \ldots, r-1$, be the unique solutions of the linear system*

$$\sum_{\mu=0}^{r-1} a_{\mu r}(-i\varepsilon_\mu)^j = \left(\frac{1}{\widehat{M_r}}\right)^{(j)}(0) \qquad (j = 0, 1, \ldots, r-1), \qquad (5.2.17)$$

*where $i = \sqrt{-1}$. Then*

$$\varphi_r(t) \equiv \sum_{\mu=0}^{r-1} a_{\mu r} M_r(t - \varepsilon_\mu) \qquad (t \in \mathbf{R}) \qquad (5.2.18)$$

*is a polynomial spline of order $r$ satisfying (5.2.12) and having support contained in $[\varepsilon_0 - r/2, \varepsilon_{r-1} + r/2]$.*

**Proof** The assertion concerning the support follows from the corresponding property of the $B$-splines. In order to show that (5.2.12) holds, the equivalent characterization of (5.2.12) by Lemma 3 will be used. For this purpose, consider the Fourier transform of $\varphi_r$, namely, (cf. Lemma 1(a)),

$$\widehat{\varphi_r}(v) = \widehat{M_r}(v) \left\{ \sum_{\mu=0}^{r-1} a_{\mu r} e^{-i\varepsilon_\mu v} \right\} =: \widehat{M_r}(v) p(v).$$

Obviously one has that $(\widehat{\varphi_r})^{(j)}(2\pi k) = 0$ for $k \in \mathbf{Z} \setminus \{0\}$ and $0 \leq j \leq r-1$ by (5.2.10). Furthermore, since

$$p^{(v)}(0) = \sum_{\mu=0}^{r-1} a_{\mu r}(-i\varepsilon_\mu)^v = \left(\frac{1}{\widehat{M_r}}\right)^{(v)}(0) \qquad (v = 0, 1, \ldots, r-1)$$

by (5.2.17), one has immediately

$$
\begin{aligned}
(\widehat{\varphi_r})^{(j)}(0) &= \sum_{s=0}^{j} \binom{j}{s} (\widehat{M_r})^{(s)}(0) \left(\frac{1}{\widehat{M_r}}\right)^{(j-s)}(0) \\
&= \left(\widehat{M_r} \cdot \frac{1}{\widehat{M_r}}\right)^{(j)}(0) \\
&= \begin{cases} 1, & j = 0 \\ 0, & 1 \leq j \leq r-1. \end{cases}
\end{aligned}
$$

This completes the proof. ∎

To solve the system of linear equations (5.2.17), the derivatives $(1/\widehat{M_r})^{(j)}(0) = (d/dv)^j((v/2)/\sin(v/2))^r|_{v=0}$ need to be known, at least for smaller values of $r$. This can be achieved with the aid of the expansion

$$\left(\frac{v/2}{\sin v/2}\right)^r = \sum_{k=0}^{\infty}(-1)^k \frac{(r-1-2k)!}{(r-1)!} t(r, r-2k)\frac{x^{2k}}{(2k)!}$$

$$(|v| < 2\pi),$$

where $t(r, k)$ are the central factorial numbers of the first kind ([164, 165], [771, Chapter 6]). Since $M_r\widehat{\ }$ is an even function, the right side of (5.2.17) vanishes for $j$ odd. Hence the imaginary unit $i$ can be cancelled out, and the solutions $a_{\mu r}$ are always real.

| r \ j | 0 | 1 | 2 | 3 | 4 | 5 |
|---|---|---|---|---|---|---|
| 2 | 1 | 0 | — | — | — | — |
| 3 | 1 | 0 | $\frac{1}{4}$ | — | — | — |
| 4 | 1 | 0 | $\frac{1}{3}$ | 0 | — | — |
| 5 | 1 | 0 | $\frac{5}{12}$ | 0 | $\frac{9}{16}$ | — |
| 6 | 1 | 0 | $\frac{1}{2}$ | 0 | $\frac{4}{5}$ | 0 |

TABLE 1: $\left(\dfrac{1}{M_r\widehat{\ }}\right)^{(j)}(0)$ for $r = 2, 3, 4, 5, 6;\ j = 0, 1, 2, \ldots, r-1.$

**Corollary 2** *Given any $r \in \mathbf{N}$, one can construct a kernel $\varphi_r \in C_{00}(\mathbf{R})$ with support in $(0, \infty)$ such that the associated generalized sampling series approximates all $f \in C^{(r)}(\mathbf{R})$ with order $\mathcal{O}(W^{-r})$, $W \to \infty$. Moreover, the kernel $\varphi$ can be chosen such that the number of samples needed for the evaluation of the series is $r+1$ at most.*

**Proof** Taking $\varepsilon > r/2$ in Theorem 3 yields a kernel $\varphi_r$ with support in $(0, \infty)$, and the generalized sampling series built up from this kernel has the required order. Furthermore, choosing, e.g., $\varepsilon_0 = r/2+1$ and $\varepsilon_{r-1} = r/2+3/2$, then the associated $\varphi_r$ has support in $[1, r+3/2]$, and one easily verifies that $\varphi(Wt - k) \neq 0$ for $r+1$ integers. $k$ at most. ∎

Observe that it is an open question whether there exists a *closed* form of the solutions $a_{\mu r}$, $\mu = 0, 1, ..., r - 1$, of (5.2.17). So far the construction can be used in actual practice only for smaller values of $r$. However, as we will see below, even the case $r = 2$ gives a pretty good upper bound for the approximation error, namely, $15 \, \|f''\|_{C(\mathbf{R})} W^{-2}$.

Let us now consider what sample values are needed in order to evaluate the series (5.2.1) in case the kernel $\varphi = \varphi_r$ is of type (5.2.18). If one sets $T_0 \equiv \varepsilon_0 - r/2$, $T_1 \equiv \varepsilon_{r-1} + r/2$, then the support of $\varphi_r$ is contained in $[T_0, T_1]$, and moreover $[T_0, T_1] \subset (0, \infty)$ provided one chooses $\varepsilon_0 > r/2$, to be assumed in the following. So to approximate $f(t)$ by the series (5.2.1) with $\varphi = \varphi_r$, sampling points $k/W$ taken exclusively from the interval $[t - T_1/W, t - T_0/W]$ will be needed. This means that the last point required lies at least $T_0/W$ time units previous to the time $t$ for which one wishes to determine $f(t)$. One can expand the interval $(t - T_0/W, t)$ by enlarging $\varepsilon_0$; this implies an enlargement of $T_0$ for $W > 0$ fixed. In this process the moments $m_r(\varphi)$ increase, however, so that the upper bound for the approximation error (5.2.13) grows. This is not surprising. In fact, it can be shown that $m_r(\varphi)$ behave like $\varepsilon_0^r$.

In the following, the kernels $\varphi_r$ for $r = 2, 3, 4, 5$ are constructed according to Theorem 3, where $\varepsilon_0$ is chosen as $r/2 + 1$ if $r$ is even, and as $r/2 + 1/2$ if $r$ is odd. The other $\varepsilon_\mu$ are taken as $\varepsilon_{\mu+1} = \varepsilon_\mu + 1$, $\mu = 0, 1, ..., r - 2$.

More specifically, if $r = 2$, $\varepsilon_0 = 2$, so that $[T_0, T_1] = [1, 4]$, then the system (5.2.17) reads, noting Table 1,

$$
\begin{aligned}
a_{02} + a_{12} &= 1, \\
-2i a_{12} - 3i a_{12} &= 0.
\end{aligned}
$$

The solution is $a_{02} = 3$, $a_{12} = -2$, giving the kernel

$$\varphi_2(t) = 3 M_2(t - 2) - 2 M_2(t - 3). \tag{5.2.19}$$

Thus if one would want to predict the signal $f(t)$ for $t \in (j/W, (j+1)/W)$, any $j \in \mathbf{Z}$, the associated sampling series $S_W^{\varphi_2} f$ would consist of three terms only, namely, those for $k = j - 3, j - 2, j - 1$ for which $t - 4/W < k/W < t - 1/W < t$. If $f'' \in C(\mathbf{R})$, then

$$\|S_W^{\varphi_2} f - f\|_C \leq 15 \|f''\|_C W^{-2}.$$

Hence $S_W^{\varphi_2} f$ enables one to predict a signal at least $1/W$ units ahead with error $\mathcal{O}(W^{-2})$. Note that $m_2(\varphi_2) \leq 30$. (This estimate for $m_2(\varphi_2)$ as well as the corresponding estimates for the kernels below were calculated numerically by using a computer.)

Now take $r = 3$, $\varepsilon_0 = 2$, so that $[T_0, T_1] = [1/2, 11/2]$. The system (5.2.17) then reads

$$
\begin{aligned}
a_{03} + a_{13} + a_{23} &= 1, \\
-2i a_{03} - 3i a_{13} - 4i a_{23} &= 0, \\
-4 a_{03} - 9 a_{13} - 16 a_{23} &= \tfrac{1}{4}
\end{aligned}
$$

which has as solutions $a_{03} = 47/8$, $a_{13} = -62/8$, $a_{23} = 23/8$. So

$$\varphi_3(t) = \frac{1}{8}\{47\, M_3(t-2) - 62\, M_3(t-3) + 23\, M_3(t-4)\},$$

the sampling series now consisting of those $k \in \mathbf{Z}$ for which $t - 11/2W < k/W < t - 1/2W < t$, thus of five terms at most. Concerning the approximation error, one has in particular that

$$\|f - S_W^{\varphi_3} f\|_C \le 54\|f'''\|_C W^{-3} \qquad (f''' \in C(\mathbf{R})),$$

since $m_3(\varphi_3)/3! \le 54$.

In case $r = 4$, $\varepsilon_0 = 3$ one has $[T_0, T_1] = [1, 8]$. By solving a system of four equations in four unknowns one can readily show that

$$\varphi_4(t) = \frac{1}{6}\{115\, M_4(t-3) - 256\, M_4(t-4)$$
$$+ 203\, M_4(t-5) - 56\, M_4(t-6)\}.$$

This time the series consists of seven terms at most, namely, those $k \in \mathbf{Z}$ for which $t - 10/W < k/W < t - 1/W < t$. In particular. if $f^{(4)} \in C(\mathbf{R})$, then the corresponding rate of approximation can, in comparison with the first examples, be improved to $\mathcal{O}(W^{-4})$ with $\mathcal{O}$ constant given by $970\|f^{(4)}\|_{C(\mathbf{R})}$. Let us finally take $r = 5$, $\varepsilon_0 = 3$ so that $[T_0, T_1] = [1/2, 19/2]$; then

$$\varphi_5(t) = \frac{1}{1152}\{36767\, M_5(t-3) - 108188\, M_5(t-4)$$
$$+ 127914\, M_5(t-5) - 70268\, M_5(t-6)$$
$$+ 14927\, M_5(t-7)\}.$$

Here nine samples will be needed, the order of approximation being $\mathcal{O}(W^{-5})$ provided $f^{(5)} \in C(\mathbf{R})$. The constant in the order is however very large: in fact $m_5(\varphi_5)/5! \le 3600$.

We conclude these considerations with a slight generalization of the kernel $\varphi_2$ of (5.2.19). If one would take $r = 2$ as above, but $\varepsilon_0 > r/2 = 1$ arbitrary and $\varepsilon_1 = \varepsilon_0 + 1$, then $[T_0, T_1] = [\varepsilon_0 - 1, \varepsilon_0 + 2] \subset (0, \infty)$, and

$$\varphi_{2,\varepsilon_0}(t) = (1 + \varepsilon_0)M_2(t - \varepsilon_0) - \varepsilon_0 M_2(t - \varepsilon_0 - 1).$$

In particular, if $\varepsilon_0 = 8$ and $t \in (j/W. (j+1)/W)$ for any $j \in \mathbf{Z}$, the sampling series based on the kernel $\varphi_{2,8}$ consists of three terms at $k = j - 9$, $j - 8$, $j - 7$, for which $k/W < t - 7/W < t$. So the kernel $\varphi_{2,8}$ allows one to predict at least $7/W$ units ahead. The moment $m_2(\varphi_{2,8})$ is, however, much larger than in the case $\varepsilon_0 = 2$, in fact $m_2(\varphi_{2,8})/2! \le 612$. This procedure can be applied to all kernels constructed above, and it would enable one to predict a signal *arbitrarily far ahead*, at least theoretically.

The choice of $\varepsilon_\mu$ made above is of course not the only possible one, but choosing $\varepsilon_{\mu+1} = \varepsilon_\mu + 1$ the sampling series based on the kernel $\varphi_r$ can be rewritten as

$$\sum_{k=-\infty}^{\infty} f\left(\frac{k}{W}\right) \varphi_r(Wt - k)$$

$$= \sum_{k=-\infty}^{\infty} \left\{ \sum_{\mu=0}^{r-1} a_{\mu r} f\left(\frac{k-\mu}{W}\right) \right\} M_r(Wt - \varepsilon_0 - k)$$

$$= \sum_{k=N_1}^{N_2} \left\{ \sum_{\mu=0}^{r-1} a_{\mu r} f\left(\frac{k-\mu}{W}\right) \right\} M_r(Wt - \varepsilon_0 - k) \qquad (5.2.20)$$

with $N_1 = [-r/2 - \varepsilon_0 + Wt] + 1$ and $N_2 = [r/2 - \varepsilon_0 + Wt] + 1$, $[x]$ denoting the largest integer $\leq x$. Representation (5.2.20) is very convenient for computational purposes.

Table 2 gives the kernels $\varphi_r$ for $r = 2, 3, 4, 5$, and $\varphi_{2,8}$ as just constructed. The entries in the second column give the number of samples needed for the evaluation of (5.2.20), whereas the third column contains an estimate for the constant $m_r(\varphi_r)/r!$ in (5.2.13). In the rightmost column, one has the best possible order of approximation which can be achieved according to Theorem 2.

| Kernels | $r$ | Samples | Order |
|---|---|---|---|
| $\varphi_2(t) = 3\,M_2(t-2) - 2\,M_2(t-3)$ | 3 | 15 | $\mathcal{O}(W^{-2})$ |
| $\varphi_3(t) = \frac{1}{8}\{47\,M_3(t-2) - 62\,M_3(t-3) + 23\,M_3(t-4)\}$ | 5 | 54 | $\mathcal{O}(W^{-3})$ |
| $\varphi_4(t) = \frac{1}{6}\{115\,M_4(t-3) - 256\,M_4(t-4) + 203\,M_4(t-5)\}$ | 7 | 970 | $\mathcal{O}(W^{-4})$ |
| $\varphi_5(t) = \frac{1}{1152}\{36767\,M_5(t-3) - 108188\,M_5(t-4) + 127914\,M_5(t-5)$ $-70268\,M_5(t-6) + 14927\,M_5(t-7)\}$ | 9 | 3600 | $\mathcal{O}(W^{-5})$ |
| $\varphi_{2,8}(t) = 9\,M_2(t-8) - 8\,M_2(t-9)$ | 3 | 612 | $\mathcal{O}(W^{-2})$ |

TABLE 2: $\varphi_r$ for $r = 2, 3, 4, 5$, and $\varphi_{2,8}$

It is worthwhile to note that the foregoing kernels have nothing in common with the familiar bell-shaped kernels normally used in approximation theory. For example, the kernel $\varphi_2$ has the piecewise polynomial representation

$$\varphi_2(t) = \begin{cases} 0, & t \leq 1 \text{ or } t \geq 4 \\ 3t - 3. & 1 < t \leq 2 \\ 13 - 5t. & 2 < t \leq 3 \\ 2t - 8. & 3 < t \leq 4. \end{cases}$$

For the other $\varphi_r$ such representations can readily deduced from (5.2.9). Note that for the evaluation of the $\varphi_r$ one should make use of the recurrence formula (5.2.11) rather than of the piecewise polynomial representation.

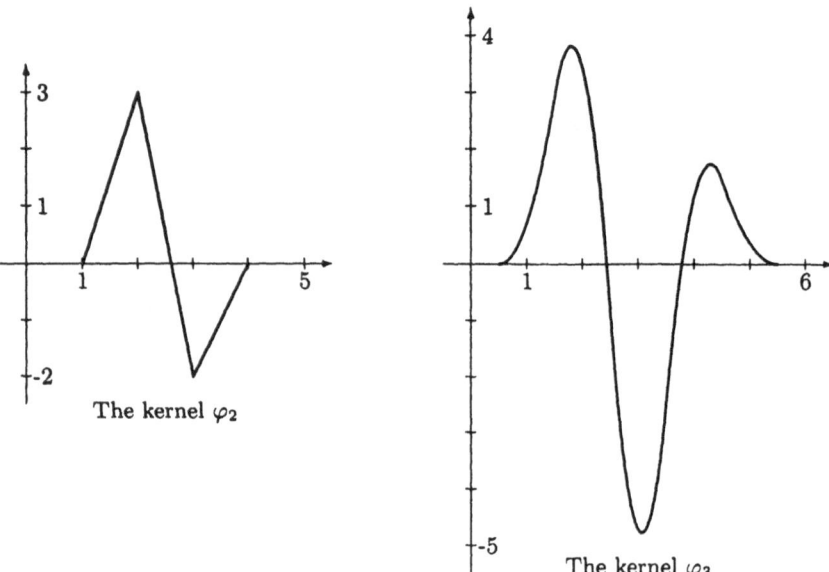

The kernel $\varphi_2$

The kernel $\varphi_3$

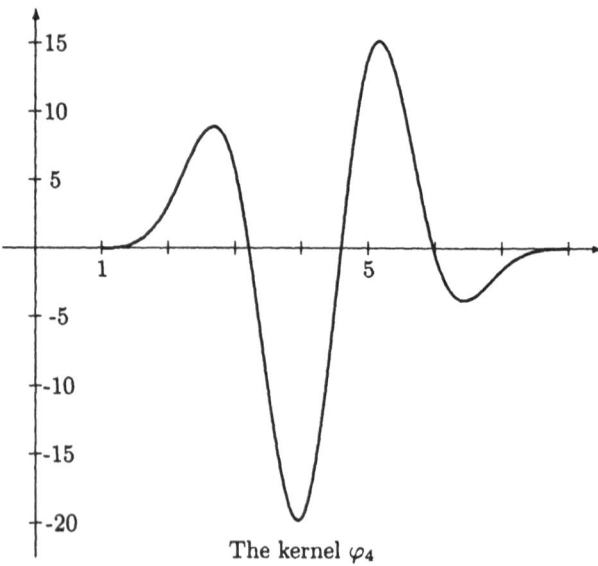

The kernel $\varphi_4$

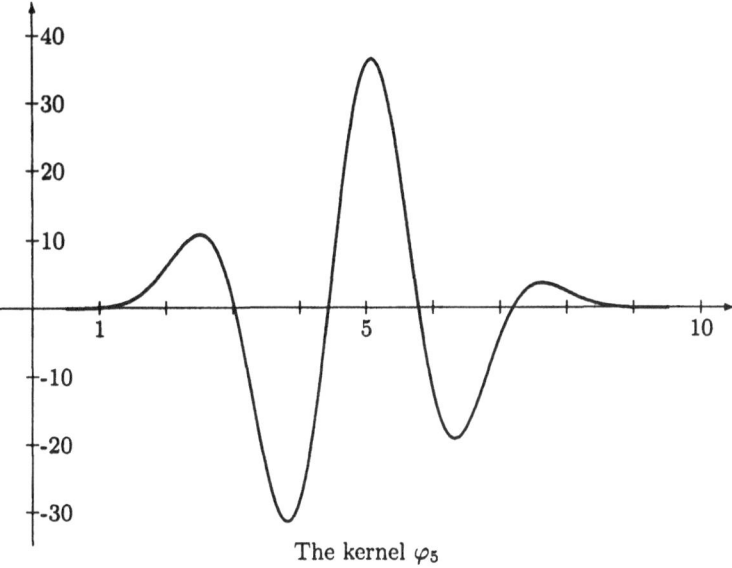

The kernel $\varphi_5$

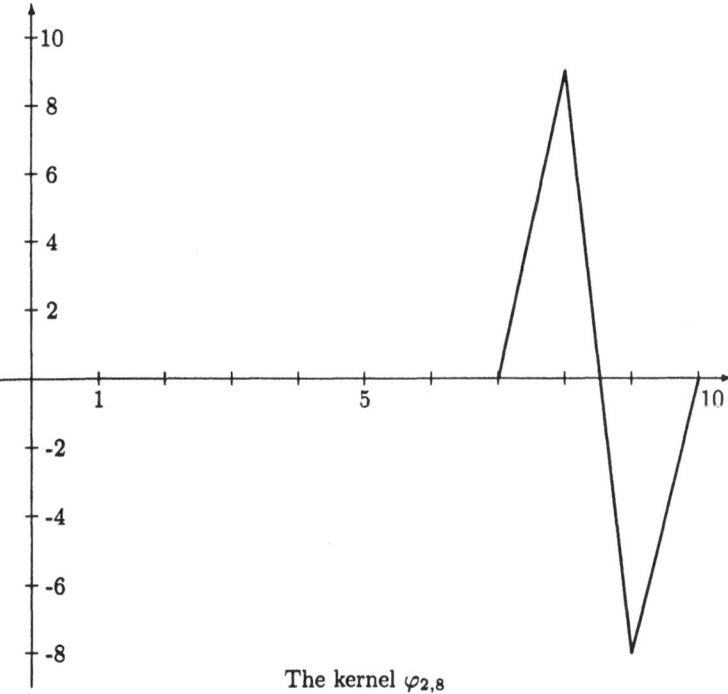

The kernel $\varphi_{2,8}$

## 5.2.3  An Inverse Result

At this point one may ask whether it is possible to construct other kernels $\varphi$ based on polynomial splines of order $r$ such that the order of approximation is better than those for the kernels constructed by the method of Theorem 1. The answer to this question is given by the following inverse-type result.

**Theorem 4** *Let $\varphi \in C_{00}(\mathbf{R})$ be any kernel consisting of piecewise polynomials of degree $r - 1$ for some $r \in \mathbf{N}$. $r \geq 2$, with knots at the points $a_0 < a_1 < \cdots < a_m$ for some $m \in \mathbf{N}_0$. Then for each $s \in \mathbf{N}_0$ with $s \leq r$ and each $\varepsilon > 0$ there exists a function $f^* \in C^{(s)}(\mathbf{R})$ such that*

$$\|f^* - S_W^\varphi f^*\|_{C(\mathbf{R})} \neq \mathcal{O}(W^{-s-\varepsilon}) \qquad (W \to \infty).$$

*Moreover, if for some $f \in C(\mathbf{R})$ there holds*

$$\|f - S_W^\varphi f\|_{C(\mathbf{R})} = o(W^{-r}) \qquad (W \to \infty), \qquad (5.2.21)$$

*then $f$ is a constant.*

The first part of this theorem shows that when using spline kernels one cannot approximate the elements of the whole class $C^{(s)}(\mathbf{R})$ with a rate better than $\mathcal{O}(W^{-s})$. In its second part the theorem states that for spline kernels of order $r$ the best possible order of approximation which can be achieved for non-constant functions $f$ is $\mathcal{O}(W^{-r})$, even if $f$ is arbitrarily smooth.

The remaining part of this section is devoted to the proof of this theorem. To this end we introduce the concept of a *modulus of continuity*. It is defined for $f \in C(I)$ and $r \in \mathbf{N}$ by

$$\omega_r(\delta; f; C(I)) \equiv \sup_{|h| \leq \delta} \| \Delta_h^r f \|_{C(I)} \qquad (\delta > 0). \qquad (5.2.22)$$

where $\Delta_h^r f$ is the difference of $f$ of order $r$, namely,

$$(\Delta_h^r f)(t) \equiv \begin{cases} \displaystyle\sum_{j=0}^r (-1)^j \binom{r}{k} f(t + jh), & t, t + rh \in I \\ 0, & \text{elsewhere} \end{cases} \qquad (h \in \mathbf{R}).$$

Note that the right-hand side of (5.2.22) can be replaced by $\sup_{0 < h \leq \delta} \| \Delta_h^r f \|_{C(I)}$ in view of the identity

$$(\Delta_h^r f)(t) = (-1)^r (\Delta_{-h}^r f)(t + rh).$$

Whereas the results of the previous sections give upper bounds for the error occurring when approximating by generalized sampling series, the next lemma gives a *lower* bound in terms of the modulus of continuity.

**Lemma 4** *Let $\varphi \in C_{00}(\mathbf{R})$ be as in Theorem 4. Then for each $f \in C(\mathbf{R})$,*

$$\omega_r(W^{-1}; f; C[0, \infty)) \le (2 + 4r/d)^r \sup_{V \ge W} \|f - S_V^\varphi f\|_{C[0,\infty)} \quad (W > 0) \quad (5.2.23)$$

*for some $d > 0$ depending on $\varphi$ only. The same result holds for $C[0, \infty)$ in (5.2.23) replaced by $C(-\infty, 0]$.*

**Proof** The assumptions upon $\varphi$ imply that for each $f \in C(\mathbf{R})$ the series $S_1^\varphi f$ considered as a function on $[0, \infty)$ is a piecewise polynomial of degree $r - 1$ having knots at the points $0 < b_1 < b_2 < b_3 < \cdots$, where the $b_\nu$ are of the form $b_\nu = a_\mu + j$ with $\mu \in \{0, 1, \ldots, m\}$ and $j \in \mathbf{Z}$. Thus, setting $b_0 = 0$, $S_W^\varphi f$ coincides with an algebraic polynomial $p_{r-1}$ of degree $r - 1$ on each interval $I_W(\nu) \equiv [b_\nu/W, b_{\nu+1}/W]$, $\nu \in \mathbf{N}_0$, $W > 0$. Hence, since $(\Delta_h^r p_{r-1})(t) = 0$ for all $t, h \in \mathbf{R}$, it follows that $(\Delta_h^r S_W^\varphi f)(t) = 0$, provided $t$ and $t + rh$ belong to the same interval $I_W(\nu)$.

Now, let $t \in \mathbf{R}$ and $0 < h \le d/2rW$, where $d > 0$ is the minimal distance between two consecutive points $b_\nu$ and $b_{\nu+1}$. There exists a $\nu_0 \in \mathbf{N}_0$ such that $t \in I_W(\nu_0)$. Let us distinguish the cases $t \le (b_{\nu_0} + b_{\nu_0+1})/2W$ and $t > (b_{\nu_0} + b_{\nu_0+1})/2W$. In the first case, $t + rh \in I_W(\nu_0)$ too, implying

$$|(\Delta_h^r f)(t)| \le |(\Delta_h^r(f - S_W^\varphi f))(t)| \le 2^r \|f - S_W^\varphi f\|_{C[0,\infty)} \quad (5.2.24)$$

in view of the considerations made above.

In the second case, choose $W_1 \equiv b_{\nu_0+1}/t$.
Then, by $(b_{\nu_0} + b_{\nu_0+1})/2W < t \le b_{\nu_0+1}/W$, one has $W \le W_1 < 2W$ and

$$\frac{b_{\nu_0+1}}{W_1} = t < t + rh \le \frac{b_{\nu_0+2}}{W_1},$$

the right-hand inequality following from

$$t + rh \le \frac{b_{\nu_0+1}}{W_1} + \frac{d}{2W} < \frac{b_{\nu_0+1} + d}{W_1} \le \frac{b_{\nu_0+2}}{W_1}.$$

Hence, $t$ and $t + rh$ belong to $I_{W_1}(\nu_0+1)$, and one obtains, as above,

$$|(\Delta_h^r f)(t)| \le 2^r \|f - S_{W_1}^\varphi f\|_{C[0,\infty)}. \quad (5.2.25)$$

Noting that $t \in [0, \infty)$ was arbitrary and that $W_1 \ge W$, the inequalities (5.2.24) and (5.2.25) can be combined to give

$$\omega_r(d/2rW; f; C[0, \infty)) \le 2^r \sup_{V \ge W} \|f - S_V^\varphi f\|_{C[0,\infty)}.$$

This proves assertion (5.2.23) since (cf. [461, (2.4)])

$$\omega_r(\delta; f; C(I)) \le (1 + \lambda^{-1})^r \omega_r(\lambda\delta; f; C(I)) \quad (\delta, \lambda > 0).$$

The case $C(-\infty, 0]$ can be treated by considering $f(-t)$ and $\varphi(-t)$. ∎

**Proof of Theorem 4.** Let $r, s$ and $0 < \varepsilon \leq 1$ be given. Since the case $s = r$ is contained in the second assertion, one can assume $s < r$. For those $s$ and $\varepsilon$ there exists a function $f^* \in C^{(s)}(\mathbf{R})$ which is $2\pi$-periodic and satisfies

$$\omega_r(\delta; f^* : C(\mathbf{R})) = \mathcal{O}(\delta^{s+\varepsilon/2}) \qquad (\delta \to 0+),$$

where the exponent of $\delta$ cannot be replaced by a larger one: see [900, pp. 77,337 ] and [461, (2.7)]. Now, if there would hold

$$\|f^* - S_W^{\varphi} f^*\|_{C(\mathbf{R})} = \mathcal{O}(W^{-s-\varepsilon}) \quad (W \to \infty),$$

then Lemma 4 implies

$$\omega_r(\delta; f^* : C[0, \infty)) = \mathcal{O}(\delta^{s+\varepsilon}) \qquad (\delta \to 0+),$$

which is a contradiction to the choice of $f^*$.

Concerning the second assertion, condition (5.2.21) yields again by Lemma 4 that

$$\omega_r(\delta; f; C[0, \infty)) = o(W^{-r}) \qquad (\delta \to 0+).$$

Hence by [461, p. 291] the function $f$ is a polynomial of degree $r-1$ on $[0, \infty)$ which must be a constant since it is bounded. The same considerations for the interval $(-\infty, 0]$ show that $f$ is a constant even on $\mathbf{R}$. ∎

## 5.2.4   PREDICTION OF DERIVATIVES $f^{(s)}$ BY SAMPLES OF $f$

The aim of this section is to show that the derivatives $f^{(s)}$ of a signal $f$ can be predicted by the derivatives $(S_W^{\varphi} f)^{(s)}$, i.e., by

$$(S_W^{\varphi} f)^{(s)}(t) = W^s \sum_{k=-\infty}^{\infty} f\left(\frac{k}{W}\right) \varphi^{(s)}(Wt - k), \qquad (5.2.26)$$

thus by samples of $f$ only, provided the kernel $\varphi$ belongs to $C_{00}^{(s)}(\mathbf{R})$ and satisfies (5.2.12) for some $r \geq s$.

**Lemma 5** *Let* $j \in \mathbf{N}_0$, $r, s \in \mathbf{N}$ *with* $j, s \leq r - 1$. *If* $\varphi \in C_{00}^{(s)}(\mathbf{R})$ *satisfies* (5.2.12), *then*

$$\sum_{k=-\infty}^{\infty} (u - k)^j \, \varphi^{(s)}(u - k) = \begin{cases} (-1)^s s!, & j = s \\ 0, & j \neq s. \end{cases} \qquad (5.2.27)$$

The proof follows by mathematical induction or by Lemma 3, noting that in view of Lemma 1(b), (c) and Poisson's summation formula,

$$(-i)^j \sum_{k=-\infty}^{\infty} (u-k)^j \varphi^{(s)}(u-k) \sim \sum_{k=-\infty}^{\infty} \left(\frac{d}{dv}\right)^j (iv)^s \hat{\varphi}(v)\Big|_{v=2\pi k} e^{i2\pi ku}.$$

(5.2.28)

The counterpart of Theorems 1 and 2 for the approximation of derivatives now reads:

**Theorem 5** *Let $r, s \in \mathbb{N}$, $s \leq r-1$, and assume that $\varphi \in C_{00}^{(s)}(\mathbb{R})$ satisfies (5.2.12).*
*a) If $f: \mathbb{R} \mapsto \mathbb{R}$ is such that $f^{(s)}(t)$ exists for some $t \in \mathbb{R}$, then*

$$\lim_{W \to \infty} (S_W^\varphi f)^{(s)}(t) = f^{(s)}(t).$$

*b) Furthermore, one has*

$$\lim_{W \to \infty} \|f^{(s)} - (S_W^\varphi f)^{(s)}\|_{C(\mathbb{R})} = 0 \qquad (f \in C^{(s)}(\mathbb{R})),$$

*and there holds the error estimate*

$$\|f^{(s)} - (S_W^\varphi f)^{(s)}\|_{C(\mathbb{R})} \leq \frac{m_r(\varphi^{(s)})}{r!} \|f^{(r)}\|_{C(\mathbb{R})} W^{-r+s}$$
$$(f \in C^{(r)}(\mathbb{R}); W > 0).$$

**Proof** Since $f^{(s)}(t)$ exists, one has the representation

$$f(u) = \sum_{\nu=0}^{s} \frac{f^{(\nu)}(t)}{\nu!}(u-t)^\nu + (u-t)^s \rho(u-t) \qquad (u \in \mathbb{R})$$

with $\rho: \mathbb{R} \mapsto \mathbb{R}$ satisfying $\lim_{u \to 0} \rho(u) = \rho(0) = 0$. Now let $\varepsilon > 0$ be given. Choose $\delta > 0$ such that $|\rho(u)| < \varepsilon$ for all $|u| < \delta$. Then it follows by (5.2.27) that

$$(S_W^\varphi f)^{(s)}(t) = W^s \sum_{\nu=0}^{s} \frac{f^{(\nu)}(t)}{\nu!} \sum_{k=-\infty}^{\infty} \left(\frac{k}{W} - t\right)^\nu \varphi^{(s)}(Wt - k)$$

$$+ W^s \sum_{k=-\infty}^{\infty} \left(\frac{k}{W} - t\right)^s \rho\left(\frac{k}{W} - t\right) \varphi^{(s)}(Wt - k)$$

$$= f^{(s)}(t) + \sum_{|k-Wt|<\delta W} (k - Wt)^s \rho\left(\frac{k}{W} - t\right) \varphi^{(s)}(Wt - k)$$

for all $W > 0$ so large that the support of $\varphi$ is contained in the interval $[-\delta W, \delta W]$. This yields part (a) since $|\rho(k/W - t)| < \varepsilon$ for all $k$ in question, implying that the absolute value of the latter series is bounded by $\varepsilon\, m_s(\varphi^{(s)})$.

Concerning the proof of (b), choose $\delta > 0$ such that $|f^{(s)}(t) - f^{(s)}(y)| < \varepsilon$ for all $|t-y| < \delta$. An application of (5.2.26) to the Taylor expansion (5.2.14) with $r = s$ yields, in view of (5.2.27) for all $W$ sufficiently large,

$$(S_W^\varphi f)^{(s)}(t) = W^s \sum_{|k-Wt|<\delta W} f\left(\frac{k}{W}\right)\varphi^{(s)}(Wt - k)$$

$$= \frac{W^s}{(s-1)!} \sum_{|k-Wt|<\delta W} \int_t^{k/W} f^{(s)}(y)\left(\frac{k}{W} - y\right)^{s-1} dy\, \varphi^{(s)}(Wt - k).$$

So one obtains again by (5.2.27) that

$$|f^{(s)}(t) - (S_W^\varphi f)^{(s)}(t)| = \left|\frac{W}{(s-1)!} \sum_{|k-Wt|<\delta W} \int_t^{k/W} \{f^{(s)}(t) - f^{(s)}(y)\}\right.$$

$$\left. \times (k - Wy)^{s-1} dy\, \varphi^{(s)}(Wt - k)\right|$$

$$< \frac{\varepsilon}{s!} \sum_{k=-\infty}^{\infty} |Wt - k|^s\, |\varphi^{(s)}(Wt - k)| \le \frac{\varepsilon\, m_s(\varphi^{(s)})}{s!}.$$

This establishes the first assertion of part (b). The proof of the error estimate is similar to that of Theorem 2, applying now (5.2.26) to (5.2.14) as it stands.   ∎

Examples of kernels satisfying the assumption of Theorem 5 are $\varphi_3$, $\varphi_4$ and $\varphi_5$ of Table 2 with $s \le 1$, $s \le 2$ and $s \le 3$, respectively. The associated sampling series $(S_W^{\varphi_r} f)^{(s)}$ again only need samples from the past. This means that these series can be used to predict the derivatives $f^{(s)}$ in terms of samples of $f$.

Let us consider the kernel $\varphi_3$ in more detail. Since the derivatives of the $B$-splines can be calculated according to the formula

$$M_r'(t) = M_{r-1}(t + 1/2) - M_{r-1}(t - 1/2) \qquad (t \in \mathbf{R}),$$

which can be easily deduced from (5.2.10) and Lemma 1(b) (see also Schoenberg [803, p. 12]), it follows that

$$\varphi_3'(t) = \frac{1}{8}\{47\, M_2(t - 3/2) - 109\, M_2(t - 5/2)$$

$$+ 85\, M_2(t - 7/2) - 23\, M_2(t - 9/2)\}.$$

Hence it follows by Theorem 5 that for $f \in C^{(3)}(\mathbf{R})$

$$\left\| W \sum_{k=-\infty}^{\infty} f\left(\frac{k}{W}\right) \varphi_3'(Wt-k) - f'(t) \right\|_{C(\mathbf{R})} = \mathcal{O}(W^{-2}), \qquad (W \to \infty).$$

This result enables one to predict the derivative $f'(t)$ in terms of a series which involves just five samples of $f$ which all lie to the left of $t-1/2W < t$.

### 5.2.5 ROUND-OFF AND TIME JITTER ERRORS

When setting up the sampling series (5.2.1) it may happen that one does not have the exact sample values $f(k/W)$ at one's disposal but only rounded values $\bar{f}(k/W)$, differing by $\eta_k \equiv f(k/W) - \bar{f}(k/W)$ with $|\eta_k| \leq \eta$, for all $k \in \mathbf{Z}$ and some $\eta > 0$. In digital signal processing this is the case when, for example, the sample values are replaced by the nearest discrete (quantized) values.

In this respect one is interested in the error occurring when $f(t)$ is approximated by the series

$$(S_W^\varphi \bar{f})(t) \equiv \sum_{k=-\infty}^{\infty} \bar{f}(k/W)\varphi(Wt - k).$$

This error can be split up in the form

$$|f(t) - S_W^\varphi \bar{f}(t)| \leq |f(t) - (S_W^\varphi f)(t)| + (Q_\eta f)(t). \tag{5.2.29}$$

The first term on the right is the error arising when the exact sample values are used, and the second term is the so-called *total round-off* or *quantization error*, defined by

$$Q_\eta f(t) \equiv |(S_W^\varphi f)(t) - (S_W^\varphi \bar{f})(t)|.$$

Now the latter one can be estimated by

$$\|(Q_\eta f)\|_{C(\mathbf{R})} = \sup_{t \in \mathbf{R}} \left| \sum_{k=-\infty}^{\infty} \eta_k \varphi(Wt - k) \right| \leq \eta m_0(\varphi).$$

meaning that the maximal error due to rounding is, apart from a constant, as large as the maximal difference $|f(k/W) - \bar{f}(k/W)|$, $k \in \mathbf{Z}$.

Hence (5.2.29) can be replaced by

$$\|f - S_W^\varphi \bar{f}\|_{C(\mathbf{R})} \leq \|f - S_W^\varphi f\|_{C(\mathbf{R})} + \eta m_0(\varphi). \tag{5.2.30}$$

where the first term on the right-hand side can be estimated further by the bounds given in Subsections 5.2.1 and 5.2.2.

It may also happen that the samples cannot be taken at the instants $k/W$ but at $(k/W + \delta_k)$, the sampled values now being $f(k/W + \delta_k)$. These errors in timing give rise to the so-called *total time jitter error*

$$(J_\delta f)(t) \equiv \left| \sum_{k=-\infty}^{\infty} f\left(\frac{k}{W}\right) \varphi(Wt - k) - \sum_{k=-\infty}^{\infty} f\left(\frac{k}{W} + \delta_k\right) \varphi(Wt - k) \right|.$$

Assuming that $|\delta_k| \leq \delta$ for all $k \in \mathbf{Z}$ and some $\delta > 0$, this error can be estimated using the mean value theorem in terms of the derivative of $f$, namely,

$$\|J_\delta f\|_{C(\mathbf{R})} \leq \sup_{k \in \mathbf{Z}} \left\{ \sup_{t \in \mathbf{R}} |f(t) - f(t + \delta_k)| \right\} \sup_{t \in \mathbf{R}} \sum_{k=-\infty}^{\infty} |\varphi(t - k)| \qquad (5.2.31)$$

$$\leq \delta\, m_0(\varphi) \|f'\|_{C(\mathbf{R})}.$$

So the jitter error essentially depends on the smoothness of $f$. For the associated approximation error it follows that for each $f \in C^{(1)}(\mathbf{R})$

$$\left\| f(\cdot) - \sum_{k=-\infty}^{\infty} f\left(\frac{k}{W} + \delta_k\right) \varphi(W \cdot - k) \right\|_{C(\mathbf{R})}$$

$$\leq \|f - S_W^\varphi f\|_{C(\mathbf{R})} + \delta\, m_0(\varphi) \|f'\|_{C(\mathbf{R})}. \qquad (5.2.32)$$

Using the concept of moduli of continuity introduced in Subsection 5.2.3 the right-hand sides of (5.2.31) and (5.2.32) could also be rewritten in terms of $\omega_1(\delta; f; C(\mathbf{R}))$ instead of $\|f'\|_{C(\mathbf{R})}$. The resulting estimates would then hold for each $f \in C(\mathbf{R})$ rather than for $f \in C^{(1)}(\mathbf{R})$ only.

The results of this section show that the sampling series $S_W^\varphi f$ are much more stable in regard to round-off or time jitter errors than is the classical Shannon series. This is due to the fact that the $S_W^\varphi$ define linear operators that are uniformly bounded with respect to $W > 0$, whereas Shannon's series does not. For round-off and jitter error in connection with Shannon's series, see [170]. also [144]. and the literature cited there.

## 5.3   Prediction of Random Signals

### 5.3.1   Continuous and Differentiable Stochastic Processes

Given a probability space $(\Omega, \mathcal{A}, \mathcal{P})$, a real-valued stochastic (random) process, namely, an $\mathcal{A}$-measurable function

$$X \equiv X(t) \equiv X(t, \omega)$$

of $\omega$ in $\Omega$ for each $t \in \mathbf{R}$, belongs to $L^2(\Omega)$, if the norm

$$\|X(t,\cdot)\|_2 \equiv \left\{ \int_\Omega |X(t,\omega)|^2 \, dP(\omega) \right\}^{1/2} \equiv \{E(|X(t)|^2)\}^{1/2} \qquad (5.3.1)$$

is finite for all $t \in \mathbf{R}$. $X$ is said to be weak sense stationary (w.s.s.) if its autocorrelation function (a.c.f.)

$$R_X(t, t + \tau) \equiv \int_\Omega X(t,\omega) \, X(t + \tau, \omega) \, dP(\omega) \qquad (5.3.2)$$

is independent of $t \in \mathbf{R}$. i.e., $R_X(t, t + \tau) = R_X(\tau)$. Note that for w.s.s. processes $R_X(\tau)$ is an even function with $\sup_{\tau \in \mathbf{R}} |R_X(\tau)| = R_X(0)$. In particular, the norm (5.3.1) is independent of $t$ and equals $\{R_X(0)\}^{1/2}$. A process $X \in L^2(\Omega)$ is called continuous in the mean (i.m.) at $t_0 \in \mathbf{R}$ if

$$\lim_{h \to 0} \left\{ \int_\Omega |X(t_0 + h, \omega) - X(t_0, \omega)|^2 \, dP(\omega) \right\}^{1/2} = 0,$$

and differentiable i.m. at $t_0 \in \mathbf{R}$ if there exists a process $X' \in L^2(\Omega)$ such that

$$\lim_{h \to 0} \left\{ \int_\Omega \left| \frac{X(t_0 + h, \omega) - X(t_0, \omega)}{h} - X'(t_0, \omega) \right|^2 \, dP(\omega) \right\}^{1/2} = 0.$$

Higher order derivatives $X^{(s)}$ are defined iteratively.

**Lemma 6** *Let $X \in L^2(\Omega)$ be a w.s.s. process and $s \in \mathbf{N}$.*
*a) The following conditions are equivalent:*
(i)      *$X$ is continuous i.m. at some $t_0 \in \mathbf{R}$,*
(ii)     *$X$ is continuous i.m. on $\mathbf{R}$,*
(iii)    *$R_X \in C(\mathbf{R})$.*
*b) The following conditions are equivalent:*
(i)      *$X^{(s)}$ exists i.m. at some $t_0 \in \mathbf{R}$,*
(ii)     *$X^{(s)}$ exists i.m. on $\mathbf{R}$,*
(iii)    *$R_X \in C^{(2s)}(\mathbf{R})$.*
*Furthermore, if $X^{(s)}$ exists, then it is w.s.s. too, and*
*$R_{X^{(s)}} = (-1)^s R_X^{(2s)}$.*

For a proof of this lemma see Splettstösser [849] or Papoulis [710, Section 9-6].

If $X$ is a w.s.s. process, then continuity (differentiability) i.m. at some point $t_0 \in \mathbf{R}$ therefore implies continuity (differentiability) i.m. for all $t \in \mathbf{R}$. So one can simply speak of a continuous (differentiable) process.

### 5.3.2  PREDICTION OF WEAK SENSE STATIONARY STOCHASTIC PROCESSES

When dealing with random signals one often uses stochastic processes which are stationary in the weak sense as a model. For the prediction of such a process $X \in L^2(\Omega)$ let us consider the sampling series

$$(S_W^\varphi X)(t,\omega) \equiv \sum_{k=-\infty}^{\infty} X\Big(\frac{k}{W},\omega\Big)\varphi(Wt-k) \qquad (t \in \mathbf{R}; \omega \in \Omega). \qquad (5.3.3)$$

It defines a family of bounded linear operators mapping $L^2(\Omega)$ into itself, satisfying

$$\|S_W^\varphi X)(t,\cdot)\|_2 = \Big\{ \sum_{k,\mu=-\infty}^{\infty} R_X\Big(\frac{k-\mu}{W}\Big)\varphi(Wt-k)\,\varphi(Wt-\mu)\Big\}^{1/2}$$

$$\leq m_0(\varphi)\{R_X(0)\}^{1/2} = m_0(\varphi)\|X\|_2.$$

For the proof of the main result of this section an auxiliary operator is needed, namely,

$$(U_W^\varphi f)(t) \equiv \sum_{k,\mu=-\infty}^{\infty} f\Big(\frac{k-\mu}{W}\Big)\varphi(Wt-k)\,\varphi(Wt-\mu)$$

$$(t \in \mathbf{R}; W > 0),$$

where $f \in C(\mathbf{R})$ and $\varphi \in C_{00}(\mathbf{R})$. $U_W^\varphi$ is a bounded linear operator from $C(\mathbf{R})$ into itself with operator norm $\|U_W^\varphi\|_{[C(\mathbf{R})]} \leq \{m_0(\varphi)\}^2$.

**Lemma 7**  *Let $\varphi \in C_{00}(\mathbf{R})$ satisfy (5.2.12). Then*

$$|f(0) - (U_W^\varphi f)(t)| \leq K\|f^{(r)}\|_{C(\mathbf{R})}W^{-r}$$

$$(f \in C^{(r)}(\mathbf{R}): t \in \mathbf{R}: W > 0) \qquad (5.3.4)$$

*for some constant $K$ independent of $f$, $t$ and $W$.*

**Proof**  Let $f \in C^{(r)}(\mathbf{R})$. Similarly as in the proof of Theorem 2 one obtains by Taylor's expansion

$$(U_W^\varphi f)(t) - f(0)$$

$$= \sum_{\nu=1}^{r-1} \frac{1}{\nu!} f^{(\nu)}(0)\Big\{ \sum_{k,\mu=-\infty}^{\infty} \Big(\frac{k-\mu}{W}\Big)^\nu \varphi(Wt-k)\varphi(Wt-\mu)\Big\}$$

$$+ \sum_{k,\mu=-\infty}^{\infty} \frac{1}{(r-1)!} \int_0^{(k-\mu)/W} f^{(r)}(y)\Big(\frac{k-\mu}{W}-y\Big)^{r-1} dy$$

$$\times \varphi(Wt-k)\varphi(Wt-\mu).$$

An application of the binomial formula to $(k - \mu)^\nu = [(k - Wt) + (Wt - \mu)]^\nu$ shows that the first double series vanishes by (5.2.12), and (5.3.4) follows by an easy estimate of the integrals. ∎

Concerning the convergence i.m. of $S_W^\varphi X$ toward $X$ one has:

**Theorem 6** *Let $\varphi \in C_{00}(\mathbf{R})$ satisfy (5.2.12) with $r$ replaced by $2r$ for some $r \in \mathbf{N}$. If $X \in L^2(\Omega)$ has a derivative i.m. of order $r$, then*

$$\{E(|X - S_W^\varphi X|^2)\}^{1/2} \le K\{E(|X^{(r)}|^2)\}^{1/2} W^{-r} \quad (t \in \mathbf{R}: W > 0) \quad (5.3.5)$$

*with $K$ independent of $X$ and $W$.*

**Proof** The left-hand side of (5.3.5) can be rewritten in terms of the operators $U_W^\varphi$ and $S_W^\varphi$ and the a.c.f. $R_X$ as

$$\begin{aligned}
E(|X - S_W^\varphi X|^2) &= (U_W^\varphi R_X)(t) - 2(S_W^\varphi R_X)(t) + R_X(0) \\
&\le |(U_W^\varphi R_X)(t) - R_X(0)| + 2|R_X(0) - (S_W^\varphi R_X)(t)|.
\end{aligned} \quad (5.3.6)$$

Since the a.c.f. $R_X$ belongs to $C^{(2r)}(\mathbf{R})$ one can apply Lemma 7 and Theorem 2 with $2r$ instead of $r$. This gives

$$E(|X - S_W^\varphi X|^2) \le K\|R_X^{(2r)}\|_{C(\mathbf{R})} W^{-2r} = K\|X^{(r)}\|_2^2 W^{-2r}$$

which is just (5.3.5). ∎

Concerning applications, the best possible order of approximation according to Theorem 6 for the kernels $\varphi_r$ of Table 2 is $\mathcal{O}(W^{-1})$ in the case of $\varphi_2$ and $\varphi_3$, as well as $\mathcal{O}(W^{-2})$ for the kernels $\varphi_4$ and $\varphi_5$.

For error estimates in terms of a suitable modulus of continuity see [175, 176]. For the approximation of w.s.s. stationary processes by the original Shannon series, see Splettstösser [850].

# 6

# Polar, Spiral, and Generalized Sampling and Interpolation

Henry Stark

## 6.1 Introduction

In this chapter we collect a number of results related to polar, spiral, and generalized non-uniform sampling. In most cases we shall not give proofs since those are often lengthy and the proofs are given in other places. A number of polar sampling theorems and their proofs are given in [858]. However, these are of limited application since they involve sampling the function at zeros of Bessel functions. A notable exception is the uniform sampling theorem which has applications in computerized tomography (CT). Another result with application in medical imaging is reconstruction from samples along spiral scans. This result was motivated by and has application in magnetic resonance imaging (MRI). A generalized reconstruction formula from non-uniform samples is given in Section 6.4. The derivation of this result is based on the powerful theory of projections onto convex sets (POCS) also known as the theory of convex projections. Other results on non-uniform sampling theory are given by Marvasti in a recent book [628] as well as in a chapter by him in this book.

## 6.2 Sampling in Polar Coordinates

### 6.2.1 SAMPLING OF PERIODIC FUNCTIONS

Let $f(t)$ satisfy the Dirichlet conditions, be periotic in $t$, with period $T$. Assume that $f(t)$ satisfies a Fourier series expansion of the form

$$f(t) = \sum_{k=-K}^{K} C_k e^{j2\pi \frac{kt}{T}}, \qquad (6.2.1)$$

i.e., $C_k = 0$ for $|k| > K$. Then we say that $f(t)$ is band-limited to highest (radian) frequency $\omega_m = \frac{2\pi K}{T}$. In that case we can reconstruct $f(t)$ from the interpolation formula

$$f(t) = \sum_{l=0}^{2K} f(lT_s) \frac{\sin[\frac{\pi}{T_s}(t - lT_s)]}{(2K+1)\sin[\frac{\pi}{T}(t - lT_s)]}, \tag{6.2.2}$$

where $T_s$, the sampling interval, is $T_s = \frac{T}{(2K+1)}$. This result is given in [858] but apparently is known in the digital signal processing community. It is a straightforward exercise, using the same technique as in [858], to show that (6.2.2) can easily be extended to

$$f(t) = \sum_{l=0}^{N-1} f(lT_s) \frac{\sin[\frac{\pi}{T_s}(t - lT_s)]}{N \sin[\frac{\pi}{T}(t - lT_s)]}, \tag{6.2.3}$$

where $N$ is any odd integer with $N > 2K + 1$. For interpolation in azimuth, the argument is $t = \theta$, $T = \pi$, $t_s = \frac{\pi}{N}$ and (6.2.3) takes the form

$$f(\theta) = \sum_{l=0}^{N-1} f\left(\frac{2\pi l}{N}\right) \frac{\sin[\frac{N}{2}(\theta - \frac{2\pi l}{N})]}{N \sin[\frac{1}{2}(\theta - \frac{2\pi l}{N})]}. \tag{6.2.4}$$

For $N$ even, the derivation of the interpolation formula is given in [864]. The result is

$$f(\theta) = \sum_{l=0}^{N-1} f\left(\frac{2\pi l}{N}\right) \frac{\sin[\frac{1}{2}(N-1)(\theta - \frac{2\pi l}{N})]}{N \sin[\frac{1}{2}(\theta - \frac{2\pi l}{N})]}. \tag{6.2.5}$$

A compact formula valid for arbitrary $N, N > 2K$, is the following:

$$f(t) = \sum_{l=0}^{N-1} f(lT_s) \frac{\sin[\frac{\pi}{T}(2K+1)(t - lT_s)]}{N \sin[\frac{\pi}{T}(t - lT_s)]}. \tag{6.2.6}$$

As before, for interpolation in azimuth, we let $t = \theta, T = 2\pi, T_s = \frac{2\pi}{N}$ and Eq. (6.2.6) takes the form

$$f(\theta) = \sum_{l=0}^{N-1} f\left(\frac{2\pi l}{N}\right) \frac{\sin[\frac{2K+1}{2}(\theta - \frac{2\pi l}{N})]}{N \sin[\frac{1}{2}(\theta - \frac{2\pi l}{N})]}. \tag{6.2.7}$$

All these results are derived in [864]. It turns out that for $N$ large, Eqs. (6.2.4) and (6.2.5) give approximately the same results except when $\theta$ is near zero or a multiple of $2\pi$. When precision at those points is needed or when $N$ is small, it is important to use the correct formula.

Equation (6.2.4) can be used to interpolate in azimuth a function $f(r, \theta)$ in two polar variables: radius $r$ and azimuth $\theta$. Thus given the sample points $f\left(r, 2\pi k/(2K+1)\right) k = 0, ..., 2K$ we obtain as a reconstruction formula

$$f(r, \theta) = \sum_{k=0}^{2K} f\left(r, \frac{2\pi k}{2K+1}\right) \sigma_k(\theta), \qquad (6.2.8)$$

where

$$\sigma_k(\theta) \equiv \sigma\left(\theta - \frac{2\pi k}{2K+1}\right) \equiv \frac{\sin[\frac{2K+1}{2}(\theta - \frac{2\pi k}{2K+1})]}{(2K+1)\sin[\frac{1}{2}(\theta - \frac{2\pi k}{2K+1})]}. \qquad (6.2.9)$$

This result is rather obvious from the earlier formulas. A somewhat cumbersome formula, of limited utility, for interpolating in both $r, \theta$ when the function is both radially and angularly band-limited is given in [858], Eq. (5.2). We repeat it for the reader's interest without proof. Thus let $f(r, \theta)$ satisfy the angular bandwidth constraint

$$f(r, \theta) = \sum_{n=-K}^{K} C_n(r) e^{jn\theta} \qquad (6.2.10)$$

and let its Fourier transform satisfy $F(\rho, \theta) = 0$ for $\rho \geq a$. Also define $Z_{ni}$ to be the $i$th zero of $J_n(\cdot)$, the $n$th-order Bessel function of the first kind; finally define $\alpha_{ni} \equiv Z_{ni}/a$ as the scaled zero associated with $Z_{ni}$. Then it is proven in [858] that

$$f(r, \theta) = \sum_{i=1}^{\infty} \sum_{k=0}^{2K} f\left(\alpha_{0i}, \frac{2\pi k}{2K+1}\right) \Phi_{0i}(r)(2K+1)^{-1}$$

$$+ 2(2K+1)^{-1} \sum_{n=1}^{K} \sum_{i=1}^{\infty} \sum_{k=0}^{2K} f\left(\alpha_{ni}, \frac{2\pi k}{2K+1}\right) \Phi_{ni}(r)$$

$$\times \cos\left[n(\theta - \frac{2\pi k}{2K+1})\right], \qquad (6.2.11)$$

where

$$\Phi_{ni}(r) \equiv \frac{2\alpha_{ni} J_n(ra)}{a J_{n+1}(\alpha_{ni}a)(\alpha_{ni}^2 - r^2)}. \qquad (6.2.12)$$

The sample points are proportional to the interlaced zeros of the first $K$ Bessel functions of the first kind. The interlacing refers to $0 < \alpha_{n1} < \alpha_{n+1,1} < \alpha_{n2} < \alpha_{n+1,2} < \cdots$ . for $n = 0, 1, \ldots$ . To use the sampling theorem, the zeros could be stored in a computer or computed by using a zero-finding algorithm [3], p. 371.

To the best of the author's knowledge there are very few instances in the real world where it would be advantageous or serendipitous for sampling to occur at zeros of Bessel functions. Of much greater interest is a polar interpolating formula that uses sample points uniformly spaced in radius and azimuth. Such a formula is given next.

## 6.2.2  A FORMULA FOR INTERPOLATING FROM SAMPLES ON A UNIFORM POLAR LATTICE

If a function $f(x)$ is band-limited to $W$, its Fourier transform, $F(u)$, satisfies

$$F(u) = 0, \qquad |u| \geq W. \tag{6.2.13}$$

For such a function, the Whittaker-Shannon sampling theorem shows that $f(x)$ is completely determined by instantaneous samples uniformly spaced with intersample period $X_s$ such that $X_s \leq 1/(2W)$. At points other than the sampling points, $f(x)$ is given by the interpolation formula

$$f(x) = \sum_{n=-\infty}^{\infty} f(nX_s) \, \text{sinc}\left(\frac{x - nX_s}{X_s}\right). \tag{6.2.14}$$

In particular if $X_s = 1/(2W)$, then

$$f(x) = \sum_{n=-\infty}^{\infty} f\left(\frac{n}{2W}\right) \, \text{sinc}\left[2W\left(x - \frac{n}{2W}\right)\right]. \tag{6.2.15}$$

Now consider a function of two variables $f(r, \theta)$. If the Fourier transform $F(\rho, \phi)$ of $f(r, \theta)$ satisfies $F(\rho, \phi) = 0$, $\rho > W$ and $f(r, \theta)$ is angularly band-limited to $K$, then we can apply Eqs. (6.2.14) and, say, (6.2.5) in succession to obtain

$$f(r, \theta) = \sum_{l=0}^{N-1} \sum_{n=-\infty}^{\infty} \tilde{f}\left(\frac{n}{2W}, \frac{2\pi l}{N}\right) \text{sinc} 2W\left(r - \frac{n}{2W}\right)$$
$$\times \frac{\sin[\frac{1}{2}(N-1)(\theta - \frac{2\pi l}{N})]}{N \sin[\frac{1}{2}(\theta - \frac{2\pi l}{N})]} \tag{6.2.16}$$

where $0 \leq \theta < 2\pi$ and $r \geq 0$ if $r, \theta$ represents polar coordinates, $N$ is assumed even and $\tilde{f}(\frac{n}{2W}, \frac{2\pi l}{N})$ is given by

$$\tilde{f}\left(\frac{n}{2W}, \frac{2\pi l}{N}\right) = \begin{cases} f\left(\frac{n}{2W}, \frac{2\pi l}{N}\right), & n \geq 0 \\ f\left(-\frac{n}{2W}, \frac{2\pi l}{N} + \pi\right), & n < 0. \end{cases} \tag{6.2.17}$$

We note that Eq. (6.2.17) is a consequence of the fact that the samples are originally collected on a bi-polar coordinate system $(\hat{r}, \Omega)$ where $-\infty < \hat{r} < \infty$ and $0 \leq \Omega < \pi$. The relation between $\hat{r}$ and $r$ is

$$\hat{r} = \begin{cases} r & \text{if } \theta = \Omega \\ -r & \text{if } \theta = \Omega + \pi. \end{cases} \tag{6.2.18}$$

Equation (6.2.16) is often used in the Fourier domain, when the image function $f(x, y)$ is space-limited. Then the following theorem applies.

**Theorem:** Let $f(x, y)$ be space-limited to $2A$. Let its Fourier transform in polar coordinates $F(\rho, \phi)$ be angularly band-limited to $K$. Then $F(\rho, \phi)$ can be reconstructed from its polar samples via

$$F(\rho, \phi) = \sum_{n=-\infty}^{\infty} \sum_{k=0}^{N-1} \tilde{F}\left(\frac{n}{2A}, \frac{2\pi k}{N}\right) \operatorname{sinc}\left[\frac{2A(\rho - n)}{2A}\right] \sigma\left(\phi - \frac{2\pi k}{N}\right),$$

(6.2.19)

where $N$ is assumed even,

$$\sigma = \frac{\sin[\frac{1}{2}(N - 1)\phi]}{N \sin[\frac{1}{2}\phi]}$$

(6.2.20)

and

$$\tilde{F}\left(\frac{n}{2A}, \frac{2\pi k}{N}\right) = \begin{cases} F(\frac{n}{2A}, \frac{2\pi k}{N}), & n \geq 0 \\ \\ F(\frac{-n}{2A}, \frac{2\pi k}{N} + \pi), & n < 0. \end{cases}$$

(6.2.21)

By an appropriate change in the sigma($\cdot$) function, Eq. (6.2.19) can be used with $N$ odd but in CT, where Eq. (6.2.19) has its major application, $N$ is usually even.

## 6.2.3 Applications in Computer Tomography (CT)

CT reconstruction by direct Fourier techniques is based on the projection theorem [808] (as is the convolution back-projection method). The theorem states that the Fourier transform of a projection is a center cross section of the Fourier transform of the image. Thus, if $f(x, y)$ represents the absorptivity function of the object with respect to a fixed coordinate system $x$-$y$ and $f_\phi(\hat{x}, \hat{y})$ represents the object in a coordinate system $\hat{x}$-$\hat{y}$ rotated from $x$-$y$ by an angle $o$, then the projection of the object at view angle $o$ is defined as

$$p_\phi(\hat{x}) \equiv \int_L f_\phi(\hat{x}, \hat{y}) d\hat{y},$$

(6.2.22)

where $L$ is the beam path. In practice $p_\phi(\hat{x})$ represents the data that are actually obtained. For each view angle $\phi$, $p_\phi(\hat{x})$ is sampled at a sequence of points $\{\hat{x}_i\}$ determined by the location of the detectors. The angle $\phi$ is then incremented by a small amount and the process is repeated. As many as 360 view angles might be used in a given run. We consider only the parallel beam geometry. i.e., for each view the $M$ X-ray sources located on the side of the object form $M$ parallel pencil beams which go through the object and are detected by $M$ detectors on the opposite side of the object. We shall ignore the fact that $p_\phi(\hat{x})$ is known only over a discrete set of points; this is done only to prevent obscuring the presentation with the details of discrete sampling.

The one-dimensional (1-D) Fourier transform (FT) of $p_\phi(\hat{x})$ is given by

$$P_\phi(u) = \int_{-\infty}^{\infty} p_\phi(\hat{x}) e^{-j2\pi \hat{x} u} d\hat{x}. \tag{6.2.23}$$

In Eq. (6.2.23), $\phi$ is held fixed and $u$ is a frequency variable that can be positive or negative. If $F(\rho, \phi)$ denotes the two-dimensional (2-D) FT of $f(x, y)$ in polar coordinates, then the projection theorem states that

$$P_\phi(\rho) = F(\rho, \phi), \qquad \rho \geq 0, \quad 0 < \phi \leq 2\pi. \tag{6.2.24}$$

The correct interpretation of Eq. (6.2.24) is that the FT of each projection is a central cross section of the FT of the object. In theory if $F(\rho, \phi)$ were known everywhere, it would be a simple matter to convert this FT from polar to Cartesian coordinates. Assuming this to have been done and denoting the corresponding 2-D Cartesian FT by $F_c(u, v)$, we can reconstruct the object function by a Fourier inversion:

$$f(x, y) = \int_{-\infty}^{\infty} \int_{-\infty}^{\infty} F_c(u, v) e^{j2\pi(ux + vy)} du\, dv. \tag{6.2.25}$$

Perhaps the most serious difficulty with the above approach is that $F(\rho, \phi)$ is not known everywhere but only on a finite discrete set of points $\{\rho_j, \phi_k\}$. The problem then becomes one of interpolating from the known values at the polar points to the values required over a rectangular Cartesian grid which allows the approximate realization of Eq. (6.2.25) via an inverse 2-D FFT routine.

Because the reconstructed image is highly sensitive to the quality of the interpolations, most of the commercial machines work in the spatial domain, using what is known as the convolution-back projection (CBP) method. In CBP, the projection function $p_\phi(\hat{x})$ is first convolved with an appropriate filter function $k(\hat{x})$ to produce a filtered projection function $g_\phi(s)$ and then $g_\phi(s)$ is back-projected to form the image $f(x, y)$. In mathematical terms this amounts to

$$g_\phi(s) = \int_{-\infty}^{\infty} p_\phi(\hat{x}) k(s - \hat{x}) d\hat{x} \tag{6.2.26}$$

(convolution-filtering) followed by

$$f(x, y) = \int_{0}^{\pi} g_\phi[r \cos(\phi - \theta)] d\phi \tag{6.2.27}$$

(back-projection).

In Eq. (6.2.27), $x = r \cos \phi$ and $y = r \sin \phi$. If $N_v$ represents the number of views, then the CBP method requires $\mathcal{O}(N_v^3)$ operations (i.e, multiplications and additions) on the data. In contrast the direct Fourier method, i.e., Eqs. (6.2.23) and (6.2.25) can produce satisfactory images using only $\mathcal{O}(N_v^2 \log N_v^2)$ operations. We now sketch out how the direct Fourier method is actually used in CT.

In step (1) we compute $P_\phi(u)$ as in Eq. (6.2.23). In practice the discrete Fourier transform (DFT) is used and is implemented using the fast Fourier transform (FFT). In step (2) Eq. (6.2.24) is used to obtain $F(p, \varphi)$ on a uniform polar lattice. The key step is step (3) in which Eq. (6.2.19) is used to interpolate $F(\rho, \varphi)$ from a polar to a Cartesian lattice. Usually we take $N = 2K + 2$ and evaluate the right-hand side of Eq. (6.2.17) at the appropriate Cartesian points $\{u_l, v_m, l, m, = \ldots, -2, -1, 0, 1, 2, \ldots\}$, where

$$\rho_{lm} \equiv \sqrt{u_l^2 + u_m^2} \geq 0,$$

$$\phi_{lm} \equiv \cos^{-1}\left[\frac{u_l}{\rho_{lm}}\right]. \tag{6.2.28}$$

This gives the set $\{F_c(u_l, v_m) = F(\rho_{lm}, \phi_{lm})\}$ from which $f(x, y)$ can be computed by a two-dimensional inverse FFT (i.e., as shown in Eq. (6.2.25) for the continuous case).

In practice we must use an approximate version of Eq. (6.2.17) since we do not know the samples at an infinite number of discrete radial points. The approximate formula is given by

$$\hat{F}(\rho, \phi) = \sum_{n=n_\rho-L_\rho}^{N_\rho+L_\rho} \sum_{k=k_\phi-L_\phi}^{k_\phi+L_\phi} \tilde{F}\left(\frac{n}{2A}, \frac{\pi k}{K+1}\right) \operatorname{sinc}\left[2A\left(\rho - \frac{n}{2A}\right)\right]$$

$$\times \sigma\left(\phi - \frac{\pi k}{K+1}\right), \tag{6.2.29}$$

where the nearest neighbors to the interpolated point are $[2a\rho] \equiv n_\rho$, $[(K+1)\phi/\pi] \equiv k_\phi$, $[x]$ means rounding $x$ to the nearest integer, and $2L_\rho + 1$ and $2L_\phi + 1$ are the total number of points involved in the interpolation in the radial and azimuth directions, respectively. $L_\rho$ and $L_\phi$ are usually determined by trial and error, i.e., by examination of the reconstructed image. Typically 15 polar points suffice to produce images comparable in quality to convolution back-projection.

In practice one or more spectral conditioning operations are applied to improve the quality of the reconstructions. The actual implementation of the direct Fourier method is discussed in several places including references [866] and [867]. Here our intent was to demonstrate that polar sampling and interpolation has important application in real problems. A block diagram of the direct Fourier method is shown in Figure 6.1.

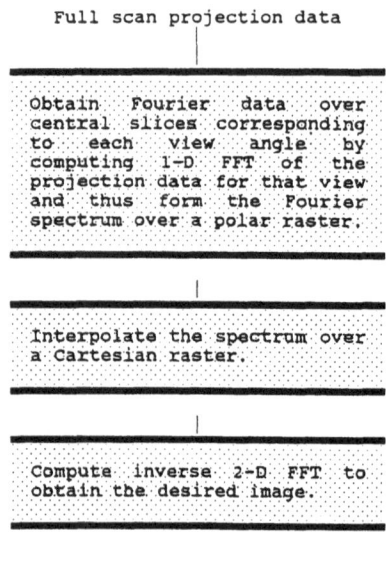

FIGURE 6.1: Direct Fourier method (DFM).

## 6.3  Spiral Sampling

### 6.3.1  LINEAR SPIRAL SAMPLING THEOREM

Research in spiral sampling and interpolation is motivated by the problem of determining optimum scanning trajectories in magnetic resonance imaging (MRI). An older, less used, name is nuclear magnetic resonance (NMR) imaging. The linear spiral sampling theorem (LSST) can be stated in the space domain or in the Fourier domain. Since its primary application is in the Fourier domain we shall state it as such.

The main result of this section is an exact interpolation theorem of the form

$$F(\rho, \phi) = \sum_j \sum_k F(\rho_{jk}, \phi_k) \Psi(\rho, \rho_{jk}) \sigma(\phi, \phi_k). \qquad (6.3.1)$$

where $F(\rho, \phi)$ is the polar Fourier transform of a space-limited function $f(x, y)$ and is assumed angularly band-limited to $K$ and $\Psi(\cdot, \cdot)$ and $\sigma(\cdot, \cdot)$ are the radial and azimuthal interpolation functions, respectively.

**Description of a Linear Spiral**

A linear spiral in the Fourier plane can be described by the equations

$$u(\varphi) \quad = \quad \alpha\varphi\cos\varphi.$$

$$v(\varphi) \;=\; \alpha\varphi\sin\varphi, \qquad\qquad 0 \le \varphi < \infty, \qquad\qquad (6.3.2)$$

where $u(\cdot)$ and $v(\cdot)$ are the horizontal and vertical displacement coordinates, respectively, $\alpha$ is a spacing parameter (determined by the extent of the object $f(x,y)$) and $\phi$ is a growing azimuthal angle. We note that for a linear spiral, the distance to the origin, $\rho(\phi)$, is given by

$$\rho(\varphi) \doteq [u^2(\varphi) + v^2(\varphi)]^{\frac{1}{2}} = \alpha\varphi. \qquad\qquad (6.3.3)$$

By defining $\phi = \varphi \bmod 2\pi$ and the revolution number $j$ as $\mathrm{int}[\varphi/2\pi]$ ($\mathrm{int}[x]$ indicates truncating $x$ to the nearest integer not larger than $x$) we can describe the sampling grid on a linear spiral scan as

$$
\begin{aligned}
u_{jk} &= \rho_{jk}\cos\phi_k, \\
v_{jk} &= \rho_{jk}\sin\phi_k, \\
\rho_{jk} &= \alpha(\phi_k + 2\pi j), \quad 0 \le \phi_k < 2\pi, \\
k &= 0, 1, 2, \ldots, N_\phi - 1 \\
j &= 0, 1, 2, \ldots \,.
\end{aligned}
\qquad (6.3.4)
$$

We let $N_\phi$ be the number of samples per revolution. Figures 6.2 and 6.3 show a linear spiral and the corresponding sampling grid for $N_\phi = 36$, respectively.

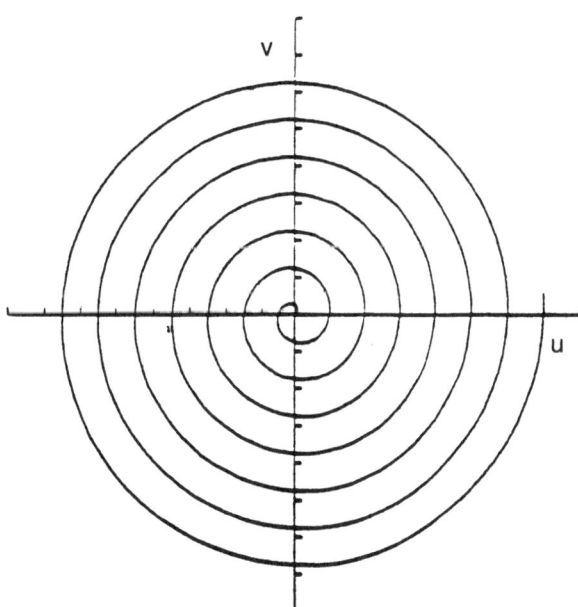

FIGURE 6.2: An example of a linear spiral scan in the $u, v$ (frequency) plane.

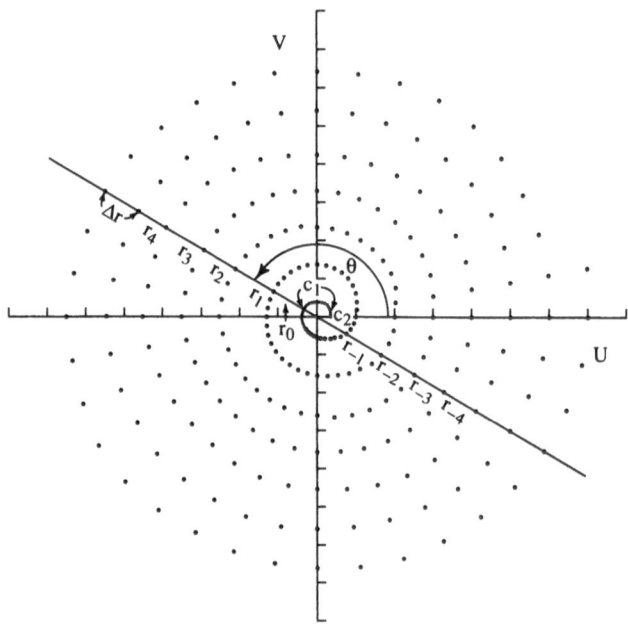

FIGURE 6.3: A spiral sampling grid for $N_\phi = 36$. The figure also shows the notation used to describe the samples along the line at angle $\theta$ from the real axis.

Spiral sampling does not yield uniformly spaced samples. This can be understood with the help of Fig. 6.3. Consider a line at some fixed angle $\theta(0 \leq \theta < \pi)$ passing through the origin. Hereafter, we will use an alternative labeling for the points along a line which is advantageous in this discussion and for computational purposes. Let $(\rho, \phi)$ be the polar coordinates of a point on the line. We define $r$ as follows:

$$r = \begin{cases} \rho, & \phi = \theta \\ -\rho, & \phi = \theta + \pi \end{cases} \qquad (6.3.5)$$
$$(0 \leq \theta < \pi).$$

Also, the samples along a line will be denoted $r_{lk}$, defined as follows:

$$r_{lk} = \begin{cases} \rho_{lk}, & l \geq 0 \\ -\rho_{(-l-1)(k+K+1)}, & l < 0 \end{cases} \qquad (6.3.6)$$

for $0 \leq k < K + 1$ and

$$r_{lk} = r_{l(k-K-1)}. \qquad (6.3.7)$$

for $K + 1 \leq k < 2K + 2$. Thus, the intercepts of the line with the spiral are labeled $\ldots, r_{-2k}, r_{1k}, r_{0k}, r_{1k}, \ldots$. The samples $r_{lk}, l \geq 0$, are located in the upper-half plane and the samples $r_{lk}, l < 0$. are located in the lower-half plane (see **Fig. 6.3**, where we omit the index $k$ as we do in what follows, since $k$ is held constant). Now, the spacings between $r_{l+1}$ and $r_l, l \geq 0$, are uniform as are the spacings between $r_l$ and $r_{l-1}$ for $l \leq -1$. However, the spacing between $r_0$ and $r_{-1}$ is different from the other spacings. It is convenient to refer to the spacing between and $r_0$ and $r_{-1}$ as the sum of two non-negative numbers $c_1 + c_2$ ($c_1 \geq 0, c_2 > 0$), where $c_1$ and $c_2$ are the actual distances of samples $r_0$ and $r_{-1}$ from the origin $u = v = 0$. Referred to the origin, the actual position of $r_{-1}$ is at $-c_2$ and the position of $r_0$ is $c_1$. Note that when $c_1 = 0$ and $c_2 = \Delta r = \Delta \rho$ the points are equally spaced, and therefore the sampling is uniform.

For a fixed $\phi_k = \theta, 0 \leq \theta < \pi$. it is not hard to see using (6.3.4) that

$$c_{1k} = \alpha \phi_k = \frac{\Delta r}{2\pi} \phi_k,$$

$$c_{2k} = c_{1k} + \frac{\Delta r}{2} \qquad (6.3.8)$$

since $\Delta r = \Delta \rho = 2\pi \alpha$, where $\alpha \leq 1/4\pi A$. Also, the following relation defines the set of non-uniform samples $\{r_{lk}\}$ along a line, for each $\phi_k$

$$r_{lk} = \begin{cases} c_{1k} + l\Delta r, & l \geq 0 \\ -c_{2k} + (l+1)\Delta r & l < 0 \end{cases} \quad \phi_k = \theta \ (0 \leq \theta < \pi) \qquad (6.3.9)$$

and

$$r_{lk} = r_{l(k-K-1)} \qquad \phi_k = \theta + \pi \qquad (0 \leq \theta < \pi). \qquad (6.3.10)$$

Thus, the reconstruction of $F(\rho, \phi)$ from its linear spiral samples is equivalent to the following interpolation problem: given the set of nonuniform sample points $\{r_{lk}\}, l = 0, \pm 1, \pm 2, \ldots, k = 0, 1, 2, \ldots, N_\phi - 1$ specified in (6.3.6), find $F(\rho, \phi)$. The theorem presented below is a solution to the problem.

**Linear Spiral Sampling Theorem (LSST)**

Let $f(x, y)$ be space-limited such that $f(x, y) = 0$ outside a disk of radius $A - \epsilon$, $0 < \epsilon < A$. centered at the origin. Let its FT in polar coordinates $F(\rho, \phi)$ be angularly band-limited to $K$. Then, $F(\rho, \phi)$ can be reconstructed from its spiral samples $F(\rho_{jk}, \phi_k)$ where

$$\rho_{jk} = \frac{\phi_k}{4\pi A} + \frac{j}{2A}, \ j = 0, 1, 2, \ldots$$

$$\phi_k = \frac{\pi k}{(K+1)}, \ k = 0, 1, 2, \ldots, 2K + 1 \qquad (6.3.11)$$

via

$$F(\rho, \phi) = \sum_{l=-\infty}^{\infty} \sum_{k=0}^{2K+1} F_1(r_{lk}, \phi_k) \Psi(r, r_{lk}) \sigma(\phi - \phi_k) \qquad (6.3.12)$$

and where the samples $F_1(r_{lk}, \phi_k)$ are obtained from $F(\rho_{jk}, \phi_k)$ as implied by (6.3.6), namely,

$$F_1(r_{lk}, \phi_k) = \begin{cases} F(\rho_{lk}, \phi_k), & l \geq 0 \\ F(\rho_{(-l-1)(k+K+1)}, \phi_{k+K+1}), & l < 0 \end{cases} \qquad (6.3.13)$$

for $0 \leq k < K + 1$, and

$$F_1(r_{lk}, \phi_k) = F_1(r_{l(k-K-1)}, \phi_{k-K-1}) \qquad (6.3.14)$$

for $K + 1 \leq k < 2K + 2$. The argument $r$ of the interpolating functions $\Psi(\cdot, \cdot)$ is obtained from (6.3.5) with $\phi = \phi_k$ as follows

$$r = \begin{cases} \rho, & 0 \leq k < K + 1 \\ -\rho, & K + 1 \leq k < 2K + 2. \end{cases} \qquad (6.3.15)$$

The interpolation function $\Psi(\cdot, \cdot)$ is given by

$$\Psi(r, r_{lk}) = B_{lk}(r) \operatorname{sinc}(2A(r - r_{lk})), \qquad (6.3.16)$$

where

$$B_{lk} = \begin{cases} \dfrac{\Gamma(2Ac_k + l)}{\Gamma(1 + l)} \dfrac{\Gamma(1 + 2A(r - c_{1k}))}{\Gamma(2A(r + c_{2k}))}, & l \geq 0 \\[4mm] \dfrac{\Gamma(2Ac'_k - l)}{\Gamma(-l)} \dfrac{\Gamma(1 - 2A(r + c_{2k}))}{\Gamma(-2A(r + c_{1k}))}, & l < 0 \end{cases} \qquad (6.3.17)$$

and $\Gamma(\cdot)$ is the standard Gamma function. Also

$$r_{lk} = \begin{cases} \dfrac{l}{2A} + c_{1k}, & l \geq 0 \\[4mm] \dfrac{l}{2A} - c'_{2k}, & l < 0 \end{cases} \qquad (6.3.18)$$

and

$$c_{1k} = \begin{cases} \dfrac{\phi_k}{4\pi A}, & 0 \leq \phi_k < \pi \\[4mm] \dfrac{\phi_k - \pi}{4\pi A}, & \pi \leq \phi_k < 2\pi. \end{cases} \qquad (6.3.19)$$

$$\begin{aligned} c_{2k} &= c_{1k} + \frac{1}{4A}, \\ c'_{2k} &\equiv c_{2k} - \frac{1}{2A}, \\ c_k &\equiv c_{1k} + c_{2k}, \\ c'_k &\equiv c_k - \frac{1}{2A}. \end{aligned} \qquad (6.3.20)$$

Finally, $\sigma(\cdot)$ is given by

$$\sigma(\phi) = \frac{\sin[\frac{1}{2}(2K+1)\phi]}{(2K+2)\sin[\frac{1}{2}\phi]}. \tag{6.3.21}$$

The proof of this theorem is given in [989]. Also in [989] are given practical approximations of Eqs. (6.3.12) and (6.3.17), and the performance of the LSST both as given here and its practical approximation.

We remark in passing that because $B_{lk}(\cdot)$ in (6.3.17) depends on $r$, the interpolating functions $\Psi(r, r_{lk})$ can behave in a manner drastically different from the sinc($\cdot$) function. The behavior of $\Psi(r, r_{lk})$ depends on the values of $c_{1k}$ and $c'_{2k}$ [980]. When $c_{1k}$ and $c'_{2k}$ have small values and $|r|$ becomes large, $\Psi(r, r_{lk})$ behaves essentially like the sinc($\cdot$) function.

### 6.3.2 RECONSTRUCTION FROM SAMPLES ON EXPANDING SPIRALS

Equation (6.3.2) described the locus of linear spirals. For linear and expanding spirals, the locus is described by

$$u(\varphi) = \alpha\varphi^\mu \cos(\varphi). \tag{6.3.22}$$

$$v(\varphi) = \alpha\varphi^\mu \sin(\varphi), \quad 0 \le \varphi < \infty. \tag{6.3.23}$$

Then, with $\rho(\varphi)$ denoting the distance to the origin, we get

$$\rho(\varphi) = [u^2(\varphi) + v^2(\varphi)]^{\frac{1}{2}} = \alpha\varphi^\mu. \tag{6.3.24}$$

where $\mu \ge 1$ and, as before, $u(\varphi)$ and $v(\varphi)$ are horizontal and vertical Cartesian coordinates. When $\mu = 1$ we obtain the linear spiral; for $\mu > 1$, we obtain an expanding spiral. The sampling grid on an expanding spiral is described by

$$u_{jk} = \rho_{jk}\cos\phi_k, \tag{6.3.25}$$

$$v_{jk} = \rho_{jk}\sin\phi_k, \tag{6.3.26}$$

$$\phi_k = \frac{2\pi k}{N_\phi}, \quad k = 0, 1, \dots, N_\phi - 1; \quad j = 0, 1, 2, \dots, \tag{6.3.27}$$

$$\rho_{jk} = \alpha(\phi_k + 2\pi j)^\mu, \quad \mu > 1. \tag{6.3.28}$$

In Eqs. (6.3.25) $\phi = \varphi \bmod 2\pi$ and $j$ is the largest integer in $\frac{\varphi}{2\pi}$ not exceeding $\frac{\varphi}{2\pi}$. It is shown in [990] that for arbitrary $\mu > 1$, it is not possible to reconstruct $F(\rho, \phi)$ (the continuous polar Fourier transform of the object function $f(x, y)$) exactly from the set of samples $\{F(\rho_{jk}, \phi_k)\}$.

At this point the reader might ask why we would even consider the non-linear case when an exact reconstruction exists for the linear spiral. The answer is, in short, that, if everything else remains the same, the non-linear spiral can reconstruct an image with higher resolution than that reconstructed by the linear spiral. This phenomenon is discussed in [990].

Before continuing, we refer to Fig. 6.4 to show how images in MRI are reconstructed from spiral samples of the Fourier transform. The spiral sampling itself is generated by the application of appropriate time-dependent gradients (see Eq. (6) in [989] and also [6]). An interpolation formula for approximating the samples on a Cartesian grid from samples on a polar grid that is valid for expanding spiral trajectories is the following:

$$
\begin{aligned}
F(\rho, \phi) &\simeq \tilde{F}(\rho, \phi) \\
&= \sum_{l=-\infty}^{\infty} \sum_{k=0}^{2K+1} F_1(r_{lk}, \phi_k) \Psi_{lk}(r) \sigma(\phi - \phi_k),
\end{aligned} \tag{6.3.29}
$$

where $F_1(r_{lk}, \phi_k)$ are the Fourier transform samples of $F(\cdot, \cdot)$ obtained using the bi-polar variable $r$ (see Eq. (6.3.5)), $\sigma(\phi)$ is given by Eq.( 6.3.21), and $\Psi_{lk}(r)$ is given by

$$
\Psi_{lk}(r) = \operatorname{sinc} \left\{ 2A \left[ T_k(r) - \frac{l}{2A} \right] \right\}. \tag{6.3.30}
$$

In Eq. (6.3.30), $T_k(r)$ is the transformation required to map radial sampling points on a non-linear spiral to uniformly spaced radial sampling points. The transformation involves the composition of two separate transformations: the first maps points on a non-linear spiral to points on a linear spiral; the second maps points on a linear spiral into a set of uniformly spaced samples. The procedure is discussed in [990] and the method for deriving $T_k(r)$ is based on the Clark-Palmer-Lawrence technique discussed in their well-known paper [224]. In reconstructing high-frequency objects, Eq. (6.3.29) proved to yield better results with expanding spirals than Eq. (6.3.12) did using uniform spirals for the same data collection times. For more on the problem of reducing data collection times see [462].

In the last few sections we dealt with the problem of interpolating from non-uniform samples to uniformly spaced Cartesian samples. The question arises: Is there an interpolation formula that deals somewhat more generally with any set of non-uniform samples? Such a formula is discussed in the next section.

# 6.4 Reconstruction from Non-Uniform Samples by Convex Projections

As seen in the earlier sections, reconstruction of functions from non-uniform-

ly spaced samples occurs in a number of practical situations. In MRI we might want to reconstruct an image from samples on a spiral scan [990]. In direct Fourier CT we wish to reconstruct the two-dimensional Fourier transform from non-uniformly spaced samples on a polar raster [867]. Examples of estimating wind velocity from non-uniformly spaced clouds from successive frames of satellite imagery, and the occurrence of non-uniform sampling in motion compensation are given in [794]. There are many other examples of naturally occurring non-uniform sampling that one could cite. Indeed a relatively large literature exists on the subject and some excellent references are given in [794] and [628]. To save time and space we refer the reader to these and other sources (e.g., [85]) for a review of non-uniform sampling reconstruction methods.

In the following subsections we derive an interpolation formula by the method of projections onto convex sets (POCS) or, equivalently, the method of convex projections. POCS has wide application in other areas as well: it has been used extensively in signal recovery problems and both the method and some applications are given in [860].

## 6.4.1 THE METHOD OF PROJECTIONS ONTO CONVEX SETS

The theory of convex projections developed by Bregman [113] and Gubin, polyak and Ralik [372] was first applied to image processing by Youla and Webb [985]. The reader unfamiliar with the method is referred to [985, 816, 983].

Here, for the reader's benefit, we furnish only the basic ideas of POCS. To begin with, assume that all the functions of interest are elements of the Hilbert space $H$ of square-integrable functions. Now consider a closed convex set $C \subset H$. The set $C$ is said to be convex if $x \in C$ and $y \in C$ imply $\alpha x + (1 - \alpha)y \in C$ for $0 \leq \alpha \leq 1$. For any $f \in H$, the projection $Pf$ of $f$ onto $C$ is the element in $C$ closest to $f$. If $C$ is closed and convex, $Pf$ exists and is uniquely determined by $f$ and $C$ from the minimality criterion

$$\|f - Pf\| = \min_{g \in C} \|f - g\|. \tag{6.4.1}$$

This rule, which assigns to every $f \in H$ its nearest neighbor in $C$, defines the (in general) non-linear projection operation $P : H \to C$ without ambiguity. In this discussion, the operator $\| \cdot \|$ is taken to be the usual $L_2$-norm.

The basic idea of POCS is as follows: Every known property of the unknown $f \in H$ will restrict $f$ to lie in a closed convex set $C_i$ in $H$. Thus, for $m$ known properties there are $m$ closed convex sets $C_i$, $i = 1, 2, \ldots, m$ and $f \in C_0 \equiv \bigcap_{i=1}^{m} C_i$. Then the problem is to find a point of $C_0$ given the sets $C_i$ and projection operators $P_i$ projecting onto $C_i$, $i = 1, 2, \ldots, m$. The convergence properties of the sequences $\{f_k\}$ generated by the recursion relation

$$f_{k+1} = P_m P_{m-1} \cdots P_1 f_k, \qquad k = 0, 1, \ldots , \qquad (6.4.2)$$

or more generally by

$$f_{k+1} = T_m T_{m-1} \cdots T_1 f_k, \qquad k = 0, 1, \ldots . \qquad (6.4.3)$$

with $T_i \equiv I + \lambda_i(P_i - I), 0 < \lambda_i < 2$, are based on fundamental theorems given by Opial [689] and Gubin et. al [372]. The $\lambda_i, 1 \ldots$, are relaxation parameters and can be used to accelerate the rate of convergence of the algorithm; $I$ is the identity operator. However, determining the ideal values of the $\lambda$'s is generally a difficult problem and for convex sets that are not linear subspaces we shall set $\lambda$'s to values somewhat arbitrarily between 1 and 2 in what follows.

## 6.4.2   ITERATIVE RECONSTRUCTION BY POCS

To avoid excessive notation we proceed in one dimension and give results in two dimensions only as needed. We are given a sequence of samples $f(x_i)$, $i = 1, 2, \ldots, N$ of an unknown real function $f(x)$ and form the $N$ sets

$$C_i = \{g(x) : g(x_i) = f(x_i), \ G(\omega) = 0, \ |\omega| > 2\pi B\}, \quad i = 1, 2, \ldots . \quad (6.4.4)$$

In Eq. (6.4.4) $B$ is the bandwidth and $G(\omega)$ is the Fourier transform of $g(x)$. In words, $C_i$ is the set of all band-limited functions of one variable whose value at the sampling point $x_i$ coincides with the value of the sampled function. Now given an arbitrary function $h(x)$, how do we find its projection $Ph \equiv g$ onto $C_i$? We leave this computation to Appendix A and present only the result

$$
\begin{aligned}
g(x) \quad = \quad & h(x) * 2B \operatorname{sinc}(2Bx) + [g(x_i) - h(x_i) * 2B \operatorname{sinc}(2Bx_i)] \\
& \times \operatorname{sinc}(2B(x - x_i))
\end{aligned}
\qquad (6.4.5)
$$

In Eq. (6.4.5) we use the notation

$$h(x_i) * 2B \operatorname{sinc}(2Bx_i) \equiv h(x) * 2B \operatorname{sinc}(2Bx)|_{x=x_i} \qquad (6.4.6)$$

and $*$ to mean ordinary convolution.) In Fig. 6.5 is shown the realization of Eq. (6.4.5) using an analog circuit. We observe that the result is intuitively satisfying; it says that the projection is obtained by first low-pass filtering $h(x)$; then using the difference between the correct sample and the sampled low-pass filtered $h(x)$ to modulate a sampling impulse; and finally adding the weighted low-pass impulse responses as a correction term to the original low-pass filtered $h(x)$.

Equations (6.4.4) and (6.4.5) can be generalized in a straightforward fashion. Thus redefining $C_i$ as

$$C_i = \{g(\mathbf{x}) \in L_2^D : g(\mathbf{x}_i) = f(\mathbf{x}_i),\ G(\omega) = \mathcal{F}\{g(\mathbf{x})\} = 0,\ \omega \notin \Omega\} \quad (6.4.7)$$

where the dimensionality $D = 1$ or $2$, we find the projection of an arbitrary $h(\mathbf{x})$ onto $C_i$ to be

$$
\begin{aligned}
g(\mathbf{x}) &\equiv P_i\, h(\mathbf{x}) \\
&= H(\mathbf{x}) * K(\mathbf{x}) + [f(\mathbf{x}_i) - (h * k)(\mathbf{x}_i)]\frac{K(\mathbf{x} - \mathbf{x}_i)}{K(0)}.
\end{aligned}
$$
$$(6.4.8)$$

In Eq. (6.4.7) we have used the notation

$$(h * k)(\mathbf{x}_i) \equiv h(\mathbf{x}) * k(\mathbf{x})|_{\mathbf{x}=\mathbf{x}_i} \quad (6.4.9)$$

and $K(\mathbf{x})$ is the kernel associated with the support region $\Omega$. The operator $\mathcal{F}$ denotes the Fourier transform operator. Thus for $D = 1$ and $\Omega = [-2\pi B, 2\pi B]$,

$$K(x) = k(x) = 2B\ \text{sinc}(2Bx). \quad (6.4.10)$$

For $D = 2$ and $\Omega$ a square region of side $4\pi B$ centered at the origin,

$$K(\mathbf{x}) = k(x, y) = (2B)^2\ \text{sinc}(2Bx)\ \text{sinc}(2By), \quad (6.4.11)$$

For $D = 2$ and $\Omega$ a circular region of radius $2\pi B$,

$$K(\mathbf{x}) = \frac{1}{2\pi} \int_0^{2\pi B} \rho J_0(\sqrt{x^2 + y^2}\rho)d\rho \quad (6.4.12)$$

$$= \frac{B}{r} J_1(2\pi r B), \quad (6.4.13)$$

where $J_n$ is the $n$th-order Bessel function of the first kind and $r = \sqrt{(x^2 + y^2)}$. With $P_i$ explicitly defined as in Eq. (6.4.7), and the relaxed projection $T_i$ being $I + \lambda(P_i - I), 0 < \lambda < 2$, the iterative reconstruction algorithm becomes

$$f_{k+1}(\mathbf{x}) = T_N T_{N-1} \cdots T_1 f_k(\mathbf{x}), \quad (6.4.14)$$

with

$$f_0 \equiv h(\mathbf{x}). \quad (6.4.15)$$

In Eq. (6.4.14), $f_k(\mathbf{x})$ represents the $k$th estimate of a band-limited function consistent with all the samples, $f_{k+1}(\mathbf{x})$ is the one-cycle improvement over $f_k(\mathbf{x})$, and $h(\mathbf{x})$ is the initial estimate which represents our best guess about the shape of the unknown function. The initial estimate $h$ can be chosen a band-limited function so that the convolution in Eq. (6.4.7) can be omitted. Since $P_i$ operating on any function always results in a band-limited function, the convolution as low-pass filtering would never be necessary in successive operations. The algorithm given in Eq. (6.4.14) converges to a point in the set $C_0 = \bigcap_{i=1}^{N} C_i$.

In addition to prior knowledge of the band-limitedness of the unknown function, we might have other prior knowledge regarding $f(\mathbf{x})$. If $C_\alpha, C_\beta, \ldots$ are closed convex sets of functions that share the prior known properties of $f(\mathbf{x})$, we can modify Eq. (6.4.14) to include this prior knowledge as

$$f_{k+1} = P_\alpha P_\beta \cdots T_N T_{N-1} \cdots T_1 f_k. \tag{6.4.16}$$

where $P_\alpha$ projects onto $C_\alpha$. etc. The use of prior knowledge usually enables faster convergence to a point in $C_0$.

## One-Step Reconstruction by POCS

Our experimental results with the iterative algorithm confirm the observation made in [794] that the convergence rate for moderate size images is slow. The question that then arises is, is it possible to project directly onto the intersection set $C_0$? The answer is yes because the $N$ constraint sets $C_i, i = 1, \ldots, N$ are very similar. Let us define the set $C_0$ explicitly as

$$C_0 \quad = \quad \{g(\mathbf{x}) \in L_2^D : g(\mathbf{x}_i) = f(\mathbf{x}_i), \quad i = 1, 2, \ldots, N$$

and

$$G(\omega) \quad = \quad \mathcal{F}\{g(\mathbf{x})\} = 0, \quad \omega \notin \Omega\} \tag{6.4.17}$$

where, as usual, $\mathcal{F}$ denotes the Fourier transform operator. The reader will note that $C_0$ defined in Eq. (6.4.17) is identical with $C_0$ defined as the intersection of the $C_i$'s, $i = 1, \ldots, N$. In Appendix B we show that the one-step projection of an arbitrary $h(\mathbf{x})$ onto $C_0$ is given by

$$P_0 h(\mathbf{x}) = q(\mathbf{x}) + \mathbf{S}(\mathbf{x})^T \mathbf{A}^{-1}(\mathbf{f} - \mathbf{q}). \tag{6.4.18}$$

where

$$
\begin{aligned}
q(\mathbf{x}) \quad &= \quad h(\mathbf{x}) * K(\mathbf{x}). & (6.4.19) \\
\mathbf{f} \quad &= \quad [f(\mathbf{x}_1) f(\mathbf{x}_2) \cdots f(\mathbf{x}_N)]^T. & \\
\mathbf{q} \quad &= \quad [q(\mathbf{x}_1) q(\mathbf{x}_2) \cdots q(\mathbf{x}_N)]^T. & \\
\mathbf{S} \quad &= \quad [K(\mathbf{x} - \mathbf{x}_1) K(\mathbf{x} - \mathbf{x}_2) \ldots K(\mathbf{x} - \mathbf{x}_N)]^T. & \\
\mathbf{A} \quad &= \quad [a_{ij}]_{N \times N}. & \\
a_{ij} \quad &= \quad K(\mathbf{x}_i - \mathbf{x}_j). \quad i, j = 1, 2, \cdots, N. & (6.4.20)
\end{aligned}
$$

and ∗ denotes convolution. Equation (6.4.20) is a formula for reconstructing a multidimensional function from $N$ arbitrary sample. As $N \to \infty$ and the sampling rate is high enough[1], the reconstructed function approaches the unknown function regardless of the initial guess $h(x)$.

It is satisfying to note that if the samples are uniformly distributed in $[0, T]$ (the one-dimensional case is assumed) so that $x_i = i\Delta x$ and $N$ is large enough, Eq. (6.4.20) yields a familiar result. In this case $\mathbf{A} = 2B\mathbf{I}$ where $\mathbf{I}$ is the identity matrix and $\mathbf{A}^{-1} = (2B)^{-1}\mathbf{I}$. Hence

$$P_0 h(x) = q(x) + \sum_{i=1}^{N} f(i\Delta x) \ \text{sinc}(2B(x - i\Delta x))$$

$$-\sum_{i=1}^{N} q(i\Delta x) \ \text{sinc}(2B(x - i\Delta x)). \tag{6.4.21}$$

But since $q(x)$ is band-limited to $B$ it follows that

$$q(x) \simeq \sum_{i=1}^{N} q(i\Delta x) \ \text{sinc}(2B(x - i\Delta x)) \tag{6.4.22}$$

for $x \in [0, T]$ and not too near the end point. Also we assume $\Delta x = T/N \leq 1/2B$.

Under these circumstances Eq. (6.4.21) reduces to

$$f(x) \simeq \sum_{i=1}^{N} f(i\Delta x) \frac{\text{sinc } 2\pi B(x - i\Delta x)}{2\pi B(x - i\Delta x)}. \tag{6.4.23}$$

The one-step projection reconstruction which gives a band-limited function matching all samples in one shot involves the inversion of the matrix $\mathbf{A}$ whose elements depend on the locations of the samples and whose dimension is the number of samples. While in some particular cases $\mathbf{A}$ may be of Toeplitz form or of block-Toeplitz form, in general $\mathbf{A}$ is not Toeplitz and its inversion requires $\mathcal{O}(N^3)$ floating point operations. This is the main drawback to the one-step projection technique. In particular when $N$ is large, the inversion of $\mathbf{A}$ becomes very sensitive to noise and the iterative algorithm, despite its slow convergence rate, may be preferable.

---

[1]A necessary condition is that $\lim N(T)/T = 2B$ where $N(T)$ is the number of samples in an interval of length $T$. For more discussion see [516].

| data sets | one step | iterative method | | iterative method with magnitude constrain | |
|---|---|---|---|---|---|
| | mean sq. err. | mse | # of iter. | mse | # of iter. |
| S1 | .0442 | .0754 | 149 | 648 | 144 |
| S2 | .00175 | .275 | 82 | .228 | 63 |
| S3 | .0473 | .0889 | 81 | 786 | 75 |
| S4 | .48 | .196 | 75 | .149 | 216 |
| S5 | .24 | .2097 | 107 | .247 | 118 |
| Average | .0655 | .136 | 88.2 | .139 | 89.9 |

TABLE 6.I: Summary of experimental results on 1-D reconstruction.

## 6.5 Experimental Results

### 6.5.1 RECONSTRUCTION OF ONE-DIMENSIONAL SIGNALS

The function $f_T(x) = \exp(-x^2/2) \cos(6\pi x)$ was used as a test signal. If we take as effective bandwidth $B$ the frequency band within the one percent magnitude level of the spectrum then $B = 6$. A total of 48 samples were drawn from the interval $[-2, 2]$, which made the average sampling rate not lower than the Nyquist rate. The signal was reconstructed in the region $[-1, 1]$. The non-uniformly spaced samples were drawn at locations which deviate from the uniformly spaced locations by independent random amounts which are distributed according to the normal distribution. The standard deviation $\sigma$ of the normal distribution was taken to be the interval between samples if they were uniformly spaced. A summary of the experimental results using five sets of samples as well as averaged values over ten sets of samples is shown in Table 6.I. Both the one-step projection method and the iterative method were tested. Since there is a random factor in the distribution of sample locations, it is possible that two or more samples might be very close. In such case the matrix **A** used in the one-step method would be ill-conditioned, which would usually result in poor reconstruction. This difficulty can be avoided by first checking the distance between sampling points. If two points are very close together, then one of the samples is discarded. Since the sample discarded gives essentially the same information as the other point, by discarding it we do not lose any information available. However the effective average sampling rate is reduced slightly.

Note that on the average one-step method (with overly close points removed) reconstructs the signal better than the iterative method. One possible explanation is that the convergence rate of the iterative method essentially goes to zero after a certain number of iterations.

Figures 6.6 and 6.7 illustrate reconstructions from two sets of samples. The last two columns in Table 6.I show that, in this case at least, the

| | iterative method without magnitude constraint | | iterative method with magnitude constraint | |
|---|---|---|---|---|
| data sets | mse | # of iterations | mse | # of iterations |
| S1 | .271 | 106 | .242 | 96 |
| S2 | .141 | 130 | .114 | 90 |
| S3 | .340 | 89 | .345 | 81 |
| S4 | .491 | 96 | .367 | 85 |
| S5 | .571 | 129 | .473 | 64 |

TABLE 6.II: Comparison of iterative method with and without magnitude constraint.

imposition of the amplitude constraint set

$$C_\alpha = \{g(x) : \ -1 \le g(x) \le 1\} \tag{6.5.1}$$

does not improve the results. This is because the average sampling rate is high enough to contain magnitude information. However when the sampling rate is reduced it is conjectured that prior-known amplitude information would become important. To test this conjecture, we reduced the number of samples to 40 points over the range $[-2, 2]$, which is 17% less than previously used, and compared the results of iterative reconstruction with and without the magnitude constraint.

From Table 6.II it is clear that the amplitude constraint improves the convergence rate when the samples are not sufficient.

### 6.5.2 RECONSTRUCTION OF IMAGES

The object we used was a 64-by-64-pixel low-pass-filtered "T." The image was low pass-filtered to 4.5 cycles per image dimension (which is 64). Each set of samples consists of 81 points drawn independently from normal populations whose means were the locations of uniformly spaced points and whose standard deviations $\sigma$ were all 64/9.

Figure 6.8 illustrates an example of reconstruction by 1) one- step method and 2) the iterative method presented in the previous section. In Table 6.III is shown the average performance of the two methods over eight sets of samples. The performance is measured in signal to error ratio. We note that the signal to error performances of the two methods are about the same.

For $64 \times 64$ images with 81 sampling points. using a fast matrix inversion algorithm from the IMSL library, the speed of the one-step method is about the same as that of the POCS iterative method. With existing matrix inversion algorithms, it seems that one-step reconstruction has a speed advantage over the iterative method only for small images or when the sampling pattern translates into a Toeplitz $\mathbf{A}$ matrix. This would be the

| | one step | POCS Iterative | |
|---|---|---|---|
| data sets | S/E (dB) | S/E (dB) | # of iterations |
| S1 | 12.56 | 12.37 | 200 |
| S2 | 14.30 | 14.28 | 95 |
| S3 | 6.75 | 6.57 | 200 |
| S4 | 7.10 | 7.05 | 200 |
| S5 | 12.60 | 12.60 | 54 |
| S6 | 13.47 | 13.74 | 37 |
| S7 | 11.64 | 11.64 | 57 |
| S8 | 15.59 | 15.59 | 37 |
| Average | 11.79 | 11.73 | 110 |

TABLE 6.III: Experimental results on reconstruction of the 64 × 64 image square, limited to 4.5 cycles/image dimension.

case for example when a uniform sampler goes dead over an interval and then resumes normal operation.

A final remark is in order. At the time the research in Section 6.4 was done the authors became aware of the work of Sauer and Allebach [794] who also attempted to derive an interpolation formula by POCS for non-uniform samples. They proposed an iterative algorithm based on two constraint sets, one involving prior knowledge of the band-limitedness of the function and the second based on the sampled data. However since isolated samples in $L_2$ spaces have zero norm, they exert no influence on the solution. To deal with their problem, the authors "smear" or average the samples over a finite support; this yields a finite norm. However this smearing has the potential for making trouble in several ways: 1) the smearing itself can reduce the resolution of the reconstruction and 2) when applied to a discrete image, the mapping is not a projection.

In a recent paper [979], there is a comparison between the method described here and the Sauer-Allebach method. Experimental results for both methods are given and the reader is referred to [979] for more details.

## 6.6   Conclusions

In this chapter we discussed polar, spiral, and general nonuniform sampling grids and the reconstruction / interpolation of functions from their samples on such grids. In medical imaging, where situations arise that make it necessary to interpolate from polar to uniformly spaced Cartesian points, we introduced two formulas for doing this exactly: Equation (6.2.16) from interpolating from samples on a uniform-in-radius, uniform-in-angle polar grid; and Eqs. (6.3.12–6.3.14) for interpolating from samples on a linear spiral.

For spirals that "grow" faster that the linear spiral, there exists no interpolation formula that will exactly interpolate from samples on points of such spirals. However approximate interpolation formulas, that work well in practice, exist and can be used efficiently in magnetic resonance imaging by direct Fourier methods.

Finally we introduced the basic ideas of the method of projections onto convex sets and derived iterative and one-shot procedures for interpolating from samples at arbitrary locations.

## Acknowledgments

The author is grateful to his students who helped in the sampling research, especially I. Paul, R.Hingorani, E. Yudilevich, H. Peng, and S-J. Yeh. Thanks are due to the National Science Foundation for funding much of this research.

**Spiral Samples**

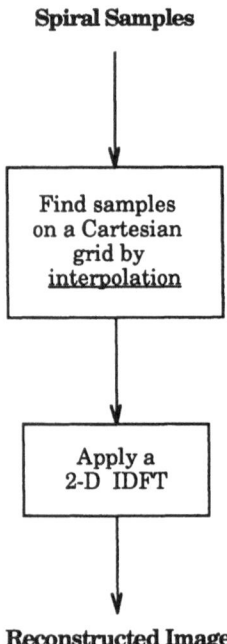

**Reconstructed Image**

FIGURE 6.4: Flowchart describing the direct Fourier reconstruction method in MRI.

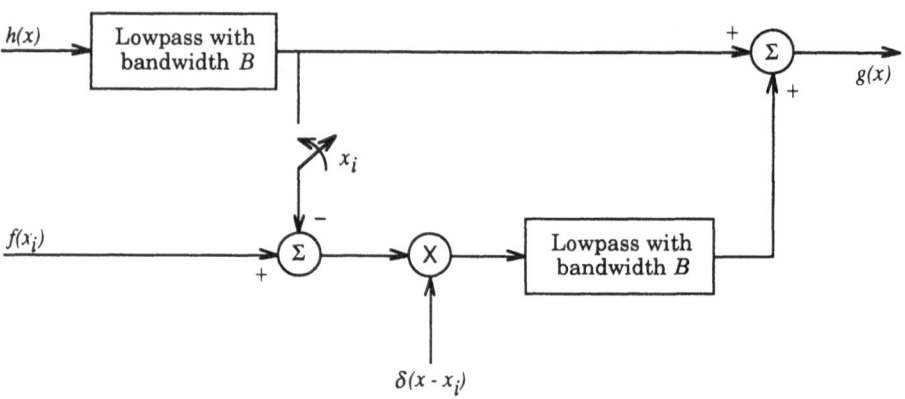

FIGURE 6.5: Analog realization of the projection onto $C_i$.

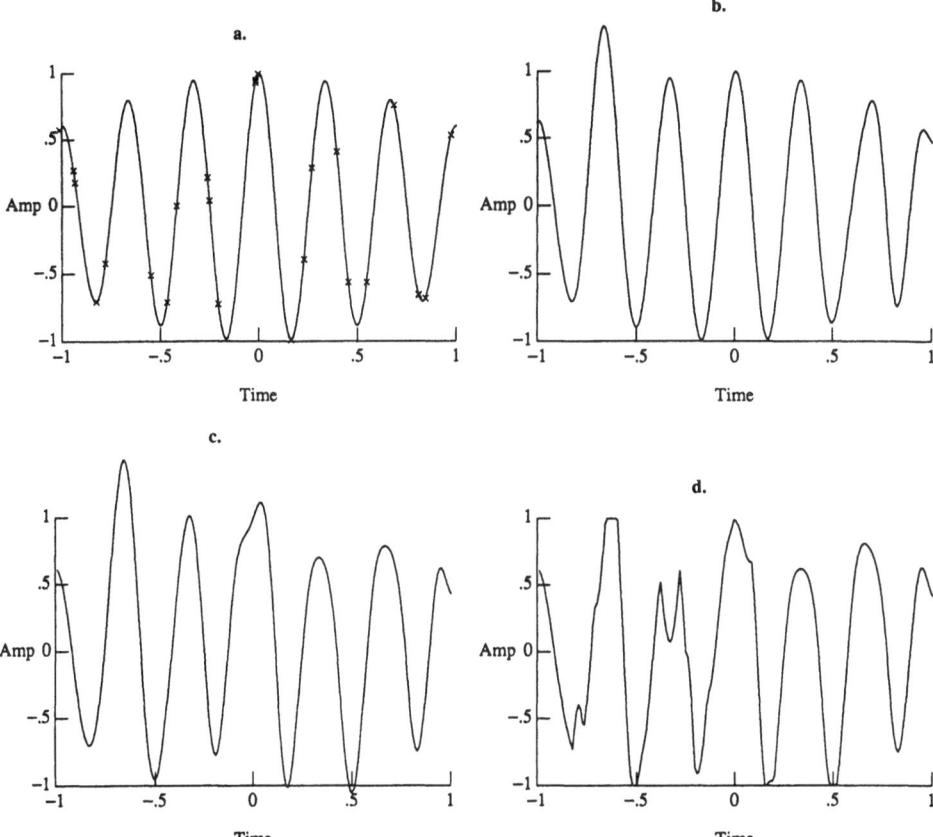

FIGURE 6.6: (a) Original function and sampling data set S1; (b) Reconstruction by one-step method from S1; (c) Reconstruction by iterative method from S1; and (d) Reconstruction by iterative method with magnitude constraint from S1.

*Henry Stark*

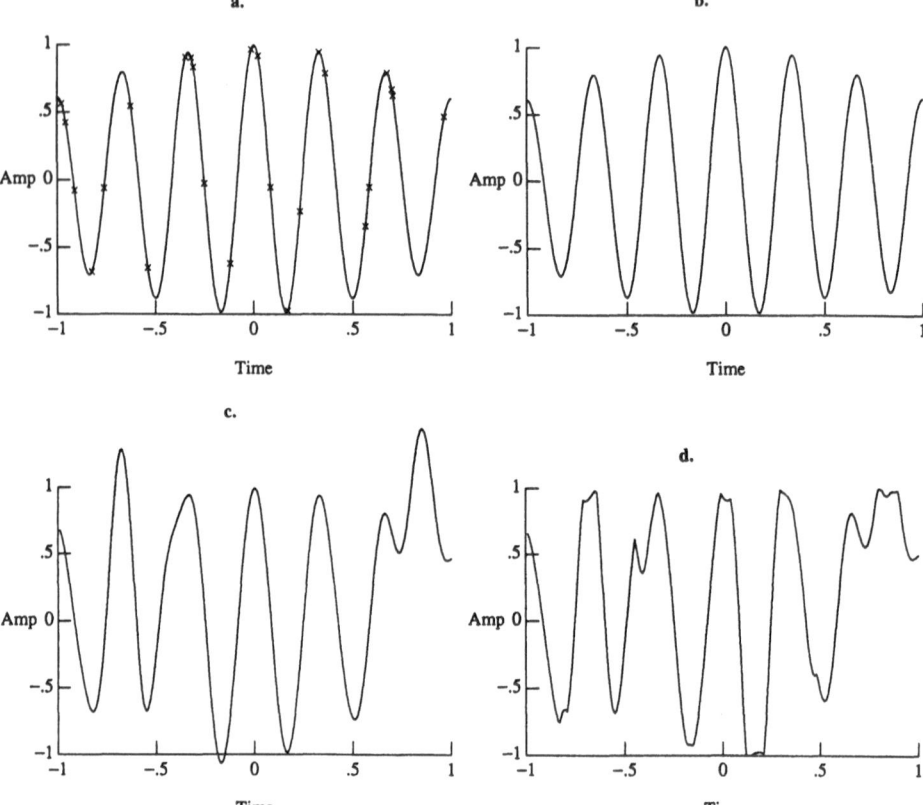

FIGURE 6.7: (a) Original function and sampling data set S2; (b) Reconstruction by one-step method from S2; (c) Reconstruction by iterative method from S2; and (d) Reconstruction by iterative method with magnitude constraint from S2.

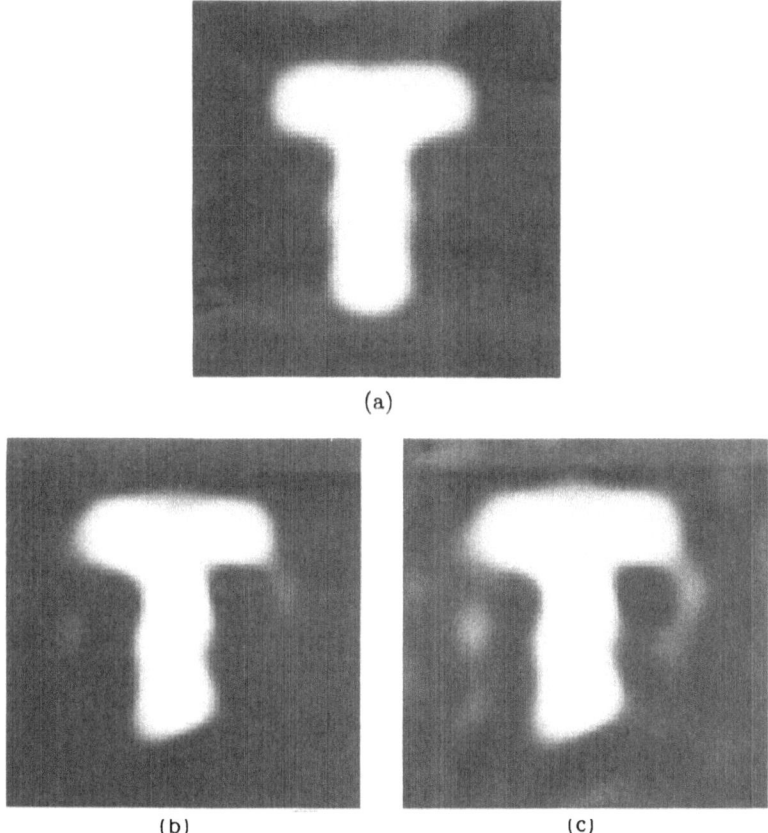

FIGURE 6.8: (a) Original image to be reconstructed; (b) Reconstruction by one-step method, S/E = 12.56 dB; (c) Reconstruction by POCS iterative method, S/E = 12.37 dB.

# Appendix A

## A.1 Derivation of Projections onto Convex Sets $C_i$

Let

$$
\begin{aligned}
C_i &= \{g(\mathbf{x}) \in L_2^D : g(\mathbf{x}_i) = f(\mathbf{x}_i), \\
G(\boldsymbol{\omega}) &= \mathcal{F}\{g(\mathbf{x})\} = 0, \ \boldsymbol{\omega} \notin \Omega\}.
\end{aligned}
\tag{A.1}
$$

The conditions that define $C_i$ can be rewritten in the frequency domain as

$$
\int_\Omega G(\boldsymbol{\omega}) e^{j\boldsymbol{\omega}\mathbf{x}_i} d\boldsymbol{\omega}/(2\pi)^D = f(\mathbf{x}_i)
\tag{A.2}
$$

and

$$
G(\boldsymbol{\omega}) = 0. \quad \boldsymbol{\omega} \notin \Omega.
\tag{A.3}
$$

Let $P$ be the projection onto $C_i$. For any $h$ in $L_2$, $g = Ph$ is the element in $C_i$ which minimizes

$$
J = \|f - g\|^2,
\tag{A.4}
$$

$\|\cdot\|$ being the $L_2$-norm. Using the method of Lagrange multipliers we let $J_\alpha$ be the auxiliary function to be minimized:

$$
J_\alpha = \|g - h\|^2 + \lambda[g(\mathbf{x}_i) - f(\mathbf{x}_i)].
\tag{A.5}
$$

Then

$$
\begin{aligned}
J_\alpha &= \int_\Omega |G(\boldsymbol{\omega}) - H(\boldsymbol{\omega})|^2 \frac{d\boldsymbol{\omega}}{(2\pi)^D} + \lambda \left[ \int_\Omega G(\boldsymbol{\omega}) e^{j\boldsymbol{\omega}\mathbf{x}_i} \frac{d\boldsymbol{\omega}}{(2\pi)^D} - f(\mathbf{x}_i) \right] \\
&= \int_\Omega [(G_R - H_R)^2 + (G_I - H_I)^2] \frac{d\boldsymbol{\omega}}{(2\pi)^D} \\
&\quad + \lambda \left\{ \int_\Omega [G_R \cos(\boldsymbol{\omega}x_i) - G_I \sin(\boldsymbol{\omega}x_i)] \frac{d\boldsymbol{\omega}}{(2\pi)^D} - f(\mathbf{x}_i) \right\}.
\end{aligned}
\tag{A.6}
$$

The imaginary term in Eq. (A.6) has been set to zero since we assume that all $h, g$ etc. are real. Using the variational principle, the quantity $J_\alpha$ is minimized when, for all $\boldsymbol{\omega} \notin \Omega$,

$$
2(G_R - H_R) + \lambda \cos(\boldsymbol{\omega}x_i) = 0
\tag{A.7}
$$

and

$$2(G_I - H_I) + \lambda \sin(\boldsymbol{\omega}\mathbf{x}_i) = 0. \tag{A.8}$$

Therefore

$$
\begin{aligned}
G &= G_R + jG_I \\
&= H_R + jH_I - \frac{\lambda}{2}[\cos(\boldsymbol{\omega}\mathbf{x}_i) - j\sin(\boldsymbol{\omega}\mathbf{x}_i)] = 0 \tag{A.9}
\end{aligned}
$$

for $\boldsymbol{\omega} \in \Omega$, and is zero for $\boldsymbol{\omega} \notin \Omega$. The function $g$ can then be obtained from the inverse Fourier transform of $G$, i.e.,

$$
\begin{aligned}
g(\mathbf{x}) &= \int_\Omega \left[ H(\boldsymbol{\omega}) - \frac{\lambda}{2}e^{-j\boldsymbol{\omega}\mathbf{x}} \right] e^{j\boldsymbol{\omega}\mathbf{x}} \frac{d\boldsymbol{\omega}}{(2\pi)^D} \\
&= \int_{-\infty}^{\infty} \left[ H(\boldsymbol{\omega}) - \frac{\lambda}{2}e^{-j\boldsymbol{\omega}\mathbf{x}} \right] W(\boldsymbol{\omega})e^{j\boldsymbol{\omega}\mathbf{x}} \frac{d\boldsymbol{\omega}}{(2\pi)^D}. \tag{A.10}
\end{aligned}
$$

where $W(\boldsymbol{\omega})$ is a window function, i.e.,

$$W(\boldsymbol{\omega}) = \begin{cases} 0, & \boldsymbol{\omega} \notin \Omega \\ 1, & \boldsymbol{\omega} \in \Omega. \end{cases} \tag{A.11}$$

We denote the inverse Fourier transform of $W(\boldsymbol{\omega})$ by $K(\mathbf{x})$ and define it as the kernel of the band-limiting system. Then the function $g$ can be written as

$$g(\mathbf{x}) = h(\mathbf{x}) * K(\mathbf{x}) - \frac{\lambda}{2}k(\mathbf{x} - \mathbf{x}_i). \tag{A.12}$$

The constant $\lambda$ is obtained from the condition $g(\mathbf{x}_i) = f(\mathbf{x}_i)$, which requires that

$$\frac{\lambda}{2} = -\frac{1}{K(0)}[g(\mathbf{x}_i) - h(\mathbf{x}) * k(\mathbf{x})|_{\mathbf{x}_i}]. \tag{A.13}$$

Using Eq. (A.13) in Eq. (A.12) gives the final result for the projection as

$$
\begin{aligned}
Ph &= g \\
&= h(\mathbf{x}) * K(\mathbf{x}) + \frac{1}{K(0)}[g(\mathbf{x}_i) - (h * k)(\mathbf{x}_i)]k(\mathbf{x} - \mathbf{x}_i). \tag{A.14}
\end{aligned}
$$

The kernel $K(\mathbf{x})$ can be expressed explicitly for various particular band-limiting systems as follows. By definition,

$$K(\mathbf{x}) = \int_{-\infty}^{\infty} W(\boldsymbol{\omega})e^{j\boldsymbol{\omega}\mathbf{x}} \frac{d\boldsymbol{\omega}}{(2\pi)^D}. \tag{A.15}$$

For the one-dimensional case $(D = 1)$ and $\omega = [-2\pi B, 2\pi B]$,

$$
\begin{aligned}
K(x) &= \int_{-2\pi B}^{2\pi B} e^{j\omega x} \frac{d\omega}{(2\pi)} \\
&= 2B \frac{\sin 2\pi Bx}{2\pi Bx} = 2B \, \text{sinc}(2Bx).
\end{aligned}
\tag{A.16}
$$

and $K(0) = 2B$. For $D = 2$ and $\Omega$ a centered square region, $\omega = \{(u, v) : u, v \in [-2\pi B, 2\pi B]\}$,

$$
\begin{aligned}
K(x, y) &= \int_{-2\pi B}^{2\pi B} \int_{-2\pi B}^{2\pi B} e^{j(ux+vy)} du \, dv \\
&= (2B)^2 \, \text{sinc}(2Bx) \, \text{sinc}(2By),
\end{aligned}
\tag{A.17}
$$

and $K(0, 0) = (2B)^2$ . For $D = 2$ and $\Omega$ a circular region of radius $2\pi B$,

$$
\begin{aligned}
K(x, y) &= \frac{1}{4\pi} \int_0^{2\pi B} \int_0^{2\pi} e^{j(x\rho\cos\theta + y\rho\sin\theta)} \rho \, d\rho \, d\theta \\
&= \frac{1}{2\pi} \int_0^{2\pi B} \rho J_0(\sqrt{x^2 + y^2}\rho) d\rho \\
&= \frac{B}{\sqrt{x^2 + y^2}} J_1(2\pi B\sqrt{x^2 + y^2}).
\end{aligned}
\tag{A.18}
$$

Here $J_n(\cdot)$ denotes the $n$th-order Bessel function of the first kind. In this case $K(0, 0) = \pi B^2$.

Again the assumption of real functions has been used so that imaginary terms are set to zero.

From algebraic manipulation of Eqs. (B.5), (B.6) and (B.2), $G$ is found to be

$$G(\omega) = G_R + jG_I = \begin{cases} H(\omega) - \sum_{i=1}^{N} \frac{\lambda_i}{2} e^{-j\omega \mathbf{x}_i}, & \omega \in \Omega \\ \\ 0, & \omega \notin \Omega. \end{cases} \quad (B.7)$$

Therefore

$$\begin{aligned} g(\mathbf{x}) &= \int_{\Omega} H(\omega) e^{j\omega \mathbf{x}} \frac{d\omega}{(2\pi)^D} - \sum_{i=1}^{N} \lambda_i K(\mathbf{x} - \mathbf{x}_i) \\ &= h(\mathbf{x}) * K(\mathbf{x}) - \frac{1}{2} \sum_{i=1}^{N} \lambda_i K(\mathbf{x} - \mathbf{x}_i). \end{aligned} \quad (B.8)$$

The constants $\lambda_1, \lambda_2, \ldots, \lambda_N$ are determined by the $N$ sampled values

$$\begin{aligned} g(\mathbf{x}_i) &= (h * K)(\mathbf{x}_i) - \frac{1}{2} \sum_{i=1}^{N} \lambda_i K(\mathbf{x} - \mathbf{x}_i) \\ &= f(\mathbf{x}_i), \quad i = 1, 2, \ldots, N. \end{aligned} \quad (B.9)$$

The system of equations of Eq. (B.9) can be written in matrix form as

$$\mathbf{q} - \frac{1}{2} \mathbf{A}\boldsymbol{\lambda} = \mathbf{f}, \quad (B.10)$$

where

$$\begin{aligned} \mathbf{f} &= [f(\mathbf{x}_1) f(\mathbf{x}_2) \cdots f(\mathbf{x}_N)]^T, \\ \mathbf{q} &= [q(\mathbf{x}_1) q(\mathbf{x}_2) \cdots q(\mathbf{x}_N)]^T, \\ \boldsymbol{\lambda} &= [\lambda_1 \lambda_2 \cdots \lambda_N]^T, \\ \mathbf{A} &= [a_{ij}]_{N \times N}, a_{ij} = K(\mathbf{x}_i - \mathbf{x}_j). \quad i, j = 1, 2, \ldots, N, \\ q(x) &= (h * k)(x). \end{aligned}$$

If the matrix $\mathbf{A}$ is not singular, then

$$\frac{1}{2}\boldsymbol{\lambda} = -\mathbf{A}^{-1}(\mathbf{f} - \mathbf{g}) \quad (B.11)$$

and, finally,

$$g = Ph = q(\mathbf{x}) - \mathbf{S}(\mathbf{x})^T \boldsymbol{\lambda} = q(\mathbf{x}) + \mathbf{S}(\mathbf{x})^T \mathbf{A}^{-1}(\mathbf{f} - \mathbf{q}), \quad (B.12)$$

where

$$\mathbf{S}(\mathbf{x}) = [K(\mathbf{x} - \mathbf{x}_1) K(\mathbf{x} - \mathbf{x}_2) \cdots K(\mathbf{x} - \mathbf{x}_N)]^T. \quad (B.13)$$

# Appendix B

## B.1 Derivation of the Projection onto the Set $C_0 \equiv \cap_i C_i$

The set $C_0$ of all band-limited functions which match the $N$ sampled values of the function $f$ at $\mathbf{x}_1, \mathbf{x}_2, \ldots, \mathbf{x}_N$ is

$$C_0 = \{g(\mathbf{x}) \in L_2^D : g(\mathbf{x}_i) = f(\mathbf{x}_i), \ i = 1, 2, \ldots, N$$

and

$$G(\boldsymbol{\omega}) = \mathcal{F}\{g(\mathbf{x})\} = 0, \ \boldsymbol{\omega} \notin \Omega\}.$$

Let the projection of an arbitrary function $h$ onto $C_0$ be denoted by $g = Ph$. Then $g$ minimizes

$$J = ||g - h||^2 = ||G - H||^2 \tag{B.1}$$

and $G = \mathcal{F}\{g\}$ satisfies the constraints

$$G(\boldsymbol{\omega}) = 0, \ \boldsymbol{\omega} \notin \Omega, \tag{B.2}$$

and

$$\int_\Omega G(\boldsymbol{\omega}) e^{j\boldsymbol{\omega}\mathbf{x}_i} \frac{d\boldsymbol{\omega}}{(2\pi)^D} = f(\mathbf{x}_i), \ \ i = 1, 2, \ldots, N. \tag{B.3}$$

As usual let $J_b$ be the auxiliary function

$$J_b = \int_\Omega |G(\boldsymbol{\omega}) - H(\boldsymbol{\omega})|^2 \frac{d\boldsymbol{\omega}}{(2\pi)^D}$$

$$+ \sum_{i=1}^N \lambda_i \left[ \int_\omega G(\boldsymbol{\omega}) e^{j\boldsymbol{\omega}\mathbf{x}_i} \frac{d\boldsymbol{\omega}}{(2\pi)^D} - f(\mathbf{x}_i) \right]. \tag{B.4}$$

The quantity $J_b$ is minimized when, for all $\boldsymbol{\omega} \notin \Omega$,

$$2(G_R - H_R) + \sum_{i=1}^N \lambda_i \cos(\boldsymbol{\omega}\mathbf{x}_i) = 0 \tag{B.5}$$

and

$$2(G_I - H_I) - \sum_{i=1}^N \lambda_i \sin(\boldsymbol{\omega}x_i) = 0. \tag{B.6}$$

# 7

# Error Analysis in Application of Generalizations of the Sampling Theorem

Abdul J. Jerri

*Dedicated to the memory of my professor*
*William M. Stone*
*of Oregon State University*

*The sampling theorem was extended to include transforms other than those of Fourier less than a decade after its popularization by Shannon. The first extension used the Bessel transform and had roots in an explicit suggestion by J.M. Whittaker in 1935. A decade later the question of physical interpretation of such an extension was addressed in relation to time-variant systems. Research in this field in the past two decades has centered around finding error bounds for truncation and aliasing errors that are incurred in the practical applications of such generalizations of the sampling theorem. This chapter is devoted to the general integral transform type of band-limited functions, along with some desirable extensions, and more importantly the analysis of the familiar "aliasing" and "truncation" errors of its applications.*

## Foreword: Welcomed General Sources for the Sampling Theorems

This chapter covers in detail a very important. and natural, extension of the well-known Shannon sampling theorem [818]. Specifically, signals are evaluated by integral transforms other than the Fourier transform. One familiar example is the use of the Hankel (or Bessel) transform in optics, where the kernel of the transform is a Bessel function instead of the usual (complex) exponential kernel of the Fourier transform. Papoulis [704] shows application of such transforms in optical systems. An overview of sampling theory applied to optics can be found in Gori's chapter in this volume.

Clearly, such analysis necessitates the development of new tools to facilitate practical applications. Of particular significance are formulas that deal with bounds on the essential errors of the sampling expansions, namely, *aliasing* and *truncation* errors. This is the major topic of this chapter.

The Shannon sampling theorem, its applications and various extensions, have been the subject of intensive research in the field of communications and almost all fields of engineering and science in the past thirty or so years. As witness to this are the hundreds of research papers that are available on the subject. Over one thousand references are listed in the bibliography of this volume.

## 7.1    Introduction—Sampling Theorems

### 7.1.1    THE SHANNON SAMPLING THEOREM—A BRIEF INTRODUCTION AND HISTORY

The statement of the sampling theorem, when introduced by Shannon, was that "If a function $f(t)$ contains no frequencies higher than W cycles per seconds (cps), it is completely determined by giving its ordinates at a series of points spaced $(\frac{1}{2W})$ seconds apart." Shannon's simple proof starts by representing such signal $f(t)$ as Fourier transform of the spectrum $F(w)$, which is band-limited to $(-2\pi W, 2\pi W)$

$$f(t) = \frac{1}{2\pi} \int_{-\infty}^{\infty} F(w)e^{-jwt}dw = \frac{1}{2\pi} \int_{-2\pi W}^{2\pi W} F(w)e^{-jwt}dw. \qquad (7.1.1)$$

Then he moved to establish the sampling series for $f(t)$.

$$f(t) = \sum_{n=-\infty}^{\infty} f\left(\frac{n}{2W}\right) \frac{\sin \pi(2Wt - n)}{\pi(2Wt - n)}, \qquad (7.1.2)$$

after writing the Fourier series for the spectrum $F(w)$ on the interval $(-2\pi W, 2\pi W)$ in (7.1.1),

$$F(w) = \sum_{n=-\infty}^{\infty} c_n e^{jwn/2W} \qquad (7.1.3)$$

allowing the term-by-term integration of this series inside the integral of (7.1.1) and recognizing that the Fourier coefficients $c_n$ in (7.1.3) are proportional to the samples of the signal as

$$c_n = \frac{1}{2W} f\left(\frac{n}{2W}\right).$$

The term by term integration is justified when the spectrum $F(w)$ is assumed to be square-integrable on $(-2\pi W, 2\pi W)$.

Since this sampling theorem is treated in detail in many places, we will be satisfied with the above simple derivation, which can still be found here as a special case of the generalized sampling theorem that we shall present soon in Subsection 7.1.2.

## The Sampling Theorems

Consider the Fourier transform of the signal $f(t)$

$$f(t) = \int_{-\infty}^{\infty} F(w)e^{iwt}dw. \tag{7.1.4}$$

Signals are transmitted on a finite band of frequency, for example, $w \in (-a, a)$. This means that $F(w)$ vanishes outside this band. Such signals are called *band-limited* to the bandwidth $a$, and we shall denote them by $f_a(t)$:

$$f_a(t) = \int_{-a}^{a} F(w)e^{iwt}dw. \tag{7.1.5}$$

This can also be written as

$$f_a(t) = \int_{-\infty}^{\infty} p_a(w)F(w)e^{iwt}dw \tag{7.1.6}$$

where

$$p_a(w) = \begin{cases} 1, & |w| \leq a \\ 0, & |w| > a \end{cases} \tag{7.1.7}$$

is the *gate function* which is the system function representation for the *ideal low-pass filter*. If we consider $f(t)$ as the voltage across one unit of resistance, then $\int_{-\infty}^{\infty} |f(t)|^2 dt$ should represent the total energy, which for all practical purposes is finite and we call the signal a finite energy signal. If we consult the Parseval equality for the Fourier transforms, we can easily see that

$$\frac{1}{2\pi}\int_{-\infty}^{\infty} |F(w)|^2 dw = \int_{-\infty}^{\infty} |f(t)|^2 dt. \tag{7.1.8}$$

Hence, for a *finite energy signal*, the right-hand side of (7.1.8) is finite and we conclude that $F(w)$ is square-integrable on $(-\infty, \infty)$, i.e., $\int_{-\infty}^{\infty} |F(w)|^2 dw < \infty$. As indicated earlier, this condition is important for facilitating the analysis of the series representation of band-limited signals, especially in regard to allowing the term-by-term integration of the Fourier series of $F(w)$ on $(-a, a)$. It should be easy to see from (7.1.8) that a band-limited signal has finite energy when $F(w)$ is square-integrable on $(-a, a)$: i.e.,

$$\int_{-\infty}^{\infty} |f_a(t)|^2 dt = \frac{1}{2\pi}\int_{-a}^{a} |F(w)|^2 dw < \infty. \tag{7.1.9}$$

## The Shannon Sampling Theorem

As was shown in (7.1.1)-(7.1.3), the Shannon sampling theorem interpolation for a band-limited signal $f_a(t)$ in terms of its samples $f_a(\frac{n\pi}{a})$ states that for the band-limited signal (7.1.5) with square-integrable $F(w)$ on $(-a, a)$ we have

$$f_a(t) = \sum_{n=-\infty}^{\infty} f_a\left(\frac{n\pi}{a}\right) \frac{\sin(at - n\pi)}{(at - n\pi)}. \qquad (7.1.10)$$

As we indicated earlier, one of the simplest proofs starts with writing the Fourier series expansion for $F(w)$ of (7.1.5) in terms of the orthogonal set of functions $\{e^{-(in\pi/a)w}\}$ on the interval $(-a, a)$; i.e.,

$$F(w) = \sum_{n=-\infty}^{\infty} c_n e^{-\frac{in\pi}{a}w}. \qquad (7.1.11)$$

The Fourier coefficients $c_n$ are obtained as

$$c_n = \frac{1}{2a} \int_{-a}^{a} F(w) e^{\frac{in\pi}{a}w} dw = \frac{1}{2a} f_a\left(\frac{n\pi}{a}\right) \qquad (7.1.12)$$

After consulting (7.1.5) we recognize the above integral as $f_a(\frac{n\pi}{a})$; the samples of the signal $f_a(t)$ in (7.1.5). The Fourier series (7.1.11), then, becomes

$$F(w) = \frac{1}{2a} \sum_{n=-\infty}^{\infty} f_a\left(\frac{n\pi}{a}\right) e^{-\frac{in\pi}{a}w}. \qquad (7.1.13)$$

To obtain $f_a(t)$ as in (7.1.5), we multiply (7.1.13) by $e^{iwt}$ and then integrate both sides from $-a$ to $a$:

$$f_a(t) = \int_{-a}^{a} F(w) e^{iwt} dw = \frac{1}{2\pi} \int_{-a}^{a} e^{iwt} \sum_{n=-\infty}^{\infty} f_a\left(\frac{n\pi}{a}\right) e^{-\frac{in\pi}{a}w} dw. \qquad (7.1.14)$$

The assumption that $F(w)$ is square-integrable on $(-a, a)$ is sufficient for exchanging the integration with the infinite summation to obtain

$$f_a(t) = \frac{1}{2a} \sum_{n=-\infty}^{\infty} f_a\left(\frac{n\pi}{a}\right) \int_{-a}^{a} e^{iw(t - \frac{n\pi}{a})} dw$$

which simplifies immediately to (7.1.10), after performing the above simple integration.

We should stress here that the importance of the Shannon sampling expansion lies not only in interpolating signals, but also in specifying the required spacing $\pi/a$ in terms of the bandwidth $a$. Such a spacing requirement plays an important role in developing the discrete Fourier transform, whose efficient algorithm is the useful fast Fourier transform (FFT). In this regard, as we cover the present chapter generalization to other integral transforms-type sampling theorem, the latter sampling plays a parallel role in specifying the spacing for the discrete version of such transforms [682, 454, 442, 456].

Next we give a very brief presentation of the main errors incurred in the practical application of the sampling series (7.1.10). This paves the way for drawing the parallel analysis of deriving bounds for such errors associated with the (Kramer–Weiss) extension, or generalization, of the Shannon sampling theorem (7.1.10, 7.1.5).

Two main errors may be involved in the practical application of the sampling expansion of signals, namely, those of $\epsilon_N$, the *truncation error* and the *aliasing error*, $\epsilon_A$. The truncation error is due to use of only a finite number $(2N + 1)$ of samples instead of the infinite number required by the series (7.1.10),

$$\epsilon_N(t) = \left| f_a(t) - \sum_{n=-N}^{N} f_a\left(\frac{n\pi}{a}\right) \frac{\sin(at - n\pi)}{(at - n\pi)} \right|. \tag{7.1.15}$$

The aliasing error, on the other hand, is due to the uncertainty of knowing exactly the bandwidth $a$ of the received signal $f(t)$ of (7.1.4) which is assumed (incorrectly) to be band-limited signal $f_a(t)$ for the sampling expansion (7.1.10) to apply to $f(t)$,

$$\epsilon_A = \left| f(t) - \sum_{n=-\infty}^{\infty} f\left(\frac{n\pi}{a}\right) \frac{\sin(at - n\pi)}{(at - n\pi)} \right|. \tag{7.1.16}$$

The task here is to find estimates for the upper bounds for each of these errors. Numerous results for the upper bound of the truncation error have been reported since the early fifties. These results are covered in our tutorial review [440]. Weiss [944] and then Brown [117], have established that

$$\epsilon_A \leq \frac{1}{\pi} \int_{|w|>a}^{\infty} |F(w)| \, dw. \tag{7.1.17}$$

Recently, Splettstösser [853] presented a very useful review of the aliasing error for the Shannon sampling theorem.

We can also speak of *time limited* signals,

$$F_T(w) = \frac{1}{2\pi} \int_{-T}^{T} f(t) e^{iwt} \, dt. \tag{7.1.18}$$

where $f(t)$ is given in (7.1.4). According to the uncertainty principle, a signal cannot be band-limited as in (7.1.5) and time limited as in (7.1.18) at the same time. In case we are forced into such a situation, we must admit an error and we usually optimize to find the shape of the signal that minimizes such an error. An obvious situation may occur when we have a time limited signal (7.1.18) but still want to use the sampling expansion (7.1.10). even though we cannot assume a band-limited function to $a$. i.e.. an uncertainty in knowing the required bandwidth $a$ for (7.1.10). In practice we assume a bandwidth $a$ and admit the *aliasing error* (as in (7.1.16)). which requires derivation of an upper bound like that of (7.1.17); hence. the results will be reported within the accuracy of such a bound.

Next we will present the extension of the Shannon sampling theorem to include more transforms with general kernel $K(x,t)$ besides the exponential Fourier $e^{ixt}$ kernel of (7.1.4) and (7.1.5).

## 7.1.2 THE GENERALIZED TRANSFORM SAMPLING THEOREM

In the last section we presented the Shannon sampling expansion (7.1.10) for the finite limit Fourier transform (7.1.5) as a representation of band-limited signals. Here we will present a generalization of the sampling theorem for functions represented by *general integral transforms* with kernel $K(x,t)$, for example, as a solution of the Sturm-Liouville problem.

Consider $f_I(t)$ as the following $\kappa$-transform of $F(x)$:

$$f_I(t) = \int_I \rho(x)K(x,t)F(x)dx \qquad (7.1.19)$$

where $F(x)$ is square-integrable on the interval $I$ and $\{K(x,t_n)\}$ is a complete orthogonal set on the interval I with respect to a weight function $\rho(x)$. Signals $f_I(t)$, as represented by the finite limit integral transform (7.1.19), are sometimes termed *transform limited* [437. 444] as opposed to *band-limited* in the case of the truncated inverse Fourier transform with its special exponential kernel $K(w,t) = e^{iwt}$. The *generalized sampling theorem* [943, 503] gives the following sampling expansion for $f_I(t)$ of (7.1.19):

$$f_I(t) = \lim_{N \to \infty} \sum_{|n| \leq N} f_I(t_n)S_n(t) \qquad (7.1.20)$$

where

$$S_n(t) = S(t,t_n) = \frac{\int_I \rho(x)K(x,t)\overline{K(x,t_n)}dx}{\int_I \rho(x)|K(x,t_n)|^2dx} \qquad (7.1.21)$$

is the *interpolation (sampling) function*. Here $\overline{K(x,t_n)}$ is the complex conjugate of $K(x,t)$.

The simplest proof parallels that for the Shannon sampling theorem in (7.1.11). (7.1.13). We write the orthogonal expansion for $F(x)$ in (7.1.19) in terms of $\overline{K(x.t_n)}$; i.e.,

$$F(x) = \sum_{n=1}^{\infty} c_n\overline{K(x,t_n)}. \qquad (7.1.22)$$

$$c_n = \frac{\int_I \rho(x)F(x,t)K(x,t_n)dx}{\int_I \rho(x)|K(x,t_n)|^2 dx} = \frac{f_I(t_n)}{\int_I \rho(x)|K(x,t_n)|^2 dx} \qquad (7.1.23)$$

after using (7.1.19) to give the value of the integral as $f_I(t_n)$. We then multiply both sides of (7.1.22) by $\rho(x)K(x,t)$ and formally integrate term by term to obtain

$$\int_I \rho(x)K(x,t)F(x)dx = f_I(t)$$

$$= \sum_{n=1}^{\infty} \frac{f_I(t_n)\int_I \rho(x)K(x,t)\overline{K(x,t_n)}dx}{\int_I \rho(x)|K(x,t_n)|^2 dx}$$

$$= \sum_{n=1}^{\infty} f_I(t_n)S_n(t)$$

after using (7.1.19) for $f_I(t)$ and (7.1.21) for the sampling function $S_n(t)$. We indicate that the same proof can be followed when $K(x,t)$ of (7.1.19) is expanded in terms of the same orthogonal functions $K(x,t_n)$. However, the shortest proof is to use Parseval's equality for the integral in (7.1.19) with the Fourier coefficients $c_n$ of (7.1.23) and $S_n(t)$ of (7.1.21) for $F(x)$ and $K(x,t)$, respectively. It is clear that the Shannon sampling expansion (7.1.10) for $f_a(t)$ in (7.1.5) is a special case of (7.1.20) corresponding to $K(x,t) = e^{ixt}$. More on the theoretical aspects of integrating the results of such general integral transforms is found in Higgins [393, p. 69], and in more detail in our monograph [453]. The emphasis here is on our own contributions over the past two decades, especially in regard to application and error analysis.

As we have mentioned, the conditions on the kernel $K(x,t)$ in (7.1.19) for this theorem are exhibited by the solutions of the Sturm-Liouville problem, which we will illustrate next for the case of $K(x,t) = J_0(xt)$, the Bessel function of the first kind of order 0. Further study and illustration of the generalized sampling theorem was done by Campbell [189], which included the case of the Legendre function kernel $P_t(x)$. Other illustrations including the associated Legendre function $P_t^m(x)$, the Gegenbauer function $C_t^\nu(x)$, the Chebyshev functions $T_t(x)$ and $U_t(x)$, and the prolate spheroidal function $P_{s_t}^m(x,\theta)$ [827] were done in detail in [434]. We also suggested using the aforementioned results in scattering problems in physics [439]. Illustrations were also done for orthogonal expansions on the infinite interval instead of the usual finite interval. This included the $L_t^\alpha(x)$-Laguerre transform on $(0,\infty)$ as found in [439], and the $H_t(x)$-Hermite (or parabolic cylinder functions) transform in [445]. Campbell [189] was the first to raise the question concerning the possibility of a relation, or equivalence, between the Shannon sampling theorem and its present generalization. This was later studied in some depth and was formulated as a few basic theorems that gave clear conditions for the equivalence between the two sampling theorems [440, 435, p. 1569].

As indicated in the tutorial review article [440, p. 1568], this extension was first suggested by Whittaker [948]. The above statement is very closely associated with that of Kramer [503] in his 1959 paper. However, this same expansion, its proof and the association of the kernel with the solution of second-order self-adjoint Sturm-Liouville problem, was considered earlier in 1957 by Weiss [943]. Weiss presented this result and, unfortunately, was satisfied with sending only a short abstract of his detailed presentation of this result. Later, this author, through his personal correspondence, received a copy of the original manuscript for Weiss' detailed presentation [943]. It may be only fair now to say that this extension had been originated by Whittaker, or even Lagrange, and was finally settled by both Kramer [503] and Weiss [943]; thus we may call it the Kramer-Weiss sampling theorem. Higgins [393, p. 69] also noted that their usual association of the general kernel in (7.1.19) with the Sturm-Liouville problem orthonormal solutions is not necessary. He points to Kak [468], who derived the Walsh sampling theorem as a special case of the above theorem, without restoring to a solution of a differential equation. We add that there are more examples of this sort including the very clear and dependable method of using complex contour integration for this general sampling expansion (7.1.20), as well as for its extension that involves the samples of the function and its derivatives [458, 440, p. 1573]. The generalized sampling theorem with more and different conditions on the transformed function $F(x)$ in (7.1.19) was addressed in [450].

### The Bessel-Type Sampling Series

Here we will illustrate the above extension of the sampling theorem for the case of the kernel being a Bessel function of the first kind of order zero. $J_0(xt)$. This will be our primary example for illustrating the necessary tools and the error bounds that we shall present in this chapter. Also. it is the most commonly used integral transform. other than the Fourier transform. with applications for instance in optics [704]. In this case, the transform (7.1.19) becomes the following finite limit $J_0$-Hankel (or Bessel) transform. where the interval $I$ is taken as $(0, b)$.

$$f_b(t) = \int_0^b x J_0(xt) F(x) dx. \qquad (7.1.24)$$

The sampling function $S_n(t)$ of (7.1.21) is

$$S(t, t_n) = S\left(t, \frac{j_{0,n}}{b}\right) = \frac{\int_0^b x J_0(xt) J_0(x \frac{j_{0,n}}{b}) dx}{\int_0^b x J_0^2(x \frac{j_{0,n}}{b}) dx}$$

$$= \frac{2 j_{0,n} J_0(bt)}{b^2 (\frac{j_{0,n}^2}{b^2} - t^2) J_1(j_{0,n})}. \qquad (7.1.25)$$

where $t_n = \frac{j_{0,n}}{b}$ and $\{j_{0,n}\}$ are the zeros of the Bessel function $J_0$; i.e., $J_0(j_{0,n}) = 0$, $n = 1, 2, \ldots$ . Some familiar properties of the Bessel functions are used to evaluate the integrals of (7.1.25). The final sampling series (7.1.20) for the finite limit $J_0$-Hankel transform becomes

$$f_b(t) = \sum_{n=1}^{\infty} f_b\left(\frac{j_{0,n}}{b}\right) \frac{2j_{0,n}J_0(bt)}{b^2(\frac{j_{0,n}^2}{b^2} - t^2)J_1(j_{0,n})}, \tag{7.1.26}$$

$$J_0(j_{0,n}) = 0, \qquad n = 1, 2, \ldots . \tag{7.1.27}$$

We note here that the weight function $\rho(x) = x$ is introduced explicitly in (7.1.24) instead of implicitly.

The *truncation error* for this generalized Bessel sampling series, as a special case of (7.1.20), can be defined in the same way as in (7.1.15):

$$\epsilon_N(t) = \left| f_b(t) - \sum_{n=1}^{N} f_b\left(\frac{j_{0,n}}{b}\right) \frac{2j_{0,n}J_0(bt)}{b^2(\frac{j_{0,n}^2}{b^2} - t^2)J_1(j_{0,n})} \right|. \tag{7.1.28}$$

In comparison to the numerous results for improving the upper bound for the truncation error (7.1.15) of the Shannon sampling expansion (or cardinal series), we find that until 1977 [440, p. 1568], there is one simple result for the generalized sampling expansion (7.1.20), attributed to Yao [974]. This was followed many years later by our practical bound [459] for the above truncation error of the Bessel sampling series where complex contour integration was the main derivation tool. This will be the subject of discussion in Subsection 7.2.2 , where we establish a lower bound for the Bessel function that is essential for the derivation, yet was not previously available in the literature.

Again, the idea of the *aliasing error* in applying the generalized sampling theorem (7.1.19)-(7.1.20) is that we are usually not sure about the exact finite interval $I$, i.e., $f(t)$ may not be "band-limited" (or "transform limited") to $(0. b)$ as $f_b(t)$, but we still apply the above Bessel sampling series (7.1.26) to its samples $f(\frac{j_{0,n}}{b})$. Such an application to $f(t)$ and not $f_b(t)$ results in the "aliasing error" $\epsilon_A$ of the sampling expansion,

$$\epsilon_A = \left| f(t) - \sum_n f(t_n)S(t, t_n) \right|, \tag{7.1.29}$$

which in the case of the $J_0$-Hankel transform becomes

$$\epsilon_A = \left| f(t) - \sum_{n=1}^{\infty} f\left(\frac{j_{0,n}}{b}\right) \frac{2j_{0,n}J_0(bt)}{b^2(\frac{j_{0,n}^2}{b^2} - t^2)J_1(j_{0,n})} \right|. \tag{7.1.30}$$

We stress again the presence (or forced use) of the samples $\{f(\frac{j_{0,n}}{b})\}$ of the "non-band-limited" $f(t)$ instead of the required band-limited samples $f_b(\frac{j_{0,n}}{b})$, which is the essence of the aliasing error.

The first practical aliasing error bound for the generalized sampling expansion was achieved in 1988 as a result of our long-standing attempts [454]. This was illustrated for $f(t)$, the $J_0$-Hankel (Bessel) transform of *monotonically decreasing* spectrum $F(w) \geq 0$, as

$$\epsilon_A(t) \leq \frac{3}{2} \int_b^\infty wF(w)dw + \left| \frac{bJ_1(bt)}{t} \right| F(b). \tag{7.1.31}$$

We note that the first (integral) term of the bound (7.1.31) is of the same form as that of the aliasing error bound of Weiss [943] and Brown [118, 117], given by (7.1.17). The method for the derivation of such an aliasing error bound (7.1.31) will be the subject of Subsection 7.2.1. This will be presented after preparing the "relatively new tools" necessary for such derivation. This will be discussed in Subsections 7.1.4 and 7.1.5.

It is now time to comment on the long delay in having the aliasing error bound of (7.1.31). As will become evident in Subsections 7.1.4, 7.1.5 and 7.2.1, use of kernels other than complex exponential (or trigonometric) kernels results in loss of many important properties and tools. This includes additivity of the exponents in multiplying Fourier exponential kernels, and the closely related periodicity of such Fourier kernels for the Fourier series expansion. Looking at the method of deriving the aliasing error bound (7.1.17) for the Shannon sampling series one feels helpless with a Bessel function kernel, where the two above essential properties of the derivation of (7.1.17) are non-existent. The solution is development of parallel tools, e.g., a *"generalized" Poisson sum formula* [454] for the Bessel transform without the usual reliance on the periodicity of the trigonometric Fourier series. This formula was derived, as we shall see in Subsection 7.1.5, with the use of our own concept of *"generalized translation"* [437, 444] that is compatible with the Hankel transform. This means that such general translation must stand as a parallel to the usual translation seen in the convolution product of the Fourier transforms. Earlier attempts did not use these tools, and the results were abandoned as they involved tedious derivation and long-winded reliance on properties of Bessel functions [940].

In the following Subsection we will discuss our first attempt [437] at giving a system interpretation for the present extension of the sampling theorems (7.1.19)-(7.1.20).

## 7.1.3  SYSTEM INTERPRETATION OF THE SAMPLING THEOREMS

In this section we will discuss the first interpretation for the generalized sampling series (7.1.20) that we suggested in 1969 [437]. It parallels that given for the Shannon sampling series (7.1.10), but it is for a *time varying* system impulse response. This is indicated by the $t$ and $\tau = t_n$ dependence in the sampling function $S(t, t_n)$ of (7.1.21) versus the time invarying (time shift dependent only) sampling function,

$$S(t, t_n) \equiv \frac{\sin(at - n\pi)}{(at - n\pi)}$$

of (7.1.10). Such analysis would necessitate the introduction [437, 444] of the *"generalized translation"* (of $t$ by $\tau$): $t\theta\tau$ instead of $t - \tau$, to be compatible with the $t, \tau$ relations of the sampling function $S(t, t_n)$ which we had adopted; writing it, then, as $S(t\theta t_n)$. The summary of the analysis of this section for a system interpretation of the Bessel-type sampling series of (7.1.26), for example, is that the signal $f_b(t)$ in (7.1.24), (7.1.26) can be considered as "the output of *band (or transform)-limited* and (time invarying) low-pass filter with a *time varying impulse* response." Such response is expressed as the (main) integral in the numerator of the sampling function $S(t, t_n)$ in (7.1.25). Again, it is this same concept of the general translation that is needed for the development of Bessel-type Poisson sum formula, which became our main tool for deriving the first aliasing error bound for general sampling expansion [454]. We mention that whereas we introduced such a generalized translation concept in 1969, Churchill [221] had a similar concept in mind for general integral transforms other than the Fourier one, which appeared in the 1972 edition of his operational calculus book [221].

### System Interpretation–Time Varying System and the Generalized Translation

Consider the following transform

$$f(t) = \int \rho(w) K(w, t) F(w) dw \equiv \kappa\{F\} \qquad (7.1.32)$$

with the Fourier-type inverse

$$F(w) = \int \rho(t) \overline{K(w, t)} f(t) dt \equiv \kappa^{-1}\{f\}. \qquad (7.1.33)$$

Here $\overline{K(w, t)}$ stands for the complex conjugate of $K(w, t)$. Also, unless otherwise indicated, the limits of integration are not finite and will be specified for the particular integral transform.

### a) Generalized translation

As we had indicated above for the generalized sampling expansion (7.1.20), when we deal with transforms of non-exponential kernels, we do not expect the usual translation "or shift" property that we are so accustomed to in dealing with the Laplace and Fourier transforms. This shift property was very important in defining the convolution products that are essential for developing the convolution theorems for the Laplace and Fourier transforms. Thus, in considering a convolution product for transforms (7.1.32)-(7.1.33), which are, in general, without exponential kernels, it is necessary to introduce a compatible transformation called a *"generalized translation"* [437, 444]. The $\tau$-generalized translation $f(t\theta\tau)$ of $f(t)$ is defined for such transforms (7.1.32)-(7.1.33) by

$$f(t\theta\tau) = \int \rho(w)F(w)K(w,t)\overline{K(w,t)}dw. \qquad (7.1.34)$$

This definition is not surprising since the translation for the Fourier transform is also the result of multiplying the transformed function by an exponential function which is of the same type as the transform kernel.

We have already indicated that the form $S(t, t_n)$ of the sampling function (7.1.20) in the Bessel-sampling series (7.1.26) represents a generalized translation which can be written now as $S(t\theta t_n)$. To illustrate this idea, we can show that the $J_0$-Hankel transform of the gate function $p_a(w)$ is $S(t) = \frac{aJ_1(at)}{t}$. To find the $\tau$-generalized translation $S(t,\tau) \equiv S(t\theta\tau) = \frac{aJ_1(a(t\theta\tau))}{t\theta\tau}$ of this function $S(t)$, we compute the $J_0$-Hankel transform of $J_0(w\tau)p_a(w)$,

$$\frac{aJ_1(a(t\theta\tau))}{t\theta\tau} = \int_0^a wJ_0(wt)J_0(w\tau)dw \equiv S(t\theta\tau)$$

$$= \frac{atJ_1(at)J_0(a\tau) - a\tau J_1(a\tau)J_0(at)}{t^2 - \tau^2}$$

$$(7.1.35)$$

as can be found with the aid of the Bessel functions properties [940].

## b) Generalized convolution product and theorem

Let $f(t)$ and $g(t)$ be the integral transforms of $F(w)$ and $G(w)$ as in (7.1.32), respectively, we define the *convolution product* $(f * g)(t)$ of $f$ and $g$ by

$$(f * g)(t) = \int \rho(\tau)g(\tau)f(t\theta\tau)d\tau = \int \rho(w)F(w)G(w)K(w,t)dw. \quad (7.1.36)$$

It is clear that this convolution product is commutative, i.e., $f*g = g*f$, and it can be shown to be associative too; i.e., $f * (g * h) = (f * g) * h$.

It should be easy now to formally state a *convolution theorem* for the general $\kappa$-transforms (7.1.32)-(7.1.33),

$$\kappa\{f * g\}(t) = F(w)G(w). \qquad (7.1.37)$$

## c) System analysis–Time varying impulse response

The *system function* $H(w,t)$ is defined as

$$H(w,t) = \frac{1}{K(w,t)}\int \rho(\tau)h(t,\tau)K(w,\tau)d\tau. \qquad (7.1.38)$$

where $h(t, \tau)$ is the *time varying impulse response*,

$$h(t, \tau) = \int \rho(w) H(w, t) K(w, t) \overline{K(w, \tau)} dw. \qquad (7.1.39)$$

These definitions are in agreement with those given in D'Angelo [243]. In this formal treatment we assume the validity of interchanging the order of integration which requires the absolute integrability of either one of the integrals.

If we consider $f(t)$ in (7.1.32) as the output of the time varying impulse response $h(t, \tau)$, and $g(t)$ as the output, it is known [991, 243] that by summing the responses $h(t_j, \tau_i) f(\tau_i) \Delta\tau$ (for all j) for the inputs $f(\tau_i)$, $i = 1, 2, \ldots$, we can. for a linear system. sum over $\tau_i$ to have the output, after considering the limit of the sum. as

$$g(t) = \int \rho(\tau) h(t, \tau) f(\tau) d\tau. \qquad (7.1.40)$$

If we substitute for $f(\tau)$ of (7.1.32) in (7.1.40). interchange the integrals. use (7.1.38) for $H(w, t)$ and assume the uniqueness of $G(w)$, the inverse of $g(t)$, we obtain:

$$G(w) = F(w) H(w, t)$$

or

$$H(w, t) = \frac{F(w)}{G(w)}. \qquad (7.1.41)$$

This is a time invarying, but it still corresponds to time varying impulse response $h(t, \tau)$. In other words, it appears that such treatment shifts the time variance of the system function to the impulse response representation.

### d) System interpretation of the sampling theorem

The generalized sampling theorem uses a finite limit integral transform, given by (7.1.19) with the sampling series

$$f_I(t) = \sum_n f_I(t_n) S(t, t_n).$$

where $\sum_n$ stands for the required infinite sum. The sampling function $S(t, t_n) \equiv S_n(t)$ is given as in (7.1.21).

As is well-known, the physical interpretation [757] for the special case of $K(w, t) = e^{iwt}$, i.e.. the Shannon sampling expansion, is that $f_I(t)$ is the output of an ideal low-pass filter with impulse response $S(t, t_n) = \sin[\pi(t - n)]/\pi(t - n)$ for $I$ as $[-\pi, \pi]$ and with the input taken to be the pulse train $\{f(t_n)\} = \{f(n)\}$.

Here we will attempt a physical interpretation for the generalized sampling series (7.1.20). In (7.1.38) consider the system function $H_1(w. t) = 1$.

with its corresponding impulse response $h_1(t, \tau)$,

$$K(w, t) = \int \rho(\tau) h_1(t, \tau) K(w, \tau) d\tau. \qquad (7.1.42)$$

Thus, $h_1(t, \tau)$ can be recognized as a generalization of the *Dirac delta function* $\delta(t - \tau)$ and as the response to $K(w, t)$ instead of $e^{iwt}$. This is in agreement with the definition of the delta function given in Zemanian [1002] for the case of the Hankel transform where $K(w, t) = J_\nu(wt)$. In keeping with the usual notation we denote $h_1(t, \tau)$ by $\delta(t, \tau)$ as the limit of the integral

$$\int_I \rho(x) K(x, t) \overline{K(x, \tau)} d\tau \equiv S(t, \tau) \int_I \rho(x) \mid K(x, \tau) \mid^2 dx \qquad (7.1.43)$$

as $I$ becomes infinite. We note here that this $S(t, \tau)$ of (7.1.43) has a similar form as that of its special case, the sampling function $S(t, t_n)$ of (7.1.21).

Next we consider the band (or transform)-limited function $f_I(t)$ of (7.1.19) as the output of a system with impulse response $h_2(t, \tau)$ and with input $f(t)$, i.e.,

$$f_I(t) = \int_I \rho(\tau) h_2(t, \tau) f(\tau) d\tau. \qquad (7.1.44)$$

Here, $f_I(t)$ of (7.1.19) is the transform of $p_I(w)F(w)$. If we let the system function be $H_2(w)$ and use (7.1.41), we obtain

$$H_2(w, t) = \frac{p_I(w)F(w)}{F(w)} = p_I(w).$$

Thus, we have a system function $H_2(w, t)$ as the time invariant gate function which corresponds to a time varying impulse response,

$$
\begin{aligned}
h_2(t, \tau) &= \int p_I(w) \rho(w) K(w, t) d\tau. \\
&= S(t, \tau) \int_I \rho(x) \mid K(x, \tau) \mid^2 d\tau. \qquad (7.1.45)
\end{aligned}
$$

Hence, the physical interpretation for the sampling expansion (7.1.20) can be given as $f_I(t)$ being the output of a low-pass filter in the sense of these general transforms. Moreover, it is associated with the time varying impulse response $h_2(t, \tau)$ of (7.1.45) that is directly related to the form $S(t, \tau)$ of the sampling function in (7.1.21) and with the pulse train samples $\{f_I(t_n)\}$ as its input.

The complete details of this analysis are found in [444, 940] and in a self-contained textbook format in [456, Sect. 3.6]. The main definitions are consistent with those given by Churchill [221] in operational mathematics and in the electrical engineering references on the subject. e.g.. D'Angelo [243], Zadeh [991] and Zemanian [1002].

## 7.1.4   SELF-TRUNCATING SAMPLING SERIES FOR BETTER TRUNCATION ERROR BOUND

In this section we will show that the self-truncating version of the Shannon sampling series, as suggested by Helms and Thomas [385], is due to an indirect use of the *hill functions (B-splines)*, well known in numerical analysis. This is done in the hope of having a tighter bound on the truncation error. To clarify the analysis, we developed an extension [444] to the discretization of the Fourier convolution product [845]. This was in the sense of extending the known result to three functions, then extending this result for the general transforms of the present sampling theorem.

**The Role of the Hill Functions (B-Splines) for a Self-Truncating Sampling Series**

Consider the Fourier transform

$$F(w) = \frac{1}{2\pi}\int_{-\infty}^{\infty} f(t)e^{-iwt}dt \qquad (7.1.46)$$

of the function

$$f(t) = \int_{-\infty}^{\infty} F(w)e^{iwt}dw.$$

The *hill function (B-spline) of order* $R + 1$, $\varphi_{R+1}(a(R+1), w)$, is well-known in numerical analysis and is defined as the $R$th fold (Fourier transform) convolution product.

$$
\begin{aligned}
\varphi_{R+1}(a(R+1).w) &= (\varphi_1(a,-) * \overset{R}{\cdots} * \varphi_1(a,-))(w) \\
&= (\varphi_1(a,-) * \varphi_R(aR,-))(w) \\
&= \int_{-\infty}^{\infty} \varphi_1(a,x)\varphi_R(aR, w-x)dx
\end{aligned}
$$
$$(7.1.47)$$

of the gate function (the hill function of order one) $p_a(w)$,

$$\varphi_1(a,w) = p_a(w) = \begin{cases} 1, & |w| < a \\ 0, & |w| > a \end{cases}.$$

As a consequence of this convolution product, $\varphi_{R+1}(a(R+1),w)$ vanishes for $|w| > a(R+1)$ and is the Fourier transform of $\eta_{R+1}(t) = [2\sin(at)/t]^{R+1}$, since the Fourier transform of $\varphi_1(a,w)$ is $\eta_1(t) = 2\sin(at)/t$. Unless otherwise indicated, $\varphi_{R+1}(w)$ will be written for $\varphi_{R+1}(a(R+1),w)$. The exact explicit form of $\varphi_{R+1}(w)$ is given in Ditkin and Prudnikov [262], with some of its properties and computations done by Segethova [809] and de Boor [249, 250], see also [441].

The following demonstrates how the hill function of higher order $\varphi_{R+1}(w)$ is used in signal analysis to make its sampling series a self-truncating one.

As we have indicated in Subsection 7.1.3, the system interpretation of the Shannon sampling series in (7.1.10) is that the signal $f_a(t)$ has passed through an ideal low-pass filter whose system function is the gate function $\varphi_1(a, w)$. In practice we can only have a finite number, $2N + 1$, of the samples in (7.1.10); hence, there is a need for an upper bound of the (truncation) error $\epsilon_N$, as in (7.1.15), resulting from the truncation of (7.1.10). To obtain a better upper bound on the truncation error that decreases faster with $N$, Helms and Thomas [385] presented the following self-truncating sampling expansion for $f(t)$ band-limited to $ra$, without mentioning the hill function:

$$f(t) = \sum_{n=-\infty}^{\infty} f\left(\frac{n\pi}{a}\right)\left[\frac{\sin\frac{aq}{m}\left(t - \frac{n\pi}{a}\right)}{\frac{aq}{m}\left(t - \frac{n\pi}{a}\right)}\right]^m \frac{\sin(at - n\pi)}{(at - n\pi)} \qquad (7.1.48)$$

where $q = 1 - r$, $0 < r < 1$. For their optimal truncation error bound, $m$ was chosen to equal approximately the optimum value $Nq\pi/e$. To further clarify the systems interpretation of Helms and Thomas' self-truncating sampling series in (7.1.48), we derived the following simple extension [444] of the discretization of the Fourier convolution product [845].

Let $f$, $g$ and $h$ be Lebesgue integrable and continuous on the real line with $f$ band-limited to $ra$, $g$ to $qa$; $r + q = 1$ and $h$ to $a$; i.e., $F(w) = 0,|$ $w |> ra; G(w) = 0, | w |> qa$; and $H(w) = 0, | w |> a$. Then

$$\int_{-\infty}^{\infty} f(\tau)g(\tau - t')h(t - \tau)d\tau$$

$$= \frac{\pi}{a} \sum_{n=-\infty}^{\infty} f\left(\frac{n\pi}{a}\right) g\left(\frac{n\pi}{a} - t'\right) h\left(t - \frac{n\pi}{a}\right). \qquad (7.1.49)$$

The proof, which will help to illustrate (7.1.48), is a straightforward one when we note that $(F * (Ge^{iwt'}))H(w)$ is the inverse Fourier transform of the integral in (7.1.49), which is band-limited to $a$,

$$\frac{1}{2\pi}\int_{-\infty}^{\infty} f(\tau)g(\tau - t')h(t - \tau)d\tau$$

$$= \int_{-a}^{a} (F * (Ge^{iwt'}))(w)H(w)e^{iwt}dw. \qquad (7.1.49a)$$

since the Fourier transform of the convolution product $(F * (Ge^{iwt'}))(w)$ is band-limited to $ra + qa = a$. If we now write the Fourier series expansion of $(F * (Ge^{-iwt'}))(w)$ and $H(w)e^{iwt}$ in terms of $\{e^{(in\pi/a)w}\}$ on $(-a, a)$, then use Parseval's equality, we obtain (7.1.49). With the choice of $H(w) = \phi_1(a, w)$, (7.1.49) becomes the Shannon sampling expansion of $f(t)g(t - t')$

as band-limited to $a$,

$$f(t)g(t - t') = \sum_{n=-\infty}^{\infty} f\left(\frac{n\pi}{a}\right) g\left(\frac{n\pi}{a} - t'\right) \frac{\sin(at - n\pi)}{(at - n\pi)}. \qquad (7.1.50)$$

To obtain a self-truncating sampling expansion for $f(t)$, we may choose $g$ to improve the convergence of the series. Helms and Thomas' choice corresponds to

$$g(t) = \left[\frac{\sin(aq/m)t}{(aq/m)t}\right]^m = \left[\frac{m}{2aq}\eta_1\left(\frac{aqt}{m}\right)\right]^m \qquad (7.1.51)$$

which when used in (7.1.50) with $t = t'$ gives (7.1.48). If we compare (7.1.49) with (7.1.48), we conclude that the self-truncating series (7.1.48), compared to the usual sampling series (7.1.10), is the result of having the transform of the signal $f(t)$ being convolved with $G(w)$, a *hill function of order* $m$, before passing through the gate function $H(w) = \varphi_1(ra, w)$. Hence the role of the higher order hill function, disguised in terms of its Fourier transform $\eta_m(t) = [2\sin(aqt/m)/t]^m$ in (7.1.48), is established, improving the estimate of the truncation error bound of this signal representation.

A special case of (7.1.49), for two band-limited functions $f$ and $g$, will be used in Subsection 7.3.4 in our development of retaining the important *interpolation property* to the Splettstösser's generalized (optimal) sampling sum of approximating theory [845].

In the next section we will develop the analysis for *"new hill"* functions associated with other known integral transforms, other than the Fourier, in an attempt to derive a *self-truncating version of the generalized sampling expansion* of such transforms (7.1.19)-(7.1.21). These new functions or tools will again facilitate the physical interpretation of the general self-truncating sampling series. It is hoped that they will aid in future research toward developing a tighter truncation error bound than the first one we derived in 1982 for the case of the Bessel-type sampling series [459].

## The General Transform Hill Functions

After introducing the concept of generalized translation in (7.1.34) and its associated convolution theorem in (7.1.37) for general transforms, we can now define *"general transform hill functions"* to serve toward a self-truncating generalized sampling series.

In the sense of the general $\kappa$-transform in (7.1.32)-(7.1.33), we define the hill function $\psi_{R+1}(w)$ paralleling that of $\varphi_{R+1}(w)$ in (7.1.47), to be the $R$th fold convolution product (7.1.36) of the (gate function) $\psi_1(w) = p_I(w)$.

$$\begin{aligned} \psi_{R+1}(w) &= (\psi_1 * \overset{R}{\cdots} * \psi_1)(w) = \int \rho(x)\psi_1(x)\psi_R(w\theta x)dx \\ &= \int \rho(t)[\xi_1(t)]^{R+1}K(w,t)dt \end{aligned} \qquad (7.1.52)$$

where $\xi_1(t) \equiv \xi_1(I, t)$ is the $\kappa$-transform of $\psi_1(w)$ which is band-limited to $I$.

Note here that the general translation for constructing the new hill function (7.1.52) is not easy to perform. We have, however, developed a method of computing the regular hill functions $\varphi_{R+1}(w)$ ($B$-splines) [441] which can be easily extended to express the $\psi_{R+1}(w)$ defined on the interval $I$ as a series of the orthogonal set $\{K(w, t_n)\}$,

$$\psi_{R+1}(w) = \sum \frac{[\xi_1(t_n)]^{R+1} K(w, t_n)}{\|K(., t_n)\|^2} \qquad (7.1.53)$$

which is clearly self-truncating. Here,

$$\|(K(., t)\|^2 = \int_I \rho(w) \mid K(w, t) \mid^2 dw.$$

Also, the hill function ($B$-spline) $\varphi_{R+1}(a(R+1), w)$ is an even function in $w$, a polynomial of degree $R$, defined on each of the $(R+1)/2$ and $(R+2)/2$ subintervals of $[0, a(R+1)]$ for $R$ odd and $R$ even, respectively, with continuous derivatives up to $R-1$ (Ditkin and Prudnikov [262]; Segethova [809]). Note that even though the discontinuities of $\varphi_1(a, w)$ are being smoothed through the Fourier convolution product that produced $\varphi_{R+1}(w)$, it is still the simple translation of this convolution that propagates these discontinuities. With the generalized translation involved in generating $v_{R+1}(w)$, one can expect a smoother hill function, as we illustrated in [442, 444] for the Hankel (Bessel) transform (7.1.54)-(7.1.55).

It is reasonable to expect that the bandwidth $J$, associated with $\psi_{R+1}(w)$, is larger than $I$ of $\psi_1(w)$. This should be investigated for each transform. In the meantime, we write $J(I, R)$ to indicate such dependence where $J = (R+1)I$ for the Fourier transform. Our illustrations for this and the preceding concepts will involve the Hankel (Bessel) transform, where our investigations indicate that $J = (R+1)I$.

Consider the $J_0$-Hankel (Bessel) transform $H_0(w)$,

$$H_0(w) = \int_0^\infty t h_0(t) J_0(tw) dt \qquad (7.1.54)$$

and its inverse $h_0(t)$

$$h_0(t) = \int_0^\infty w H_0(w) J_0(wt) dw \qquad (7.1.55)$$

where $J_0$ is the Bessel function of the first kind of order zero. The above transforms are symmetric; hence, there is no confusion in referring to either one as the transform. $\xi_1(t) = (a J_1(at))/t$ is the $J_0$-Hankel transform of (the gate function) $\psi_1(w)$. The $\tau$–generalized translation of $(a J_1(at))/t$ is given by (7.1.35).

In (7.1.49) we developed the discretization of the Fourier convolution product for three functions to give a clearer physical interpretation of the self-truncating (Shannon) sampling series of Helms and Thomas [385]. Here we shall present the discretization of a similar convolution product (7.1.34) associated with the general transform, (7.1.32)-(7.1.33) which will be illustrated for the $J_0$-Hankel transforms (7.1.54)-(7.1.55). This will provide us with the simplest proof of the generalized sampling theorem and, more importantly, the development of "generalized" self-truncating sampling series as we shall present in (7.1.60).

The convolution product, (7.1.34) for the generalized $\kappa$-transforms in (7.1.32)-(7.1.33) can be discretized in a manner similar to that for the (Fourier) convolution product of (7.1.49). This is so, because when $f(t)$ and $g(t)$ are band-limited to $(0,a)$, i.e., $F(w) = G(w) \equiv 0.\ w > a$, then

$$
\begin{aligned}
\int_0^\infty \rho(\tau)f(\tau)g(t\theta\tau)d\tau &= \sum_n \frac{f_a(t_n)g_a(t\theta t_n)}{\| K(.t_n) \|_2^2} \\
&= \int_0^a \rho(w)F(w)G(w)K(w,t)dw. \quad (7.1.56)
\end{aligned}
$$

The proof is straightforward. taking into consideration the complete orthogonal set $\{K(w,t_n)\}$ on $I \equiv (0,a)$, and parallels that of (7.1.49). The case of $J_0(wt)$, the Bessel function of the first kind of order zero. and $G(w) = \psi_1(w) = p_a(w)$ gives the Bessel sampling expansion for $f_a(t)$,

$$
\begin{aligned}
f_a(t) &= \sum_{n=1}^\infty f_a(t_{0,n}) \frac{aJ_1(t\theta t_{0,n})}{t\theta t_{0,n}} \cdot \frac{1}{\frac{a^2}{2}J_1^2(j_{0,n})} \\
&= \sum_{n=1}^\infty f_a(t_{0,n}) \frac{2t_{0,n}J_0(at)}{a(t_{0,n}^2 - t^2)J_1(j_{0,n})} \quad (7.1.57)
\end{aligned}
$$

where $at_{0,n} = j_{0,n}$ is the $n$th zero of $J_0(x)$. This is a special case of the generalized sampling theorem (7.1.20) that we seek to self-truncate.

### A Self-Truncating Generalized Sampling Series

In parallel to (7.1.49) and (7.1.49a), we present the following generalization of (7.1.56) to three functions $f(t).\ g(t)$ and $h(t)$ which are band-limited to $rl,\ qI$ and $I$, respectively, such that $f(t)g(t)$ is band-limited to $I$,

$$
\begin{aligned}
&\int f(\tau)g(\tau\theta t')h(t\theta\tau)d\tau \\
&= \int_I (F * (GK(w,t')))(w)H(w)K(w,t)dw \\
&= \sum \frac{f(t_n)g(t_n\theta t')h(t\theta t_n)}{\|K(.,t_n)\|_2^2}. \quad (7.1.58)
\end{aligned}
$$

This is a band-limited function to $I$, where the condition $f(t)g(t)$ band-limited to $I$ is necessary for the convolution product $(F * (GK(w, t'))(w)$ in (7.1.58) to have the same support $I$ as that of $H(w)$, in order to express both of them in terms of the orthogonal functions $\{K(w, t_n)\}$ on $I$. For the particular transform the relation between $r$ and $q$, resulting in $f(t)g(t)$ being band-limited to $I$, should be investigated. For the Fourier transform $r + q = 1$, our preliminary investigation indicates the same is true for the $J_0$-Hankel transform.

In the case of $H(w) = \upsilon_1(w)$ we have the generalized sampling expansion, parallel to (7.1.50), for $f(t)g(t\theta t')$ as band-limited to $I$,

$$f(t)g(t\theta t') = \sum \frac{f(t_n)g(t_n\theta t')h(t\theta t_n)}{\|K(, t_n)\|_2^2}. \tag{7.1.59}$$

With the choice of a fast decaying $g(t\theta t')$ for a particular $t' = t_0$, e.g., $[\xi_1(qa, m; t\theta t_0)]^m$, we obtain a *self-truncating sampling series* for $f(t)$ as band-limited to $rI$. We stress again that the dependence of this choice on $m$ should be investigated, for the particular transform at hand, in order to ensure that $f(t)$ is band-limited to $rI$. Also, we can still choose an optimal value for $m$ during the evaluation of the truncation error bound, where typically complex integration is employed. For the case of $J_0$-Hankel transform with $t_0 = 0$ and $I = (0, a)$, we use $\xi_1(t) = \frac{aJ_1(at)}{t}$ to obtain the following *self-truncating Bessel sampling series* for $f(t)$, as band-limited to $ra$:

$$f(t) = \sum f(t_{0,n}) \frac{2t_{0,n}J_0(at)}{a(t_{0,n}^2 - t^2)} \cdot \frac{1}{J_1(j_{0,n})} \cdot \left[ \frac{tJ_1(\frac{qj_{0,n}}{m})}{t_{0,n}J_1(\frac{qat}{m})} \right]^m \tag{7.1.60}$$

where we have used

$$g(t) = [\xi_1(t)]^m = \left[ \frac{qaJ_1(\frac{qat}{m})}{mt} \right]^m \qquad \frac{qat}{m} \neq j_{1,n}, \tag{7.1.61}$$

noting that $a$ for $\xi_1$ is replaced by $\frac{qa}{m}$. We may stress again that other choices of the self-truncating factor $g(t\theta t_n)$, along with the optimal value of $m$, should be investigated. However, the choice here clearly demonstrates the role of the high-order *general hill function* $\psi_m(w)$ in the disguised form of its transform $[\xi_1(t)]^m$.

In Subsection 7.2.2 we will present the derivation of the first practical truncation error bound of the Bessel sampling series (7.1.57). There we will also refer to the role of $\xi_1^{R+1}(I, t)$, the $\kappa$-transform of the generalized hill function $\psi_{R+1}(w)$, of (7.1.52) in improving the truncation error bound via the above self-truncating sampling series (7.1.60).

## 7.1.5 A NEW IMPULSE TRAIN—THE EXTENDED POISSON SUM FORMULA

We will use the impulse train approach, as in Papoulis [707], to derive the *extended Poisson sum formula* [454] for the present general integral transforms and the associated sampling series. Such an approach will enable us to avoid our usual dependence on the periodicity of the Fourier (trigonometric) series that is often used in signal analysis. The extended Poisson sum formula will be our tool for approaching and deriving the *aliasing error bound* of the general sampling series (7.1.20) that we shall present in Section 7.2.1. To arrive at such a necessary formula, we need to establish an *impulse train* compatible with the general transform like the Hankel transform (7.1.54)-(7.1.55).

It is instructive to first introduce *the Poisson sum formula* with its simple derivation and show its importance in reducing the infinite Fourier integral to a finite limit one [707, 393, 456]. This is followed by deriving the general Poisson-type sum formula which is illustrated for the Hankel (Bessel) transform.

Let $y(t)$ be the Fourier transform of $Y(w)$,

$$y(t) = \int_{-\infty}^{\infty} Y(w)e^{iwt}dw. \tag{7.1.62}$$

$$Y(w) = \frac{1}{2\pi}\int_{-\infty}^{\infty} y(t)e^{-iwt}dt. \tag{7.1.63}$$

Consider $\bar{y}(t)$, the superposition of all the translations of $y(t)$, by $nT$,

$$\bar{y}(t) = \sum_{n=-\infty}^{\infty} y(t+nT) \tag{7.1.64}$$

which is, of course, periodic with period $T$. Now we write the Fourier series expansion of $\bar{y}(t)$,

$$\bar{y}(t) = \sum_{n=-\infty}^{\infty} y(t+nT) = \frac{1}{T}\sum_{k=-\infty}^{\infty} c_k e^{\frac{i2\pi k}{T}t}. \tag{7.1.65}$$

$$c_k = \int_{-T/2}^{T/2} \bar{y}(t)e^{\frac{-i2\pi k}{T}} dt = \int_{-T/2}^{T/2}\sum_{n=-\infty}^{\infty} y(t+nT)e^{\frac{-i2\pi kt}{T}}dt. \tag{7.1.66}$$

If we exchange the integration and summation, make the simple change of variables $x = t + nT$, and use the periodicity of $e^{-i2\pi kt/T}$, the integral in (7.1.66) becomes an infinite integral,

$$c_k = \int_{-\infty}^{\infty} y(t)e^{\frac{-i2\pi kt}{T}} dt = Y\left(\frac{2\pi k}{T}\right). \tag{7.1.67}$$

Hence, (7.1.65) and (7.1.67) give the usual *Poisson sum formula* for the Fourier transform,

$$\sum_{n=-\infty}^{\infty} y(t + nT) = \frac{1}{T} \sum_{n=-\infty}^{\infty} Y\left(\frac{2\pi k}{T}\right) e^{\frac{i2\pi nt}{T}t}. \qquad (7.1.68)$$

Higgins [393] remarked that this simple form was given (without proof) by Gauss "in a note written sometime between 1799 and 1813." With this method we can see clearly how the samples $Y(\frac{2\pi k}{T})$ as an "infinite integral" of $y(t)$ in (7.1.67) can be expressed as the "finite integral" of $\bar{y}(t)$ in (7.1.66).

$$Y\left(\frac{2\pi k}{T}\right) = \int_{-\infty}^{\infty} y(t) e^{\frac{-i2\pi kt}{T}} dt = \int_{-T/2}^{T/2} \bar{y}(t) e^{\frac{-i2\pi kt}{T}} dt. \qquad (7.1.69)$$

This is very important when we attempt to approximate the infinite integral, which is now a finite integral, by the discrete Fourier transform and its fast algorithm. the fast Fourier transform. According to the above method, obtaining (7.1.67) from (7.1.66) depends entirely on the direct application of the *periodicity*, and hence, the *translated replicas* from all the equal intervals (with length T) of the real line $(-\infty, \infty)$ to the basic finite interval $(-T/2, T/2)$. If we are to move to other orthogonal expansions, which in general are not periodic, like that of Fourier-Bessel series expansion, we have to dispense with the periodicity and should be content with some "generalized" form of repetition. In anticipation of such difficulty, we present another well-known approach [707] for obtaining (7.1.68) that more closely follows the direction of our general development. Here we use the *impulse train* as defined by the divergent Fourier series,

$$\sum_{n=-\infty}^{\infty} \delta(t + nT) = \frac{1}{T} \sum_{n=-\infty}^{\infty} e^{\frac{i2\pi nt}{T}}, \qquad (7.1.70)$$

and convolve it with $y(t)$ to obtain $\bar{y}(t)$ and the *Poisson sum formula* (7.1.68).

$$\begin{aligned}
\bar{y}(t) &\equiv \sum_{n=-\infty}^{\infty} y(t + nT) \\
&= y(t) * \sum_{n=-\infty}^{\infty} \delta(t + nT) = y(t) * \frac{1}{T} \sum_{n=-\infty}^{\infty} e^{\frac{i2\pi nt}{T}} \\
&= \frac{1}{T} \sum_{n=-\infty}^{\infty} \int_{-\infty}^{\infty} y(t) e^{\frac{i2\pi n}{T}(t-x)} dx \\
&= \frac{1}{T} \sum_{n=-\infty}^{\infty} e^{\frac{i2\pi nt}{T}} \int_{-\infty}^{\infty} y(x) e^{\frac{-i2\pi nx}{T}} dx \\
&= \frac{1}{T} \sum_{n=-\infty}^{\infty} Y\left(\frac{2\pi n}{T}\right) e^{\frac{i2\pi nt}{T}}.
\end{aligned}$$

Such an approach will be very useful in establishing the generalization of (7.1.68) for other integral transforms like the Hankel (Bessel) transform [454]. As we see next, this necessitates the introduction of a *"generalized"* form of translation for the specific transform that we introduced in (7.1.34). There, we used it for the system interpretation of the generalized sampling series in Subsection 7.1.3. This concept was also used in Subsection 7.1.4 in our attempt to obtain a self-truncating version (7.1.59) of the sampling series (7.1.20), and in particular the Bessel one in (7.1.60).

### Impulse Trains for General Transforms

For simplicity, we will limit our discussions here to Fourier type symmetric transforms (7.1.32), (7.1.33) that we have already used in the previous sections, along with their generalized translation (7.1.34), and their associated convolution theorem in (7.1.37), (7.1.36). The illustrations will be for the Fourier-Bessel series.

Consider the $\kappa$-transform in (7.1.32) and also the equations that follow [(7.1.33)-(7.1.36)].

Our main example is a Hankel (Bessel) transform in conjunction with Fourier-Bessel series. This will exemplify the contrast of the *"non-periodic"* nature of the Fourier-Bessel series expansion with that of the *"periodic"* Fourier (trigonometric) series.

Consider again the $J_0$–Hankel (Bessel) transform $H_0(w)$, in (7.1.54) and its inverse in (7.1.55). The above transforms are symmetric; hence, there is no confusion in referring to either one as the transform. Recall that $\xi_1(t) = (aJ_1(at))/t$ is the $J_0$-Hankel transform of (the gate function)

$$\psi_1(w) = \begin{cases} 1, & 0 < w < a \\ 0, & |w| > a. \end{cases}$$

Consider, then, the $\tau$-translation of $(aJ_1(at))/t$ in (7.1.35). The same can be done for $J_m(x)$ [444], but we stay with $J_0(x)$ to simplify the first illustration. The zero subscript in $H_0(w)$ and $h_0(t)$ was used to specify the $J_0$-Hankel transforms, but in the sequel it shall be dropped for simplicity. The complete details of the following analysis are found in [454].

Now we introduce the *"generalized delta function"* $\delta(w\theta x)$ for the transform (7.1.32)-(7.1.33)

$$\int \rho(t)K(t,w)\overline{K(t,x)}dt \equiv \delta(w\theta x). \qquad (7.1.71)$$

Because of the weight function $\rho(t)$ used in the definition of the above general transforms, to use this delta function to locate the impulses in the impulse train ( similar to what is done with $\delta(w-x)$ of the special case of Fourier transforms) it should be written as $\delta(w\theta x) = \delta(w-x)/\rho(w)$; i.e.,

$$\int \rho(w)F(w)\delta(w\theta x)dw = \int \rho(w)F(w)\frac{\delta(w-x)}{\rho(w)} = F(x). \qquad (7.1.72)$$

Next we introduce the impulse train in a manner similar to that of the Fourier trigonometric series, but with the sense of this general translation, and, of course, without the usual periodicity of the Fourier series.

Consider the orthogonal set $\{K(w, t_n)\}$ on the interval $I$. We define the *generalized impulse train*, via the following divergent series at $\{c_m\}$, as

$$X(w) = \sum_m d_m \delta(w \theta c_m) \equiv \sum_n \frac{K(w, t_n)}{\|K(., t_n)\|_2^2} \qquad (7.1.73)$$

where

$$\|K(., t_n)\|_2^2 = \int \rho(w) \mid K(w, t) \mid^2 dw \qquad (7.1.74)$$

and where the locations $\{c_m\}$ for the impulses are to be determined for the particular transform of interest. As we shall illustrate for the case of the Fourier-Bessel series, and in contrast to the case of the usual trigonometric series, $d_m$ in (7.1.73) may change sign and in general $\mid d_m \mid$ is not uniform.

In this case of the $J_0$-Hankel transform we have the following formal divergent Fourier-Bessel series expansion on $(0, b)$, whose natural (*non-periodic*) extension defines an impulse train, $X(w)$:

$$X(w) = \sum_{m=0}^{\infty} d_m \delta(w \theta 2mb) = \frac{2}{b^2} \sum_{n=1}^{\infty} \frac{J_0(\frac{wj_{0,n}}{b})}{J_1^2(j_{0,n})}, \quad 0 < w < \infty,$$

$$J_0(j_{0,n}) = 0, \qquad n = 0, 1, 2 \ldots . \qquad (7.1.75)$$

The location of the impulses $\{w_m = c_m = 2mb\}$ was verified numerically and can also be supported by using an asymptotic expansion (Watson [940]) with $\mid d_m \mid$ decreasing. It is also shown that $d_0 = 1, \mid d_1 \mid = 1, d_1, d_2, d_5, d_6, \ldots$, etc., are negative, whereas $d_3, d_4, d_7, d_8, \ldots$, etc., are positive, as illustrated in Fig. 7.1.

As we had remarked earlier for (7.1.72), this impulse train (7.1.75) can now be written in terms of the usual (Fourier type) delta functions as

$$X(w) = \sum_{m=0}^{\infty} d_m \delta(w \theta 2mb) = d_0 \frac{\delta(w)}{w} + \sum_{m=1}^{\infty} d_m \frac{\delta(w - 2mb)}{w}$$

$$= \frac{2}{b^2} \sum_{n=1}^{\infty} \frac{J_0(\frac{wj_{0,n}}{b})}{J_1^2(j_{0,n})}. \qquad (7.1.76)$$

Considering the integral of the Fourier coefficients in the right hand side of (7.1.76), we find that

$$\int_0^b wX(w)J_0\left(\frac{wj_{0,n}}{b}\right) dw = d_0 \int_0^b w\frac{\delta(w)}{w} J_0\left(\frac{wj_{0,n}}{b}\right) dw = d_0 = 1$$

This necessitates assigning $d_0 = 1$,

$$X(w) = \frac{\delta(w)}{w} + \sum_{m=1}^{\infty} d_m \frac{\delta(w - 2mb)}{w} = \frac{2}{b^2} \sum_{n=1}^{\infty} \frac{J_0\left(\frac{wj_{0,n}}{b}\right)}{J_1^2(j_{0,n})}. \qquad (7.1.77)$$

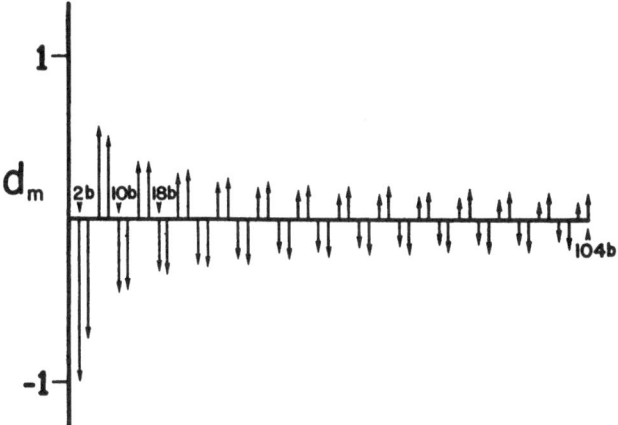

FIGURE 7.1: The impulse train of the $J_0$-Bessel series, its spacing at $2mb$, and the decreasing amplitude $\mid dm \mid, m = 0, 1, 2, \ldots, 52$. [From Jerri [454], Courtesy of *Appl. Analysis J. 26*, 199-221,1988].

If we formally take the Hankel transform of both sides of (7.1.77), we have

$$x(t) = \int_0^\infty w \frac{\delta(w)}{w} J_0(wt) dw + \sum_{m=1}^\infty dm \int_0^\infty w \frac{w-2mb}{w} J_0(wt) dw$$

$$= \frac{2}{b^2} \sum_{n=1}^\infty \frac{\int_0^\infty w J_0(wt) J_0(w j_{0,n}/b) dw}{J_1^2(j_{0,n})},$$

$$x(t) = 1 + \sum_{m=1}^\infty dm J_0(2mbt) = \frac{2}{b^2} \sum_{n=1}^\infty \frac{\delta(t\theta \frac{j_{0,n}}{b})}{J_1^2(j_{0,n})}; \qquad (7.1.78)$$

i.e., the result is another impulse train in $t$-space, where the impulses are located at $\{t_n = j_{0,n}/b\}$. We may remark again that for the purpose of locating these impulses we write $\delta(t\theta j_{0,n}/b) = \delta(t - j_{0,n}/b)/t$ . If we accept the numerical or asymptotic result that $d_1, d_2, d_5, d_6, \ldots$, etc., are negative, while $d_3, d_4, d_7, d_8, \ldots$ are positive, we may use (7.1.78) at $t = 0$, where the right-hand side vanishes, to have a basic relation between these $d_n$,

$$1 - [\mid d_1 \mid + \mid d_2 \mid + \mid d_5 \mid + \mid d_6 \mid + \cdots] + d_4 + d_7 + d_8 + \cdots = 0. \quad (7.1.79)$$

It was verified numerically, with good accuracy, that this equality (7.1.79) is satisfied when all these terms $\mid dm \mid$ are normalized in terms of $\mid d_1 \mid = 1$.

Figure 7.1 illustrates the locations $\{2mb\}$ and the variation of $d_m$ for $m = 1$ to 52, where we used 40 terms in the series of the right-hand side of (7.1.77). In the same figure we note the relative constancy of the normalized impulses $\mid d_m/d_1 \mid$, $m \neq 0$ as $m$ becomes large. This means that we can write the impulse train (7.1.77) in the $w$-space as

$$
\begin{aligned}
X(w) &= \frac{\delta(w)}{w} - \frac{\delta(w - 2b)}{w} + \sum_{m=2}^{\infty} d_m \frac{\delta(w - 2mb)}{w} \\
&= \frac{2}{b^2} \sum_{n=1}^{\infty} \frac{J_0(\frac{w j_{0,n}}{b})}{J_1^2(j_{0,n})}.
\end{aligned} \tag{7.1.80}
$$

For future reference we will use $X(w)$ for the impulse train of the left–hand side of Eq. (7.1.80), whereas $X_s(w)$ will stand for the above Bessel series.

We note here that the impulses of $X(w)$ repeat at $\{2mb\}$ with equal spacing of $2b$, whereas the impulses of $x(t)$ in (7.1.78) repeat at $\{ j_{0,n}/b \}$ which are not equally spaced, except asymptotically where $j_{0,n+1} - j_{0,n} \sim \pi$ for large $n$. This would also mean that in contrast to the symmetric and equal spacing of the Fourier series in both $w$ and $t$, we have here two differ-ent spacings for the samples. Hence, we should expect *two versions of the extension of the Poisson sum formula* (7.1.68) for the Hankel transforms. We will concentrate in the following section on the first version with equal spacing in $w$ for starting with the impulse train $X(w)$ in (7.1.80). We will give only a very brief presentation of the second version in the $t$-space at the end of the section.

### The Extended Poisson Sum Formula for the Fourier-Bessel Series

In parallel to the development of using (7.1.70) to derive the usual Poisson sum formula (7.1.68), we will introduce $\bar{H}(w)$, the *superposition* of all the *"generalized translations"* (7.1.34) of $H(w)$ in (7.1.54). This can be obtained by convolving $H(w)$ (in the sense of the convolution product (7.1.36)) with the impulse train $X(w)$ in (7.1.80) to have $\bar{H}(w) = (H * X)(w)$.

$$
\bar{H}(w) = H(w) * \left[ \frac{\delta(w)}{w} + \sum_{m=1}^{\infty} d_m \frac{\delta(w - 2mb)}{w} \right], \quad |d_1| = 1.
$$

$$
\bar{H}(w) = \int_0^{\infty} x H(w \theta x) \frac{\delta(x)}{x} dx + \sum_{m=1}^{\infty} d_m \int_0^{\infty} x H(w \theta x) \frac{\delta(x - 2mb)}{x} dx
$$

$$
\bar{H}(w) = H(w) + \sum_{m=1}^{\infty} d_m H(w \theta 2mb). \qquad | d_1 |= 1. \tag{7.1.81}
$$

Before we convolve $H(w)$ with the Bessel series $X_s(w)$ of the right-hand side of (7.1.80) we should mention that the sought $(H * X_s)(w)$ is again a

$J_0$-Bessel series.

$$(H * X_s)(w) = H * \frac{2}{b^2} \sum_{n=1}^{\infty} \frac{J_0(\frac{w j_{0,n}}{b})}{J_1^2(j_{0,n})}$$

$$= \frac{2}{b^2} \sum_{n=1}^{\infty} \frac{\int_0^{\infty} x H(w\theta x) J_0(\frac{x j_{0,n}}{b}) dx}{J_1^2(j_{0,n})}$$

$$= \frac{2}{b^2} \sum_{n=1}^{\infty} \frac{h(\frac{j_{0,n}}{b}) J_0(\frac{w j_{0,n}}{b})}{J_1^2(j_{0,n})}.$$

This series obviously vanishes at the end point $w = b$, whereas in general there is no good reason for the above $\bar{H}(w)$ in (7.1.81) to vanish at $w = b$. To be certain about this point, we have verified it numerically for a number of Hankel transforms pairs [454], as we shall see in the upcoming sections. For this reason, we will seek the same type of Bessel series on $(0, b)$ for the new function $\tilde{H}(w) \equiv \bar{H}(w) - \bar{H}(b) = (H * X)(w) - (H * X)(b)$ which vanishes at $w = b$. This requires convolving $H(w)$ with both sides of (7.1.80), then evaluating the resulting expression at $w$ and $b$ to have

$$\tilde{H}(w) = \bar{H}(w) - \bar{H}(b) = (H * X)(w) - (H * X)(b)$$

$$= (H * X_s)(w) - (H * X_s)(b)$$

$$= H(w) * \frac{2}{b^2} \sum_{n=1}^{\infty} \frac{J_0(\frac{w j_{0,n}}{b})}{J_1^2(j_{0,n})} \Big|_{w=b}^{w}$$

$$= \frac{2}{b^2} \sum_{n=1}^{\infty} \frac{\int_0^{\infty} x H(w\theta x) J_0(\frac{x j_{0,n}}{b}) dx}{J_1^2(j_{0,n})} \Big|_{w=b}^{w}$$

$$= \frac{2}{b^2} \sum_{n=1}^{\infty} \frac{h(\frac{j_{0,n}}{b}) J_0(\frac{w j_{0,n}}{b})}{J_1^2(j_{0,n})}. \tag{7.1.82}$$

If we combine this result with the expression for $\bar{H}(w)$ in (7.1.81), we obtain the desired *Bessel type Poisson sum formula* in $w$-space as the *first version*.

$$\tilde{H}(w) = \bar{H}(w) - \bar{H}(b) = \sum_{m=0}^{\infty} d_m [H(w\theta 2mb) - H(b\theta 2mb)],$$

$$\tilde{H}(w) = H(w) - H(b) + \sum_{m=1}^{\infty} d_m [H(w\theta 2mb) - H(b\theta 2mb)],$$

$$\tilde{H}(w) = \bar{H}(w) - \bar{H}(b) = \frac{2}{b^2} \sum_{n=1}^{\infty} \frac{h(\frac{j_{0,n}}{b}) J_0(\frac{w j_{0,n}}{b})}{J_1^2(j_{0,n})}. \tag{7.1.83}$$

As a relatively new result (7.1.83) was verified numerically with a number of $J_0$-Hankel transforms pairs directly in (7.1.83). We present it in the discussion following Eq. (7.1.88). and indirectly via verifying the new upper bound for the aliasing error of the Bessel sampling series in Subsection 7.2.1. whose derivation we based primarily on (7.1.83).

We emphasize here that even though we wrote $\delta(w\theta x) = \frac{\delta(w-x)}{x}$ for the purpose of locating the impulses (or sample values), it is essential in the above derivation of (7.1.82)-(7.1.83) that we use $H(w\theta x)$ with its generalized translation for the process of convolving. This is necessary to bring about the main feature of a Poisson type sum formula, which is to involve the samples $h(\frac{j_{0,n}}{b})$ of the (infinite limit) integral transform of $H(w)$ in (7.1.83). This $h(\frac{j_{0,n}}{b})$ will be in the position of the required finite integral of the Fourier-Bessel coefficients of $\tilde{H}(w) - \tilde{H}(b)$. the modified (or aliased) $H(w)$ on $(0, b)$. The $\tilde{H}(w)$ in (7.1.83) can be considered as an "aliased" version of $H(w)$ because of the use of the (infinite) integral of $h(\frac{j_{0,n}}{b})$ for the Fourier coefficients of the "Fourier-Bessel series" for $\tilde{H}(w)$ on $(0, b)$. This means that if the series on the right-hand side of (7.1.83) is to be used to approximate the original (infinite) transform $H(w)$ on $(0, b)$, there will be an aliasing error $\mathcal{E}_A$ in the general Fourier series expansion,

$$
\begin{aligned}
\mathcal{E}_A &= |H(w) - \tilde{H}(w)| \\
&= |H(b) - \sum_{m=1}^{\infty} d_m [H(w\theta 2mb) - H(b\theta 2mb)]|.
\end{aligned} \tag{7.1.84}
$$

We note again that the whole reason behind this aliasing error is our use of the samples $h(\frac{j_{0,n}}{b})$ of the (infinite) Hankel transform,

$$
h(t) = \int_0^{\infty} wH(w)J_0(wt)dw,
$$

as the Fourier-Bessel series coefficients for $H(w)$ on $(0, b)$. This is instead of $h_b(\frac{j_{0,n}}{b})$, the samples of "the correct" finite (or band-limited) integral,

$$
h_b(t) = \int_0^b wH(w)J_0(wt)dw. \tag{7.1.85}
$$

So in the case of $H(w)$ in (7.1.55) vanishing identically for $w \geq b$, the aliasing error $\mathcal{E}_A$ in (7.1.84) must vanish. This means that the sum of the generalized translations of $H(w)$.

$$
H(w\theta 2mb) = \int_0^{\infty} th(t)J_0(wt)J_0(2mbt)dt, \tag{7.1.86}
$$

in (7.1.84) must vanish. This is easily satisfied if we are to interpret the translated argument $w\theta 2mb$ of $H(w\theta 2mb)$ to be outside the interval $(0, b)$. We can easily see from (7.1.86) that, indeed, $w\theta 2mb = 2mb\theta w$ and that $w\theta 2mb = 2mb$ at $w = 0$. Also, with $w\theta 2mb > b$, and for a class of monotonically decreasing $H(w)$, we can compare $|H(w\theta 2mb)|$ to $|H(w)|$ on $(0, b)$,

$$
|H(w\theta 2mb)| \leq |H(w)|. \qquad w \in (0, b).
$$

Another way of showing that $H(w\theta 2mb)$ vanishes for $h_b(t)$ of (7.1.85) is to consult for another simple interpretation of the delta function. First we write the integral representation for $h(t)$ inside the integral of (7.1.86), then exchange the two integrals to have

$$
\begin{aligned}
H(w\theta 2mb) &= \int_0^\infty t \left[ \int_0^\infty xH(x)J_0(xt)dx \right] J_0(wt)J_0(2mbt)dt \\
&= \int_0^\infty xH(x) \left[ \int_0^\infty J_0(xt)J_0(wt)J_0(2mbt)dt \right] dx
\end{aligned}
$$

$$(7.1.87)$$

where the inside integral defines $\delta(x-w\theta 2mb)/x$ as is required for $H(w\theta 2mb)$ on the left-hand side. On the other hand, if $h(t)$ is band-limited to $(0,b)$ as in (7.1.85), we expect all the translations $H(w\theta 2mb)$ to vanish, which can be seen after we modify $h(t)$ in the above integrals to $h_b(t)$, where $H(x) \equiv 0$ for $x > b$ in (7.1.87),

$$
H(w\theta 2mb) = \int_0^\infty t \left[ \int_0^b xH(x)J_0(xt)dx \right] J_0(wt)J_0(2mbt)dt.
$$

and interchange the two integrals,

$$
H(w\theta 2mb) = \int_0^b xH(x) \left[ \int_0^\infty J_0(xt)J_0(wt)J_0(2mbt)dt \right] dx. \qquad (7.1.88)
$$

So, if we invoke the interpretation of the inner integral as $\frac{\delta(x-w\theta 2mb)}{x}$ and that $w\theta 2mb > b$ for $w \in (0,b)$, the above double integral vanishes.

These simple results and interpretations of the generalized translations will offer some support for our ultimate goal of deriving an *aliasing error bound* for the Bessel-type sampling series that we shall present in Subsection 7.2.1.

The extended Poisson type sum formula (7.1.83) was checked numerically with a number of $J_0$-Hankel transform pairs including $h(t) = \frac{1}{t}e^{-at}$ with its Hankel transform $H(w) = \frac{1}{\sqrt{w^2+a^2}}$, which decays very slowly, and has the added advantage of having its generalized translation $H(w\theta x)$ in a closed form in terms of $K$, the complete elliptic function of the first kind (Erdelyi et al. [282, p. 14, Eq. (17)]),

$$
H(w\theta x) = \frac{2}{\pi}K(2(xw)^{\frac{1}{2}}k^{-1})k^{-1}; \qquad k = \sqrt{(x+w)^2 + a^2}. \qquad (7.1.89)
$$

## A Second Version of the Extended Poisson Sum Formula

As we remarked earlier, we can expect two kinds of Poisson sum formulas, one version like (7.1.83) for the Fourier-Bessel series in the $w$-space, and a *second version* in the transform *t*-space, which is what we shall illustrate briefly here.

For this purpose, and in parallel to the above treatment for obtaining (7.1.83), we will convolve $h(t)$ with the impulse train $x(t)$ of (7.1.78) to have $\bar{h}(t)$.

$$
\begin{aligned}
\bar{h}(t) &= h(t) * x(t) = h(t) * \left[ \frac{2}{b^2} \sum_{n=1}^{\infty} \frac{\delta(t\theta \frac{j_{0,n}}{b})}{J_1^2(j_{0,n})} \right] \\
&= h(t) * \sum_{m=0}^{\infty} d_m J_0(2mbt), \\
\bar{h}(t) &= \frac{2}{b^2} \sum_{n=1}^{\infty} \frac{h(t\theta \frac{j_{0,n}}{b})}{J_1^2(j_{0,n})} = \sum_{m=0}^{\infty} d_m \int_0^{\infty} \tau h(t\theta\tau) J_0(2mb\tau) d\tau. \\
\bar{h}(t) &= \frac{2}{b^2} \sum_{n=1}^{\infty} \frac{h(t\theta \frac{j_{0,n}}{b})}{J_1^2(j_{0,n})} = \sum_{m=0}^{\infty} d_m H(2mb) J_0(2mbt). \qquad (7.1.90)
\end{aligned}
$$

We note here that the superposition $\bar{h}(t)$, of all the generalized translations by $j_{0,n}/b$ of $h(t)$, on the left-hand side of (7.1.90) lacks the original (non-translated) term $h(t)$. Also, more importantly, the right-hand side does not lend itself to a typical Fourier-Bessel series interpretation, which is the essence of any Poisson type sum formula. Thus, we will leave (7.1.90) for now and concentrate our efforts and applications for the *first version* in (7.1.83). As was pointed out to us very recently by L.L. Campbell. the series (7.1.90) is a Schlomilch series. but it should be noted that it has a different form of coefficients (Watson [940, p. 618]).

In concluding this section we restate that we have illustrated here the development of an *extended Poisson type sum formula* (7.1.83) *for the $J_0$-Hankel (Bessel)* transform. This is a first step away from the well-known Poisson sum formula which is restricted to Fourier (trigonometric) series. The extension of (7.1.83) to other Hankel transforms and the more general transforms should be clear after the above rather formal illustration [454].

It is emphasized again that the motivation for following this avenue of extending the Poisson sum formula to general orthogonal expansions and transforms was the belief that it may lead to a reasonable approach for deriving an upper bound for the aliasing error incurred in applying the generalized sampling theorem, which it did. This will be the main subject of Subsection 7.2.1. Of course, it is also one of the main topics of this chapter: establishing error bounds for the *"generalized sampling series"* (7.1.20).

# 7.2 Error Bounds of the Present Extension of the Sampling Theorem

In Subsections 7.1.2-7.1.5 we presented the generalized sampling theorem (7.1.20) and discussed new tools that enable us to give this series its first physical interpretation in terms of a system with time varying impulse response. These tools are also instrumental in deriving bounds for the two basic errors, namely, *aliasing* and *truncation errors*.

## 7.2.1 THE ALIASING ERROR BOUND

This section is devoted to a clear illustration of deriving the aliasing error for the generalized sampling expansion (7.1.20). This is done with the aid of the *"extended type Poisson sum formula"* for the general transforms (7.1.32)-(7.1.33) that we covered in detail for the case of the Hankel transform (7.1.83). The Bessel function kernel represents a sudden departure from the familiar exponential and/or periodic kernels of the usual Fourier and Laplace transforms. As such, it represents a challenge that demands new tools of analysis, which we tried to define and derive [437, 444, 454, 440] in Subsections 7.1.5 and 7.1.3. These tools include the *generalized translation* (7.1.34), (7.1.35), its *associated convolution theorem* (7.1.36), a *general type hill function* (7.1.52) for tighter truncation error bound, the *general transform impulse train* (7.1.73), (7.1.75) and its associated *extended (Bessel type) Poisson sum formula* (7.1.83). The latter sum formula (7.1.83) was vital to our development of the first aliasing error bound, for the general type Bessel sampling series, in (7.1.26).

**Aliasing Error Bound for the $J_0$-Bessel Sampling Expansion**

Aliasing error in application of this sampling theorem is due largely to uncertainty about the exact finite interval $I$, i.e., $f(t)$ may not be "band-limited" to $(0, b)$ as $f_b(t)$ in (7.1.24), but we still apply the above sampling series (7.1.26) to its samples $f(\frac{j_{0,n}}{b})$. Such an application to $f(t)$ and not $f_b(t)$ results in the "aliasing error", $\epsilon_A$, of the sampling expansion,

$$\epsilon_A = \mid f(t) - \sum_n f(t_n)S(t, t_n) \mid. \tag{7.2.1}$$

which in the case of the $J_0$-Hankel transform becomes

$$\epsilon_A = \left| f(t) - \sum_{n=1}^{\infty} f\left(\frac{j_{0,n}}{b}\right) \frac{2j_{0,n} J_0(bt)}{b^2(\frac{j_{0,n}^2}{b^2} - t^2) J_1(j_{0,n})} \right|. \tag{7.2.2}$$

Note that we used $\mathcal{E}_A$ in (7.1.84) for the aliasing error of the general orthogonal series expansion. We stress again the presence of the samples $\{f(\frac{j_{0,n}}{b})\}$

of the *"non-band-limited"* $f(t)$,

$$f(t) = \int_0^\infty \omega J_0(\omega) F(\omega) \; d\omega. \tag{7.2.3}$$

in (7.2.1) instead of the required band-limited $f_b(\frac{j_{0,n}}{b})$ which is the essence of the aliasing error in (7.2.2). It is also the presence of the samples $f(\frac{j_{0,n}}{b})$ that makes the link with *the extended Poisson sum formula* (7.1.83), where we involve the samples of the *"non-band-limited"* $f(t)$ of (7.2.3). In order to produce the sampling series in (7.2.2) from the right-hand side of (7.1.83), we replace $h(t)$ by $f(t)$ and $H(\omega)$ by $F(\omega)$ in (7.1.83), then multiply both sides by $\omega J_0(\omega t)$ and integrate from 0 to $b$ to have

$$\int_0^b \omega \tilde{F}(\omega) J_0(\omega t) \; d\omega = \frac{2}{b^2} \sum_{n=1}^\infty \frac{f(\frac{j_{0,n}}{b}) \int_0^b \omega J_0(\omega t) J_0(\omega \frac{j_{0,n}}{b}) \; d\omega}{j_1^2(j_{0,n})}$$

$$= \sum_{n=1}^\infty f(\frac{j_{0,n}}{b}) \frac{2 j_{0,n} J_0(bt)}{b^2 (\frac{j_{0,n}}{b} - t^2) J_1(j_{0,n})}. \tag{7.2.4}$$

If we use this result in (7.2.2) and invoke the expression for $\tilde{F}(\omega)$ from the left-hand side of (7.1.83), we have

$$\epsilon_A = \left| \int_0^\infty \omega J_0(\omega t) F(\omega) \; d\omega \right.$$

$$\left. - \int_0^b \omega J_0(\omega t) \left\{ \sum_{m=0}^\infty d_m [F(\omega \theta 2mb) - F(b\theta 2mb)] \right\} \; d\omega \right|$$

$$= \left| \int_0^\infty \omega J_0(\omega t) F(\omega) \; d\omega \right.$$

$$- \int_0^b \omega J_0(\omega t) F(\omega) \; d\omega \; + \; \frac{b J_1(bt)}{t} F(b)$$

$$\left. - \sum_{m=1}^\infty d_m \int_0^b \omega J_0(\omega t) [F(\omega \theta 2mb) - F(b\theta 2mb)] \; d\omega \right|$$

$$= \left| \int_b^\infty \omega J_0(\omega t) F(\omega) \; d\omega \; + \; \frac{b J_1(bt)}{t} F(b) \right.$$

$$\left. - \sum_{m=1}^\infty d_m \int_0^b \omega J_0(\omega t) [F(\omega \theta 2mb) - F(b\theta 2mb)] \; d\omega \right|. \tag{7.2.5}$$

Just as we noted that the aliasing error $\mathcal{E}_A$ for the Fourier-Bessel series in (7.1.84) vanishes when $H(\omega)$ vanishes identically for $\omega \geq b$, we also note that the aliasing error $\epsilon_A$ in (7.2.2) must vanish when the sampled function $f_b(t)$ is band-limited as in (7.1.24),

$$f_b(t) = \int_0^b \omega F(\omega) J_0(\omega t) \; d\omega,$$

i.e., $f_b(t)$ is the Hankel transform of $F(\omega)$ where $F(\omega) \equiv 0$ for $\omega \geq b$. This is easily seen in (7.2.5). Since the first (integral) term clearly vanishes, the second term vanishes when $F(b) = 0$. The third (sum) term also vanishes since we have already shown for the vanishing of $\mathcal{E}_A$ the aliasing error of the Fourier–Bessel series in (7.1.84) that the translations $F(\omega \theta 2mb)$ vanish when $F(\omega) \equiv 0$ for $\omega \geq b$, $m \geq 1$.

To establish our first example of an upper bound for the aliasing error (7.2.5) of the Bessel type sampling series, we will consider a limited, though reasonable, class of functions $F(\omega) \geq 0$ which is *monotonically decreasing*. This would allow us the following inequality:

$$F(\omega \theta 2mb) \leq F(\omega) \quad \text{on } (0, b) \tag{7.2.6}$$

since we have already interpreted in Subsection 7.1.5 that $\omega \theta 2mb \geq b$ on $(0, b)$. This result will help in providing control on the upper bound for the third (sum) in (7.2.5).

$$\epsilon_A \;\; \leq \;\; \int_b^\infty \omega \mid J_0(\omega t) \mid F(\omega) \; d\omega \;\; + \;\; \left| \frac{b J_1(bt)}{t} \right| F(b)$$

$$+ \;\; \int_0^b \omega \left| \sum_{m=1}^\infty \left| d_m \int_0^b \omega J_0(\omega t)[F(\omega \theta 2mb) - F(b \theta 2mb)] \right| \right| d\omega. \tag{7.2.7}$$

With $d_1 = 1$, $\mid d_m \mid < 1$ for $m > 1$, $\mid J_0(\omega t) \mid \leq 1$, and the result in (7.2.6) we have

$$\omega \mid d_m J_0(\omega t)[F(\omega \theta 2mb) - F(b \theta 2mb)] \mid \;\; \leq \;\; (\omega + 2mb)F(\omega + 2mb) \tag{7.2.8}$$

since for our special class of functions that allowed (7.2.6), $F(\omega \theta 2mb)$ and $F(\omega + 2mb)$ are comparable on $(0, b)$. If we use (7.2.8) for the third (sum) term of (7.2.7) we have

$$\int_0^b \omega \sum_{m=1}^\infty \mid d_m J_0(\omega t)[F(\omega \theta 2mb) - F(b \theta 2mb)] \mid \; d\omega$$

$$\leq \;\; \int_0^b \sum_{m=1}^\infty (\omega + 2mb)F(\omega + 2mb) \; d\omega$$

$$\leq \;\; \int_{2b}^{3b} x F(x) \; dx + \int_{4b}^{5b} x F(x) \; dx + \int_{6b}^{7b} x F(x) \; dx + \cdots$$

$$\leq \;\; \frac{1}{2} \int_b^\infty x F(x) \; dx \tag{7.2.9}$$

after using the fact that the sum of the left out areas (under $xF(x)$ on the intervals $(b, 2b), (3b, 4b), (5b, 6b), \ldots,$ etc.), is larger than those considered by the above integrals.

With the result (7.2.9) used in (7.2.7), *the upper bound for the aliasing error* of the sampling expansion of $f(t)$, the Hankel transform of the monotonically decreasing $F(\omega) \geq 0$ , becomes

$$\epsilon_A \leq \frac{3}{2} \int_b^\infty \omega F(\omega) \; d\omega \; + \left| \frac{bJ_1(bt)}{t} \right| F(b) \qquad (7.2.10)$$

after using again $\mid J_0(\omega t) \mid \leq 1$ inside the first integral of (7.2.7).

We note that the first (integral) term of the bound (7.2.10) is of the same form as that of the aliasing error bound of Weiss [944] and Brown [118, 117].

$$\epsilon_A \leq \frac{1}{\pi} \int_{|\omega|>b}^\infty |F(\omega)|d\omega, \qquad (7.2.11)$$

Brown [117] (see also Higgins [393]) and Papoulis [707] gave a simple example to show that the multiple constant $1/\pi$ in his upper bound (7.2.11) was optimal in the sense that the maximum error for that example was the same as the bound in (7.2.11). Here we give a parallel example for the Bessel sampling expansion case to show that the bound in (7.2.10), though not the optimal one, is very close.

Consider the following function:

$$f_{2a}(t) = \frac{J_0(at)J_1(at)}{t} = \int_0^{2a} \frac{\omega}{a\pi} \cos^{-1}\left(\frac{\omega}{2a}\right) J_0(\omega t) \; d\omega,$$

which is band-limited to $2a$ and should be sampled as

$$f_{2a}\left(\frac{j_{0,n}}{2a}\right) = \frac{2aJ_0(\frac{aj_{0,n}}{2a})J_1(\frac{aj_{0,n}}{2a})}{j_{0,n}} \qquad (7.2.12)$$

to be correctly used in the Bessel sampling series (7.1.26). However, if we incorrectly sample this function at $j_{0,n}/a$ due to our estimated assumption of it being band-limited to $a$ instead of $2a$, all the samples will vanish; i.e..

$$f_{2a}(\frac{j_{0,n}}{2a}) = a\frac{J_0(j_{0,n})J_1(j_{0,n})}{j_{0,n}} = 0$$

since $J_0(j_{0,n}) = 0$, and the resulting sampling series in (7.2.2) vanishes, which results in maximum aliasing error.

$$\epsilon_A \; = \; |f_{2a}(t) - 0| \; = \; \left| \frac{J_0(at)J_1(at)}{t} \right|.$$

The maximum of this error is $0.5a$ occurring at $t = 0$. For this same example, our new aliasing error bound in (7.2.10) gives

$$\frac{3}{2} \int_a^\infty \omega F(\omega) \; d\omega \; + \; \left| \frac{a J_1(at)}{t} \right| F(a)$$

$$= \; \frac{3}{2} \frac{1}{a\pi} \int_a^{2a} \omega \cos^{-1} \left( \frac{\omega}{2a} \right) \; d\omega \; + \; \left| \frac{a J_1(at)}{t} \right| \cdot \frac{1}{a\pi} \cos^{-1} \left( \frac{1}{2} \right)$$

$$= \; \frac{3a}{\pi} \left[ \frac{\pi}{12} + \frac{\sqrt{3}}{8} \right] + \frac{1}{3} \left| \frac{J_1(at)}{t} \right| \; = \; 0.457a + \frac{1}{3} \left| \frac{J_1(at)}{t} \right|.$$

This has a maximum value of $0.623a$ occurring at $t = 0$, which is not very much higher than $0.5a$, the actual maximum of the error involved.

In addition to the above example, we have tried the following first three functions of the class of monotonically decreasing $F(\omega) \geq 0$

| | $F(\omega)$ | $f(t)$ |
|---|---|---|
| 1. | $e^{-a\omega}/\omega$ | $(t^2 + a^2)^{-1/2}$ |
| 2. | $e^{-\omega}$ | $(t^2 + 1)^{-3/2}$ |
| 3. | $e^{-\omega^2}$ | $(1/2)e^{-t^2/4}$ |
| 4. | $F(\omega) = \begin{cases} \frac{1}{a\pi} \cos^{-1}(\omega/2a), & \omega < 2a \\ 0, & \omega > 2a, \end{cases}$ | $J_0(at)J_1(at)/t$ |

Their respective aliasing error bound from (7.2.10) is

$$1. \quad \epsilon_A \leq \frac{3}{2a} e^{-ab} + \frac{b}{2} e^{-ab}$$

$$2. \quad \epsilon_A \leq \frac{3}{2a} e^{-b}(b+1) + \frac{b^2}{2} e^{-b}$$

$$3. \quad \epsilon_A \leq \frac{3}{4} e^{-b^2} + \frac{b^2}{2} e^{-b^2}$$

$$4. \quad \epsilon_A \leq \frac{3a}{4\pi} [\sin(2\theta) - 2\theta \cos(2\theta)] + \frac{b^2}{2a\pi} \theta,$$

$$\text{where} \quad \theta = \cos^{-1} \left( \frac{b}{2a} \right).$$

These upper bounds were then compared with the actual aliasing error bound of the sampling series in (7.2.2) for three different values of truncating the integral at $\omega = b$ where we used a 21-term partial sum for the sampling series in (7.2.2). These comparisons are illustrated in Table 7.I.

In summary, this is the first attempt to extend the Poisson sum formula to other transforms besides the Fourier transform for the main purpose of developing an upper bound on the aliasing error of the associated general orthogonal expansions and the generalized sampling expansion. To this end, we have illustrated these concepts along with their requisite tools, e.g., the impulse train for the kernel $J_0(\frac{j_{0,n} x}{b})$ . Our work continues as we try to

| b | $F(\omega)$ | Actual $\epsilon_A$ of (7.2.2) | The upper bound of $\epsilon_A$ as in (7.2.10) |
|---|---|---|---|
| 1 | $e^{-\omega}/\omega$ | .515 | .736 |
| 1 | $e^{-\omega}$ | .914 | 1.290 |
| 1 | $e^{-\omega^2}$ | .312 | .460 |
| 1 | $F(\omega) = \begin{cases} \frac{1}{\pi}\cos^{-1}(\omega/2) & \omega < 2 \\ 0 & \omega > 2 \end{cases}$ | .500 | .623 |
| 2 | $e^{-\omega}/\omega$ | .221 | .338 |
| 2 | $e^{-\omega}$ | .616 | .880 |
| 2 | $e^{-\omega^2}$ | .0147 | .0504 |
| 1.5 | $F(\omega) = \begin{cases} \frac{1}{\pi}\cos^{-1}(\omega/2) & \omega < 2 \\ 0 & \omega > 2 \end{cases}$ | .242 | .453 |
| 5 | $e^{-\omega}/\omega$ | .01672 | .02695 |
| 5 | $e^{-\omega}$ | .0729 | .145 |
| 5 | $e^{-\omega^2}$ | – – –† | – – –† |
| 1.75 | $F(\omega) = \begin{cases} \frac{1}{\pi}\cos^{-1}(\omega/2) & \omega < 2 \\ 0 & \omega > 2 \end{cases}$ | .096 | .320 |

† within the accuracy of the computations.

TABLE 7.I: Illustration of the actual (sampling) aliasing error bound in (7.2.2) and its present upper bound of (7.2.10). (From Jerri [454], 1988: courtesy of *J. Appl. Analysis*).

develop those for other type Bessel series expansions and more general orthogonal expansions.

The importance of the present analysis of the aliasing error, associated with the Fourier-Bessel series expansions, stems from the need for the numerical computation of Hankel transforms or Fourier-Bessel series, where both the aliasing and truncation errors may be incurred. One recent application, which deals with such aliasing as a major error, is that of using the Hankel transform for the acoustic problem of computing wavefield due to a monochromatic point source in a stratified medium [660] (see also references [15.16.19] and [21] given in [660]). In this reference, it was pointed out clearly that the error involved is analogous to that of the discrete Fourier transform.

## 7.2.2    THE TRUNCATION ERROR BOUND

Numerous improved upper bounds for the truncation error of the Shannon sampling expansion are available in the literature [440, Sect. VI]. However, until 1982 there was only one truncation error bound associated with the generalized sampling expansion, which is attributed to Yao [974]. As we shall see in (7.2.18), this bound did not show the usual simple dependence

on the truncation limit $N$. Thus, a more computationally feasible bound
was needed and is the subject of this section [459].

The reason behind the delay in obtaining such a result was the absence of
a lower bound for the general (non-trigonometric) kernel of the given transform. This was necessary for estimating an upper bound of the complex
contour integral that is used for expressing, and finally deriving, the truncation error bound. For our present illustration of the $J_m$-Bessel sampling
series,

$$f(t) = \sum_{n=1}^{\infty} f(t_{m,n}) \frac{2t_{m,n} J_m(t)}{a(t_{m,n}^2 - t^2) J_{m+1}(at_{m,n})};$$

$$J_m(at_{m,n}) = 0, \quad n = 1, 2, \ldots: \quad t_{m,n} = \frac{j_{m,n}}{a}$$

where

$$f(t) = \int_0^a x J_m(xt) F(x) \, dx.$$

We needed a lower bound for $J_m(z)$, which, unfortunately, did not then
exist in the literature. Thus we had to derive it, or at least make a good
practical estimate of it as our starting point [459]. It became abundantly
clear that to derive truncation error bounds for other general sampling
series, we needed to keep a lower bound for the transform's kernel in mind.
This was, of course, if the complex contour integration method was used,
which, seemed the most feasible for a practical form of the truncation error
bound in that it parallels those available for the Shannon sampling series.

The truncation error bound $\epsilon_T$ (sometimes written $\epsilon_N$) for the generalized sampling series in (7.1.20), is defined as

$$\epsilon_T \equiv |f(t) - f_N(t)| = \left| \sum_{|n|>N} f(t_n) S(t, t_n) \right|. \tag{7.2.15}$$

This is defined in parallel to that of the Shannon sampling series (7.1.10)
as

$$\epsilon_T \equiv |f(t) - f_N(t)| = \left| \sum_{|n|>N} f\left(\frac{n\pi}{a}\right) \frac{\sin(at - n\pi)}{(at - n\pi)} \right|. \tag{7.2.16}$$

The special case of (7.2.15) for the Bessel series is given in (7.1.28) where
$\epsilon_T$ was written as $\epsilon_N(t)$.

One of the earliest, but elementary, truncation error bound for the Shannon sampling expansion (7.2.16) was derived by Helms and Thomas [385],

$$|\epsilon_T(t)| \leq \frac{4M}{\pi^2 N(1-r)},$$

$$M = \max|f(t)| \quad \text{for all } t; \quad -\infty < t < \infty. \tag{7.2.17}$$

For this, they employed complex contour integration and used a smaller finite bandwidth $ra$; $0 < r < 1$, instead of the required $a$ in (7.2.16). Numerous improved truncation and other relevant error bounds for (7.2.16) can be found in [440, Sect. VI]. Prior to 1982, the only known attempt for a truncation error bound of the generalized sampling expansion is due to Yao [974] (for more details see [440, p. 1586]). His method and notation, which uses the concept of *"reproducing kernel Hilbert space,"* can be simplified if we realize that the sampling functions $\{S(t, t_n)\}$ in (7.2.15) and (7.1.20) constitute an orthogonal set for transforms with Fourier type symmetric inverse (7.1.19), (7.1.32)-(7.1.33). This is the case for the Hankel transform (7.1.54)-(7.1.55) illustrated here. Yao's result for the case of $K(x, t) = J_m(xt)$ and with $a = 1$ for the (band-limited) $J_m$-Hankel transform in (7.2.14) becomes

$$| \epsilon_T(t) | \leq \left[ \int_0^t t f^2(t) \, dt - 2 \sum_{i=1}^{N} \frac{f^2(j_{m,i})}{J_{m+1}^2(j_{m,i})} \right]^{\frac{1}{2}}$$

$$\times \left[ \frac{1}{2} \{ J_m^2(t) + J_{m+1}^2(t) \} - 2 \sum_{i=1}^{N} \frac{j_{m,i}^2 J_m^2(t)}{(j_{m,i}^2 - t^2)} \right]^{\frac{1}{2}}.$$

$$(7.2.18)$$

In comparison with existing bounds for the Shannon sampling expansion, e.g., (7.2.17), which are given as more simple and direct functions of the truncation limit $N$, this bound is complicated. Thus, a more computationally friendly bound is needed and is the subject of the next section.

## Truncation Error Bound for the $J_m$–Bessel Sampling Expansion

Here we present a more computationally feasible bound for the $J_m$–Bessel sampling expansion. This can be improved and the method extended to other sampling expansions [459].

$$| \epsilon_T(t) | \leq \frac{2^{3/2} K}{\sqrt{\pi}} \left[ \ln \frac{1}{1 - 2r} \right]^{\frac{1}{2}} \left| \frac{J_m(t)}{J_m(j'_{m,N})} \right|$$

$$\times \left[ \frac{1}{| j'_{m,N} + t |} + \frac{1}{| j'_{m,N} - t |} \right] \qquad (7.2.19)$$

where the sampling series (7.2.13) is truncated at $N$, with the bandwidth $a$ replaced by $ra$, $0 < r < 1$, and

$$J'_m(j'_{m,s}) = 0, \quad s = 1, 2, \ldots,$$

$$| f(z) | \leq K \left[ \frac{e^{r|y|} - 1}{2 | y |} \right]^{\frac{1}{2}}. \qquad (7.2.20)$$

$$K = \int_0^r x^2 F^2(x) \ dx. \tag{7.2.21}$$

The method employs complex contour integration, which parallels that for deriving the error bound (7.2.17) for the Shannon sampling expansion. The present result (7.2.19), however, requires the derivation of the lower bound for $J_m(j'_{m,n} + iy)$ at any $j'_{m,n}$, the zeros of $J'_m(x)$,

$$|J_m(j'_{m,s} + iy)| \geq J_m(j'_{m,s})I_m(y) \geq \frac{1}{\sqrt{2\pi}}J_m(j'_{m,s})\frac{e^y}{\sqrt{y+1}} \tag{7.2.22}$$

where $I_m(y)$ is the modified Bessel function of order $m$. This allows us to consider a truncation error for any $N$ which is not necessarily large, the condition for the then existing asymptotic lower bounds [940, p. 584]

$$|J_\nu(z)| \geq C_2 \frac{e^{|y|}}{\sqrt{|z|}} \tag{7.2.23}$$

where $z$ is on the line joining $A_N - i\infty$ to $A_N + i\infty$, provided that $N$ exceeds a value dependent on $\nu$. The choice of $j'_{m,s}$ is to maximize $J_m(j'_{m,s})$ and hence improve the bound (7.2.19). In deriving the truncation error (7.2.19) for the Bessel sampling expansion (7.2.13), we consider a rectangular contour $C = C_1 + C_2 + C_3 + C_4$, where $C_1$ and $C_3$ are the vertical right and left sides that extend from $j'_{m,N+1} - i\infty$ to $j'_{m,N+1} + i\infty$ and $-j'_{m,N+1} - i\infty$ to $-j'_{m,N+1} + i\infty$, respectively. $C_2$ and $C_4$ are the horizontal lines. For $f(t)$ in (7.2.14), we consider the contour integral

$$\left| \int_C \frac{f(z) \ dz}{(z-t)J_m(z)} \right| \tag{7.2.24}$$

which vanishes along $C_2$ and $C_4$ as $y \to \mp\infty$, since it can be shown [458], using the Schwartz inequality, that for $f(z)$ in (7.2.14) (with $t$ replaced by $z$), with bandwidth $ra$ instead of $a$,

$$|f(z)| \leq K \left[ \frac{e^{r|y|} - 1}{2|y|} \right]^{\frac{1}{2}}, \quad 0 < r < 1. \tag{7.2.25}$$

and that $|J_m(z)|$ is bounded as in (7.2.22). Now it is easy to see that the truncation error (7.2.15) can be expressed in terms of the integrals along $C_1$ and $C_3$ as

$$\begin{aligned} |\epsilon_T(t)| &= |f(t) - S_N(t)| \\ &= \frac{1}{2\pi}|J_m(t)| \left| \int_{C_1, C_3} \frac{f(z) \ dz}{(z-t)J_m(z)} \right|, \end{aligned} \tag{7.2.26}$$

which, after evaluating the residues at $z = t$, $z = t_{m,n}$ and using bounds (7.2.20) and (7.2.22), yields the desired truncation error bound (7.2.19).

**Truncation Error for Other Sampling Expansions**

The method of the last section can be extended to other sampling series, provided, of course, that a lower bound for the kernel $K(x, t)$ is available. Another possibility is to derive the truncation error bound when the samples $f(t_n)$ and their derivatives $f'(t_n)$ are involved [458]. It is expected that such error bound may have the usual $N^{-1}$ dependence. It should be stressed that the work here represents only a starting point in finding practical truncation error bounds and efforts should be continued to derive more efficient and tighter bounds, as has been done for the Shannon sampling expansion [440, Sect. VI]. Efforts should be also aimed toward improving the convergence of the generalized sampling series (7.1.20) and introducing a tighter truncation error bound. This may, for example, consist of convolving the sampling function with higher degree weight of related form. In the frequency domain, this is equivalent to multiplying the gate function

$$p_a(x) = \begin{cases} 1 & |x| < a \\ 0 & |x| > a \end{cases}$$

by a high order general hill function $\psi_{R+1}(w)$ [444], see also (7.1.52) asso-

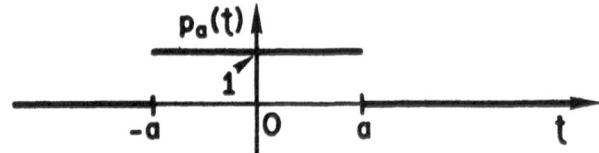

FIGURE 7.2: The gate function

ciated with the general kernel $K(x, t)$, instead of the known hill functions (*B*-spline) associated with the Fourier transform. Some references to related work may be found in the bibliography.

## 7.3    Applications

In this section we will present two examples of our early applications of the generalized sampling theorem. The first is in optics (Sect. 7.3.1) and the second is for facilitating the solution of a boundary-value problem in heat and mass transfer (Sect. 7.3.3). The third topic is related more to the error analysis and concerns the Gibbs' phenomenon for general orthogonal expansion associated with the present sampling series (Sect. 7.3.2). The section is concluded with varied applications, or illustrations, of the generalized sampling theorem that reflect our own experience in utilizing the tools that we presented in Subsections 7.1.3-7.1.5. The emphasis here is mainly on our own work.

## 7.3.1   Optics—Integral Equations Representation for Circular Aperture

We will briefly illustrate the importance of the new concepts and tools, developed here, for the analysis of optical systems of circular aperture. Recall the *"generalized convolution theorem"* in (7.1.37) where

$$(f * g)(t) = \int \rho(\tau)g(\tau)f(t\theta\tau)d\tau = \int \rho(w)F(w)G(w)K(w.t)dw$$

with its associated concept of the *generalized translation* $f(t\theta\tau)$ in (7.1.34). As we reported in [440, p. 1589], Barakat [29] presented a direct application of the Shannon sampling theorem to optical diffraction (see also Gabor [315]). He used the Shannon sampling theorem for the *point spread function for a slit aperture,*

$$f(t) = \frac{1}{2} \int_{-2}^{2} F(w)e^{iwt}dw \tag{7.3.1}$$

where $F(w)$ is the *transfer function* and where the factor $1/2$ enters in order that $f(0)$ be unity,

$$f(t) = \sum_{n=-\infty}^{\infty} f\left(\frac{n\pi}{2}\right) \frac{\sin(2t - n\pi)}{(2t - n\pi)}. \tag{7.3.2}$$

He then extended his analysis to a square aperture where the Shannon sampling theorem in two dimensions [440, p. 1571] was used.

For a *circular aperture* with rotationally symmetric point spread function, he used the $J_0$-Bessel sampling series for

$$f(t) = \int_{0}^{2} wG(w)J_0(tw)dw, \tag{7.3.3}$$

$$f(t) = 4\sum_{n=1}^{\infty} \frac{j_{0,n}f(\frac{j_{0,n}}{2})J_0(2t)}{(j_{0,n}^2 - 4t^2)J_1(j_{0,n})}. \tag{7.3.4}$$

Barakat then stated that the sampled (band-limited) function $f(t)$ in (7.3.1) satisfies the homogeneous *Fredholm integral equation* [451],

$$f(t) = \frac{2}{\pi} \int_{-\infty}^{\infty} \frac{\sin 2(t - \tau)}{t - \tau} f(\tau)d\tau. \tag{7.3.5}$$

For his proof, he multiplied both sides of (7.3.2) by $\frac{\sin m(t-y)}{t-y}$ and integrated over $(-\infty, \infty)$, using the following Hardy's integral,

$$\int_{-\infty}^{\infty} \frac{\sin m(y-x)}{(y-x)} \frac{\sin \mu(y-z)}{(y-z)} dy = \frac{\pi \sin \mu(z-x)}{(z-x)}, \qquad 0 < \mu \le m \qquad (7.3.6)$$

inside the series of (7.3.2) to obtain the Fredholm integral equation in (7.3.5).

We remark here that (7.3.5) would follow easily from (7.3.1) via the use of the (Fourier) convolution theorem. Here $f(t)$ in (7.3.1) is band-limited with a bandwidth of 2. Then (7.3.5) is saying no more than that it had passed through an ideal low-pass filter of the same bandwidth, as seen from its impulse response $\frac{\sin 2(t-\tau)}{\pi(t-\tau)}$. So, the signal should come out unaltered as an output $f(t)$ on the left side of (7.3.5). This removes the necessity of Hardy's integral. Moreover, this integral is itself a consequence of the (Fourier) convolution theorem as applied to band-limited functions. As we have in (7.3.6), an input band-limited signal (with bandwidth $\mu \le m$) passing through an ideal low-pass filter with a larger or equal bandwidth $m \ge \mu$, so it should come out unharmed as the output on the right side of (7.3.6).

These preliminaries bring us to our main point: that in the absence of the new tools (7.1.34)-(7.1.37), developed here, Barakat suggested as possible, with obviously correct intuition, but admitted at the same time that he was unable to derive a similar Fredholm integral equation [29] for the $f(t)$ of the Bessel sampling expansion in (7.3.4).

This is now easily accessible via the new generalized convolution theorem (7.1.36) with a homogeneous Fredholm integral equation for $f(t)$ in (7.3.4),

$$f(t) = \int_0^\infty \frac{2J_1(2(t\theta\tau))}{t\theta\tau} f(\tau) dt. \qquad (7.3.7)$$

The proof is simple and is in line with that given above for $f(t)$ in (7.3.2). Here $f(t)$ of (7.3.3) has a bandwidth of 2. As such, we see the system interpretation from (7.1.40) for such $f(t)$ as passing through an ideal low-pass filter. This is seen in the first integral of (7.1.36), with $S(t) = \frac{2J_1(2t)}{t}$ as the $J_0$-Hankel transform of the gate function

$$p_a(w) = \begin{cases} 1, & 0 < w < a \\ 0, & w > a \end{cases} \qquad (7.3.8)$$

for $a = 2$. Indeed with the same attitude and simple method we were able to derive the following "generalization of the Hardy's integral" [447] to have it associated with the above $J_0$-Hankel transform:

$$\int_0^\infty \frac{yJ_1(a(y\theta x))}{y\theta x} \frac{J_1(b(y\theta z))}{y\theta z} dy = \frac{J_1(b(z\theta x))}{z\theta x}. \qquad 0 < b < a. \quad (7.3.9)$$

where $\frac{J_1(a(y\theta x))}{y\theta x}$ stands for the "generalized" $J_0$-Hankel type translation of $\frac{J_1(ay)}{y}$, as given in (7.1.35). The proof of this extension (7.3.9) to the Hardy's integral needs no more than a simple application of the new generalized convolution theorem (7.1.36). We should first have in mind the generalized translation of the impulse response $aJ_1(\frac{a(t\theta \tau)}{t\theta \tau})$ along with the systems interpretation for time varying systems that we had presented in Subsection 7.1.3. The generalized Hardy's integral can be seen to parallel that of (7.3.6). In the integral on the left side of (7.3.9) we have a system with an ideal low-pass filter of bandwidth $a$. Such a system is probed by a band-limited (or transform limited) signal of narrower band width $b < a$. Thus it should come out unaltered as the output to give us the right-hand side of (7.3.9). Other modes of proof are available (see [447]).

We add that the generalized convolution theorem (7.1.36), (7.1.37) and the generalized sampling theorem (7.1.19)-(7.1.21) should be instrumental in developing general Hardy-type integrals that are associated with general transforms characterized by the kernel $K(x,t)$ of (7.1.19) and not only our illustration of the Bessel function kernel $J_0(xt)$. Also, it should be possible to develop homogeneous Fredholm integral equation representation for such general $\kappa$-transform representation (7.1.19) of the generalized sampling theorem (7.1.19)-(7.1.20).

## 7.3.2 The Gibbs' Phenomena of the General Orthogonal Expansion—A Possible Remedy

In this section we attempt to show how the new tools developed in the previous Subsections 7.1.3-7.1.5 (and used in Subsections 7.2.1-7.2.2), for the systems interpretation and the error analysis of the generalized sampling theorem can now be used to help in finding a remedy for the *Gibbs' phenomenon* of the general orthogonal expansion. Such an orthogonal expansion, as we have seen, is the backbone to deriving the generalized sampling expansion (7.1.20)-(7.1.21). This error is well studied for the trigonometric Fourier series. Here we will try to shed some light, even if it is in terms of inquiries, on this error for the general (*non-periodic*) orthogonal expansion such as the Fourier-Bessel series expansion. A brief analysis of this Gibbs' phenomenon, for general orthogonal series expansion, is found in Gottlieb and Orszag [366].

In our analysis of the Shannon sampling theorem and its present generalization we had to depend heavily on the Fourier trigonometric (periodic) series and the *general orthogonal expansion*, respectively. The latter general expansion, unfortunately, is not necessarily periodic, a property that all regular users of the typical Fourier analysis would miss very much.

As we have already mentioned in this chapter, the lack of the periodicity property presented the greatest difficulty in our (and others) attempts to give a physical interpretation of the generalized sampling theorem. This

difficulty became even more evident when deriving explicit error bounds of
such generalized sampling expansion, as was discussed in Subsections 7.2.1
and 7.2.2. Of course, such an attempt had to be preceded by developing the
new tools of Subsections 7.1.3-7.1.5, i.e., the *"generalized translation"* and
its *"associated convolution theorem,"* the *"new impulse train"* (see Fig. 7.1)
and the new *"extended Poisson sum formula."* These tools were the preliminaries in compensating for the absence of the periodicity that characterized the trigonometric Fourier series and the "related" shift property of
the (exponential) Fourier transform representation of the Shannon sampling theorem.

The analysis of the errors, e.g., truncation and aliasing, of the Shannon
sampling theorem, was not far from the analysis of the truncation and
aliasing errors of the Fourier series often used for deriving the sampling expansion. Both of these errors are related to the *Gibbs' phenomenon*, which
represents the error resulting from the "smooth" trigonometric functions
attempt in approximating a *"not so smooth"* functions with jump discontinuities at the end points, or in the interior, of interval of the expansion.
The Gibbs' phenomenon does not seem to be well-known in the case of approximating such discontinuous functions by a *general Fourier orthogonal
expansion* in terms of the smooth, but not necessarily periodic functions
[380, 366]. A clear example is the Fourier-Bessel series expansion.

After having developed new tools *compatible* with such an orthogonal
expansion in Sections 7.1 and 7.2, it is time to bring attention to their
possible use for the analysis of the Gibbs' phenomenon in such a general
orthogonal series expansion. Our attempt here is to draw a parallel analysis
to what is known for the Fourier trigonometric series and integrals and to
point out the place of the new tools in such analysis. We also suggest an
*improvement* to the existing remedy for the Gibbs' phenomenon of the
trigonometric Fourier analysis. This is in line with using a *self-truncation
factor* with the help of *high order hill functions* of (7.1.47), such as we
employed in the last section for a tighter bound on the truncation error
of the Shannon sampling expansion. After a possible remedy is suggested
for the Gibbs' phenomenon of the general orthogonal expansions, we also
suggest for it the use of high order *general hill functions* (7.1.52) to affect
a parallel improvement to that of the trigonometric series. The detailed
analytic treatment of these concepts for Fourier series and integrals are
found in Appendix A.

The simplest example for illustrating the Gibbs' phenomenon in Fourier
analysis is that of the signum function in Fig. 7.3,

$$\text{sgn}(t) = \frac{t}{|t|} = \begin{cases} 1, & t > 0 \\ -1, & t < 0 \end{cases} \tag{7.3.10}$$

with its clear jump discontinuity at $t = 0$, and its well-known "stubborn"
Gibbs' phenomenon close to $t = 0$. Indeed it can be shown analytically

[380, 456] that the error amounts to about 9% of the size of the jump discontinuity. More importantly, this error is independent of the size of the truncation of the Fourier integral. (For a detailed analysis, see the discussion after Eq. (A.9) and Eq. (A.24) in Appendix A for the Fourier integral and series, respectively). We note that the analysis for the signum function can be applied to other familiar functions with jump discontinuities such as the gate function of Fig. 7.4, which can be expressed in terms of the signum function as $p_a(t) = \frac{1}{2}[\text{sgn}(t + a) - \text{sgn}(t - a)]$.

The next important example is that of the Fourier series representation of a *square wave* of unit amplitude on $(-\pi, \pi)$,

$$f(t) = \begin{cases} 1, & 0 < t < \pi \\ -1, & -\pi < t < 0, \end{cases} \qquad (7.3.11)$$

and its very clear discontinuities at $t = 0$, $t = \mp\pi$ as shown in Fig. 7.5. Here we draw a parallel in our illustration to the Gibbs' phenomenon in case the above two functions of (7.3.10) and (7.3.11) are represented by more general forms, i.e., the Hankel integral transform and the Fourier-Bessel series, respectively.

Due to lack of space we shall be very brief, concentrating our efforts on illustrating the Gibbs' phenomenon for the Fourier-Bessel series approximation of the square wave on $(0,1)$ with its (interior) jump discontinuity at $x = \frac{1}{2}$ (see Figs. 7.9–7.11)

$$f(t) = \begin{cases} 1, & 0 < t < \frac{1}{2} \\ 0, & \frac{1}{2} < t < 1. \end{cases} \qquad (7.3.12)$$

This is to be compared with the familiar Gibbs' phenomenon of the Fourier (trigonometric) series of the same square wave function of (7.3.11) on the interval $(-\pi, \pi)$ with its (interior) jump discontinuity at $x = 0$. A *remedy* for the latter Gibbs' phenomenon will also be illustrated and a parallel one suggested for the former one of the Fourier-Bessel series.

To save space and make for a clearer comparison, we have included in our following illustrations the square wave function, the $N$th partial sum $S_N(t)$ of its Fourier series, and $S_N(t)$ after applying the $\sigma_n$-averaging remedy for reducing the Gibbs' phenomenon.

The basic details of the analytical background treatment for the Gibbs' phenomenon, of the Fourier (trigonometric) integral as well as series representation of functions, are relegated to Appendix A. For a more complete treatment (with ample illustrations) of the Gibbs' phenomenon, we refer the reader to our book [456, Chap. 4, Sect. 4.1.6]; see also references [380, 363].

There is also an illustrated suggestion for improving the existing prescribed remedy [380] by using high order hill functions [456]. The related windowing effect is also discussed in the same book [456, Sect. 4.1.6 E].

Consider $S_N(t)$, the $N$th partial sum of the (trigonometric) Fourier series

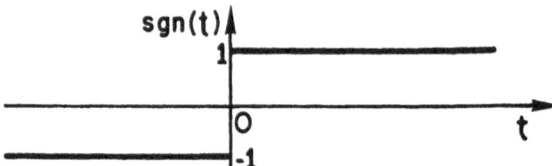

FIGURE 7.3: The signum function sgn(t). (Most of the figures and their analysis here are from Jerri [456], *Integral and Discrete Transforms with Applications and Error Analysis*, 1992, courtesy of Marcel Dekker Inc.).

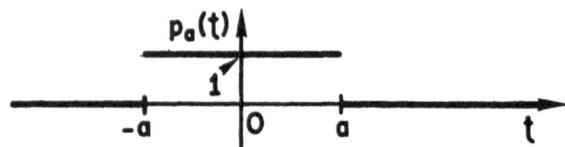

FIGURE 7.4: The gate function $p_a(t)$.

expansion of the square wave in (7.3.11), on the interval $(-\pi, \pi)$,

$$S_N(t) = \frac{4}{\pi} \sum_{n=1}^{N} \frac{1}{(2n-1)} \sin(2n-1)t. \qquad (7.3.13)$$

Figure 7.6 shows the inset of the Gibbs' phenomenon near the jump discontinuity at $t = 0$ (and also at $-\pi, \pi$). Figures 7.7 and 7.8 show that raising the number of terms $N$ from 5 to 10, then to 40 did not make any difference, in so far as the overshoot near $t = 0$ persists and is still a good percentage of the size of the jump there.

In parallel to this we noticed that $S_N(x)$, the $N$th partial sum of the $J_0$-Bessel-Fourier series expansion on $(0.1)$ of the square wave in (7.3.12), shows the same type of Gibbs' phenomenon near the jump discontinuity at $x = \frac{1}{2}$. This is illustrated for $S_{20}(x)$ in Fig. 7.9, and where doubling $N$ to 40 in Fig. 7.10, and even to a 100 in Fig. 7.11 only improved the

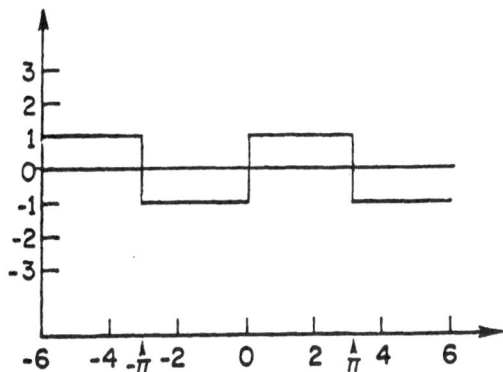

FIGURE 7.5: The square wave function (period $2\pi$).

series approximation of the function, but did not affect the same type of the overshoots and undershoots near the jump discontinuity at $t = \frac{1}{2}$. Figure 7.10, then, shows clearly a similar size first overshoot to that of Fig. 7.8 of the trigonometric Fourier series, which is about 10% of the jump size at $t = \frac{1}{2}$. For a jump discontinuity at an end point, see [366].

It is this Gibbs' phenomenon of Fig. 7.10, as a possible representative for other orthogonal expansions. that we used as an example in order to direct the attention to its possible remedy. Again, with the tools developed here, the known analytical treatment for the case of Fourier series (in Appendix A) and the analysis in [380], it should not be difficult to justify analytically this parallel result of $\sim 10\%$ overshoot for the Fourier-Bessel series near the jump discontinuity at $t = \frac{1}{2}$ in Fig. 7.11. In both Figs. 7.9 and 7.10 we notice an undershoot at $x = 0$, which may resemble a Gibbs' phenomenon. It is not the case however, since the $(J_0)$-Bessel-Fourier series used here represents an even function. Indeed Fig. 7.11 indicates that such an undershoot decreased for $S_{100}(x)$ whereas the overshoot at the discontinuity $x = \frac{1}{2}$ persisted.

Of course, the Fourier-Bessel series is not periodic: therefore, there has not been any interest in looking beyond the interval of the series expansion like $(0,1)$ in the above case of the square wave in (7.3.12). However recalling the relatively new generalized impulse train (7.1.75) of Fig. 7.1, we do have a

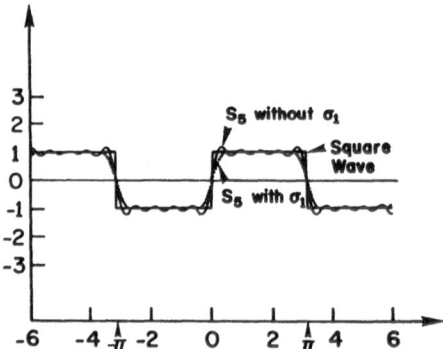

FIGURE 7.6: The square wave, its Fourier series approximation and Gibbs' phenomena and the $\sigma_1$ (averaging) remedy, $N = 5$. (From Jerri [456]).

purpose and interest in looking beyond this basic interval of the orthogonal expansion, even when we are expecting no more than just a "noisy-like *repetition*" of some kind. Figure 7.12 illustrates how the Fourier-Bessel series extends the representation of the square wave in (7.3.12) to the larger interval (0,4), which is beyond (0,1), the basic interval of the orthogonal expansion, and where the above mentioned "some type of repetition" on (1,4) is clearly indicated. We also looked at the expansion on an even larger interval of (0,20), where the *"repetition"* could be associated with the character of the repetition of the impulse train in (7.1.75) of Fig. 7.1.

### A Possible Remedy for the Gibbs' Phenomena—$\sigma$-Averaging

The usual known remedy for the Gibbs' phenomenon of the Fourier analysis like that of Fig. 7.3, is to replace the function with its abrupt jump discontinuity by "another" one which is continuous in a small neighborhood of that discontinuity. Indeed the function is replaced by a straight line in that neighborhood as shown in Fig. 7.13 for the signum function.

Such a process is seen as an averaging of the function, which is called the $\sigma_1$-*averaging* and is attributed to Lancos (see [380]). Details of the basic analysis of this process are found in Appendix A. In Fig. 7.7 we have the $S_{10}(x)$ Fourier series approximation of the square wave function in (7.3.11), with its jump discontinuity and its Gibbs' phenomenon as well as the $S_{10}(x)$

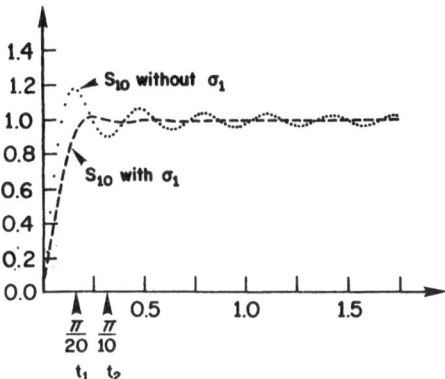

FIGURE 7.7: The Gibbs' phenomena (of the square wave) with $N = 10$. Also its $\sigma_1$ (averaging) remedy.

of the $\sigma_1$-averaging of the square wave. We see clearly the reduction of the Gibbs' phenomenon for the latter (averaged) case as well as a much better approximation where the function is continuous.

More important to us in the analysis of the Gibbs' phenomenon for the Fourier-Bessel series is that such $\sigma_1$-averaging is equivalent to convolving the square wave function with the gate function, a process that we can easily perform for the Fourier-Bessel series with the newly developed tools of Subsections 7.1.4-7.1.5 which are very compatible with such purpose. Also, equivalently. this $\sigma_1$ process means multiplying the Fourier coefficients $c_n = \frac{4}{\pi} \cdot \frac{1}{2n-1}$ of (7.3.13) by $\frac{\sin(2n-1)\frac{\pi}{2N}}{(2n-1)\frac{\pi}{2N}}$ (as Fourier coefficients of the Fourier series expansion of $Np_{\pi/2N}(t)$ on $(-\pi, \pi)$).

In the hope of improving on such $S_{N,\sigma_1}$-averaging process we have attempted to repeat it as $S_{N,\sigma_1^2}$, $S_{N,\sigma_1^3}$, ..., $S_{N,\sigma_1^m}$ (or $\sigma_1$, $\sigma_2$, $\sigma_3$, ..., $\sigma_m$: $\sigma_m \equiv \sigma_1^m$) averagings. Figure 7.8 illustrates the $S_{40}$ approximation without averaging, the $S_{40,\sigma_1}$, and $S_{40,\sigma_2}$ where we note that the $\sigma_2$-averaging is a better approximation with about completely removed Gibbs' phenomenon at $t = 0$. However, it has one drawback: it required almost double the rise time than that of the $\sigma_1$-averaging. Higher order averagings of $\sigma_3$ and $\sigma_6$. even with $S_{80}$, are illustrated in Fig. 7.14. It is observed that there seems to be an optimal order of such high averagings, where we see in Fig. 7.14 that the $S_{80,\sigma_3}$ is the best. $S_{80,\sigma_6}$. though extremely smooth. does suffer

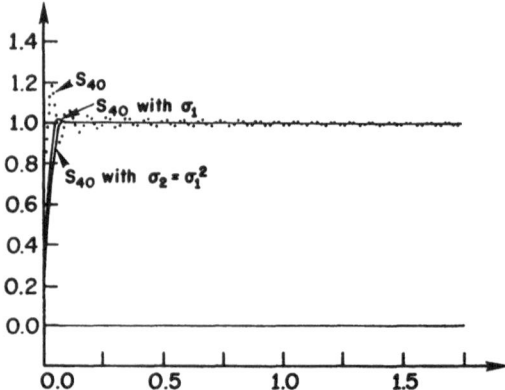

FIGURE 7.8: The Gibbs's phenomenon with $N = 40$. Also its remedies of one ($\sigma_1$) and two ($\sigma_2$) averagings.

from a longer rise time compared to the two former ones.

We again note that the $\sigma_1^m = \sigma_m$-averaging is equivalent to convolving the function $m$ times with $Np_{\pi/2N}(t)$, which in turn will introduce the self-truncating factor,

$$\sigma_m = \left[ \frac{\sin(2n-1)\frac{\pi}{2N}}{(2n-1)\frac{\pi}{2N}} \right]^m .$$

to the Fourier coefficient of $S_N(t)$ in (7.3.13). Here, again, comes the role of the *hill function* of order $m$, as the Fourier transform of $\sigma_m$, where it is making a *self-truncating Fourier* series in a manner similar to making the self-truncating Shannon sampling series. So, for our treatment of the Gibbs' phenomenon of the Fourier–Bessel series, we can employ the *generalized hill function* in (7.1.52) to reduce the Gibbs' phenomenon, or to make the $J_0$-Bessel orthogonal expansion (7.3.19) a self-truncating one. We hope, of course, that this can be applied to other general orthogonal expansions.

In all our practical computations we had to approximate the infinite Fourier series of $G_S(f)$,

$$G_S(f) = \sum_{k=-\infty}^{\infty} c_k e^{-2\pi i k T f}, \qquad -\frac{1}{2T} < f < \frac{1}{2T}. \qquad (7.3.14)$$

$$c_k = T \int_{-T/2}^{T/2} G_S(f) e^{2\pi i k T f} df \qquad (7.3.15)$$

by an $N$ term partial sum. Here, $\omega = 2\pi f$.

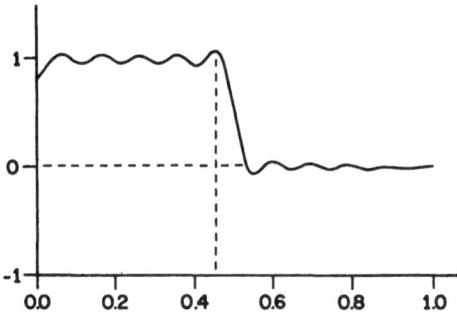

FIGURE 7.9: The square wave function and the "Gibbs' phenomenon" of its ($J_0$-Bessel)-Fourier series approximation $S_N(x)$, $N = 20$. (See also [366].)

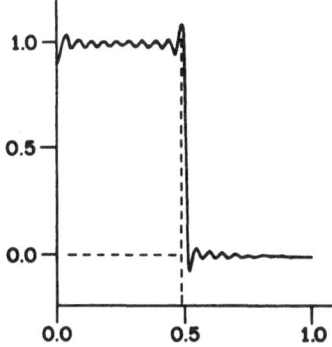

FIGURE 7.10: The square wave function and the clear "Gibbs' phenomenon" of its ($J_0$-Bessel)-Fourier series approximation $S_N(x)$, $N = 40$. (See also [456].)

In the same manner we limit the infinite Fourier integrals of $G(f)$,

$$G(f) = \int_{-\infty}^{\infty} g(t)e^{-j2\pi ft}dt \qquad (7.3.16)$$

and

$$g(t) = \int_{-\infty}^{\infty} G(f)e^{i2\pi ft}df \qquad (7.3.17)$$

to finite time limits $-b$ to $b$ and finite frequency limits $-a$ to $a$ respectively. This means that we had to assume an *"almost" time limited* function for $G(f)$ and an *"almost" band-limited* function for $g(t)$.

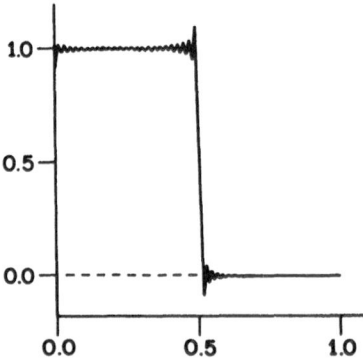

FIGURE 7.11: The square wave function and "definite" "Gibbs' phenomenon" of its ($J_0$-Bessel)-Fourier series approximation $S_N(x), N = 100$.

In all cases of truncation. we actually employed the gate function,

$$p_a(t) = \begin{cases} 1, & |t| \leq a \\ 0. & |t| > a \end{cases} \qquad (7.3.18)$$

with its abrupt discontinuities at $t = \mp 1$. Such a particular case of truncation is viewed as the result of applying a window of the gate function type $p_a(f)$ to $G(f)$, whose effect on the transform $g(t)$ appears as wiggles around $\mp b$, the ends of its support (or period). This effect is well-known as the *"windowing effect,"* and its "apparent" remedy is to increase the bandwidth $a$. However, in many situations $a$ may be fixed, thus something has to be done about the gate function and its *"troublesome"* sharp discontinuities. This means that we have to choose a window that dies out slowly around its truncation edges $\mp a$. Indeed the analysis (or art) of constructing such windows, is very intensive [380], [456, 941] for the (trigonometric) Fourier analysis. However, to our knowledge, there is very little, if any, analysis for the more general transforms or orthogonal series expansion.

It is easily shown [380, 456] (see Appendix A) that the wiggles of the windowing effect are strongly related to the Fourier transform of the given window. With the tools we have already developed in the Subsections 7.1.3-7.1.5, a "similar" statement for the Hankel transform or Fourier-Bessel series representation is in order.

In the case of the Fourier series (7.3.14). typical truncation to $N$ terms amounts to multiplying the infinite sequence of its Fourier coefficients $c_k$ by the gate function $p_{NT}(kT)$. The resulting discontinuity of $c_k$ has its effect on the transform $G_S(f)$ appearing as wiggles, or ripples, around $\mp 1/T$. the ends of the period. These ripples can be explained in terms of the

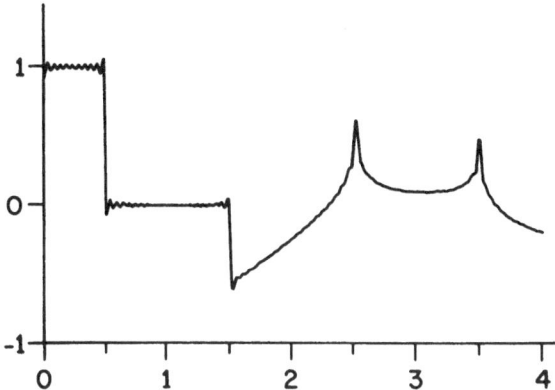

FIGURE 7.12: The $(J_0$-Bessel)-Fourier series approximation $S_N(x)$ of the square wave function on $(0,1)$, and "its some type of repetition" on $(0,4)$. See also Fig. 7.1.

Fourier transform of the truncated Fourier coefficients $c_k p_{NT}(kT)$, which is the convolution product $G_S(f) * \frac{\sin 2\pi NTf}{\pi f}$ where $\frac{\sin 2\pi NTf}{\pi f}$ is the (Fourier) transform of the gate function $p_{NT}(t)$.

Consider the analysis of the truncated Fourier-Bessel series of a function $H_S(x)$,

$$H_S(x) = 2\sum_{n=1}^{\infty} c_n \frac{J_0(x j_{0,n})}{J_1^2(j_{0,n})} \qquad (7.3.19)$$

defined on $(0,b)$, which is, of course, not periodic. The question here is whether we expect ripples around $x = b$ when we truncate the above Fourier-Bessel series (7.3.19) to $N$ terms (see [366]). There is no doubt that such truncation is the result of multiplying the coefficients $c_n$ by a gate function $p_N(x)$, i.e., we use $c_n p_N(t)$ instead of $c_n$ in (7.3.19). In parallel to the above Fourier analysis, this should correspond to "convolving" the $H_S(x)$ of (7.3.19) with the $J_0$-Hankel transform $S(x) = \frac{N J_1(Nx)}{x}$ of the gate function $p_N(t)$. Of course, this convolution product is the generalized one as defined in (7.1.36), with its generalized translation (7.1.34) replacing the usual translation of the Fourier transform. It is clear that $S(x)$ is an oscillating function since $J_1(Nx)$ is, and that it is decaying with zeros at $x_n = \frac{j_{1,n}}{N}$. So, intuitively, and in parallel to the above Fourier analysis with the oscillating $\frac{\sin 2\pi NTf}{\pi f}$, we also expect wiggles around $x = b$. However, the reality is a little different, since the Hankel convolution product does not involve a *"mere"* translation of $x$ in $S(x)$ but a *"generalized translation"*

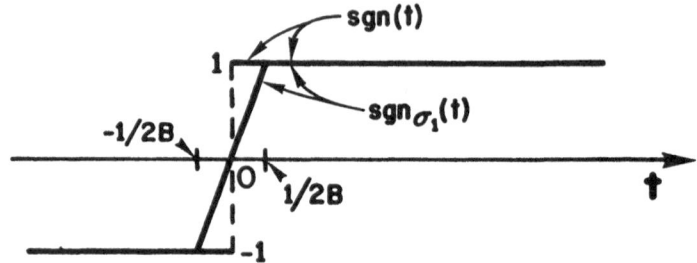

FIGURE 7.13: A possible remedy for the Gibbs' phenomenon—approximation of sgn($t$) by a continuous function $\text{sgn}_{\sigma_1}(t)$.

$S(x\boldsymbol{\theta}\tau)$ like in (7.1.35).

$$
\begin{aligned}
S(x\boldsymbol{\theta}\tau) &= \frac{NJ_1(N(x\boldsymbol{\theta}\tau))}{x\boldsymbol{\theta}\tau} \\
&= \frac{NxJ_1(Nx)J_0(N\tau) - N\tau J_1(N\tau)J_0(Nx)}{x^2 - \tau^2}.
\end{aligned}
\tag{7.3.20}
$$

The tentative computations and illustrations [442, 437, 454] show that for relatively large $\tau$, the generalized translation seems to have a smoothing effect. As to the Fourier–Bessel series representation $H_S(x)$ beyond the interval $(0, b)$, the form may be described as some kind of noisy repetition (Fig. 7.12) with extended intervals related to locations of the pulses in the impulse train of Fig. 7.1.

It is hoped that these brief notes offer a clear presentation of the Gibbs' phenomenon, and its possible remedies, of the Fourier series and the general orthogonal expansion of functions with jump discontinuities as illustrated in Figs. 7.5, 7.7, 7.14 and 7.9.

As we mentioned at the beginning of this section, we shall have most of the detailed analytical treatment of the typical Fourier (trigonometric) analysis for Appendix A.

## 7.3.3  BOUNDARY-VALUE PROBLEMS

One of our earliest applications of generalized sampling expansions (7.1.20), outside of communications or optics, was to facilitate the solution of a boundary-value problem in a plug flow [457]. Such a problem is concerned with determining the effect of axial conduction on the temperature field of a fluid (in a laminar flow) in a tube. For such a problem, for the temperature $T(r, z)$: $0 < r < 1, 0 < z < \infty$, the finite Hankel transform (or Bessel-

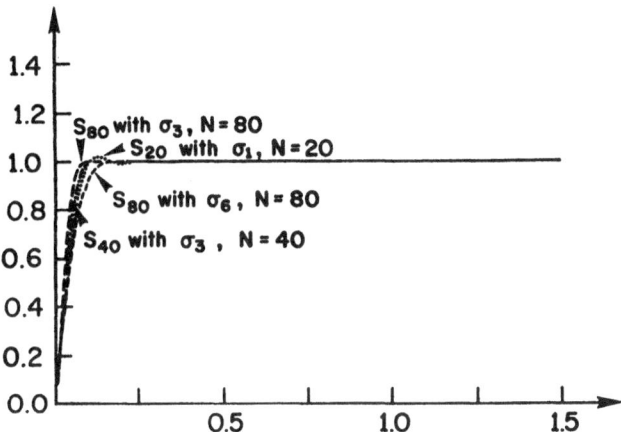

FIGURE 7.14: Higher order $\sigma_n$-averagings for reducing the Gibbs' phenomenon. The rise time for $\sigma_3$ is reduced by increasing $N$, $N = 20, 40, 80$.

Fourier series) is used for the radial variation fields $T_1(r, z)$ for $z < 0$ and $T_2(r, z)$ for $z > 0$. For complete details of the analysis and the improvement in the numerical solution see [457]. Among a number of the necessary auxiliary conditions. the following two are important compatibility conditions (at the point of entrance $z = 0$ to the environment of $z > 0$):

$$T_1(r. 0) = T_2(r, 0) \qquad (7.3.21)$$

and

$$\frac{\partial T_1}{\partial z}(r. 0) = \frac{\partial T_2}{\partial z}(r, 0) \qquad (7.3.22)$$

for matching the temperature and its gradient at the entrance point $z = 0$. As expected, such conditions involve Bessel series on both sides of each equation, but here the (different) Fourier-Bessel coefficients are the unknowns. For example, condition (7.3.21) becomes

$$\sum_{n=1}^{\infty} \frac{2D(\lambda_{1,n})J_0(\lambda_{1,n}r)}{J_0^2(\lambda_{1,n})} = 1 + \sum_{n=1}^{\infty} \frac{2C(\lambda_{0,n})J_0(\lambda_{0,n}r)}{J_1^2(\lambda_{0,n})}. \qquad (7.3.23)$$

$$J_0(\lambda_{0,n}) = 0. \qquad n = 1, 2. \ldots$$

$$J_1(\lambda_{1,n}) = 0. \qquad n = 1, 2. \ldots$$

where $D(\lambda_{1,n})$ and $C(\lambda_{0,n})$ are the Fourier-Bessel coefficients of the Bessel series solution as related to $T_1$ and $T_2$, respectively. Before our recognition of the possible use of the generalized Bessel-type sampling theorem, the infinite number of coefficients $\{D(\lambda_{1,n})\}$ and $\{C(\lambda_{0,n})\}$ had to be matched [4]. However, since we know that such coefficients are indeed the samples of a finite limit Hankel transform (or as the Fourier coefficients of the Fourier-Bessel series) we can, with a simple operation, change each of the infinite series in (7.3.23) to only one term $C(\lambda)$ and $D(\lambda)$ as the interpolations of $C(\lambda_{0,n})$ and $D(\lambda_{1,n})$, respectively. With the forms in (7.3.23) and keeping in mind the generalized sampling theorem, we can multiply both sides of (7.3.23) by $rJ_0(\lambda r)$ and integrate term by term from 0 to 1 to have

$$\sum_{n=1}^{\infty} D(\lambda_{1,n})S_2(\lambda,\lambda_{1,n}) = \frac{J_1(\lambda)}{\lambda} + \sum_{n=1}^{\infty} C(\lambda_{0,n})S_1(\lambda,\lambda_{0,n}). \qquad (7.3.24)$$

Now we recognize the above two series as *Bessel sampling series expansions* for $C(\lambda)$ and $D(\lambda)$:

$$C(\lambda) = \sum_{n=1}^{\infty} C(\lambda_{0,n})S_1(\lambda,\lambda_{0,n}) \qquad (7.3.25)$$

and

$$D(\lambda) = \sum_{n=1}^{\infty} D(\lambda_{1,n})S_2(\lambda,\lambda_{1,n}). \qquad (7.3.26)$$

Equation (7.3.24) becomes a much simpler relation between only two functions $C(\lambda)$ and $D(\lambda)$,

$$D(\lambda) - C(\lambda) = \frac{J_1(\lambda)}{\lambda}. \qquad (7.3.27)$$

This replaces the infinite matching of the coefficients $\{C(\lambda_{0,n})\}$ and $\{D(\lambda_{1,n})\}$ in (7.3.23) before employing the sampling theorem. The same thing was done for the series resulting from the second compatibility condition (7.3.22) of matching the temperature gradient at the entrance point $z = 0$. This analysis using the generalized sampling theorem resulted in simpler and more accurate numerical computations for this plug flow boundary value problem, or a conjugate boundary-value problem of Graetez type [4]. For numerical results, see Figs. 1–3 in [457].

## 7.3.4 OTHER APPLICATIONS AND SUGGESTED EXTENSIONS

In this section we will outline some of our present and suggested future research in relation to sampling theorems. This research includes the retaining of the interpolation property [448] of the generalized (optimal) sampling sum of approximation theory [845] as we shall discuss in more detail in this section. There is also our continued attempt for completing the picture, i.e., sampling a function and its derivatives, for most of the remaining, well-known general integral transforms that are associated with the generalized sampling theorem [437]. This includes the Legendre, associated Legendre, the Gegenbauer, Laguerre, Hermite and the Jacobi transforms [434, 437, 440].

Apart from the applications for the communications engineer, we found applications of the Shannon and the generalized sampling for civil, mechanical and chemical engineers. The first relates to the determination of the spacing between the railroad cross-ties [446, 510], where the Shannon sampling theorem was used and the result agreed with the long-standing railroad data. We are considering presently this problem of support in the case of thin cylindrical shells with circular supports, where the Hankel transform is used. It should be clear from our analysis (in Sections 7.1.2 and 7.1.3) how the Bessel sampling series (7.1.26) will play a similar role in determining the spacing of the circular supports. When the circular supports, for spherical domes or shells, are considered, a Legendre sampling series, or other generalized sampling series with closely related radial kernels, may be used for the spacing of such arcs or supports.

The Shannon sampling theorem and its generalization in this chapter played a role in the development of our research interest, namely, the development of a *modified iterative method* for solving *non-linear problems* [455, 449, 460] (see also more relevant references in the recent tutorial article on the subject [455]). The familiarity with the properties of band-limited functions as well as with the Shannon sampling theorem was helpful in developing this method for solving non-linear problems with rectangular geometry. When the same non-linear problem was considered for a cylindrical shape, the role of the transform-limited functions, presented in Subsections 7.1.2–7.1.3, became very evident. This is, of course, apart from the necessity of Bessel sampling series. Details of the brief history of this development, especially as it relates to the band (or transform)-limited functions and their corresponding sampling series, is presented in second part of this section.

## Interpolation for the Generalized Sampling Sum of Approximation Theory

In his treatment of the error bound for the aliasing error of not necessarily band-limited functions, Splettstösser [845] replaced the si (or $\sin x / x$) function, of Shannon sampling expansion [818, 440] by well-known functions in approximation theory.

He then illustrated his work with a number of such known functions. He showed better convergence, and hence a tighter aliasing error bound than the existing ones. He remarked, however, that such better expansions lack the *interpolation property*, which is very important for the series analysis of signals. Consider the function

$$f(t) = \frac{1}{\sqrt{2\pi}} \int_{-\infty}^{\infty} F(w)e^{iwt}dw \tag{7.3.28}$$

which is Lebesgue integrable ($f \in L(R)$), continuous ($f \in C(R)$) and has a Lebesgue integrable Fourier transform $F(w)$, $F(w) \in L(R)$ on the real line $R$,

$$F(w) = \frac{1}{\sqrt{2\pi}} \int_{-\infty}^{\infty} f(t)e^{-iwt}dt. \tag{7.3.29}$$

Let $G(w)$ be the Fourier transform of $g(t)$ with the same conditions above: the convolution product $(f * g)(t)$ of $f(t)$ and $g(t)$ is

$$\begin{aligned} (f * g)(t) &= \frac{1}{\sqrt{2\pi}} \int_{-\infty}^{\infty} f(u)g(t-u)du \\ &= \frac{1}{\sqrt{2\pi}} \int_{-\infty}^{\infty} e^{iwt} F(w)G(w)dw, \end{aligned} \tag{7.3.30}$$

with $F(w)G(w)$ as its Fourier transform. Splettstösser employed an important known tool in approximation theory concerning the discretization of the above convolution product. It states that if, in addition to $f, g \in L(R) \cap C(R)$, they are also band-limited to $\pi W$, i.e., $F(w) = G(w) = 0$ for all $|w| > \pi W, W > 0$, then

$$\frac{1}{\sqrt{2\pi}} \int_{-\infty}^{\infty} f(u)g(t-u)du = \frac{1}{\sqrt{2\pi}W} \sum_{k=-\infty}^{\infty} f\left(\frac{k}{W}\right) g\left(t - \frac{k}{W}\right). \tag{7.3.31}$$

This result was stated in Subsection 7.1.4, as a special case of our extension of (7.1.49) to the product of three functions, to help with the derivation of the self-truncating sampling series in (7.1.48). Moreover, with the help of the new tool of the generalized convolution theorem in (7.1.36), we were able to extend these results, associated with only the Fourier transforms, to the result (7.1.56) of the general transforms (7.1.32)-(7.1.33) associated with the generalized sampling theorem (7.1.20)-(7.1.21). We shall rely on such new tools in extending our following results. of retaining the interpolation property to the Fourier transform development of Splettstösser, to the general integral transforms of (7.1.32)-(7.1.33).

For not necessarily band-limited functions, Splettstösser established that for $f \in C(R) \cap L(R)$ and $F \in L(R)$, then for each $t \in R$,

$$f(t) = \lim_{W \to \infty} \sum_{k=-\infty}^{\infty} f\left(\frac{k}{W}\right) \text{si}\{\pi(Wt - k)\}, \quad \text{si } x \equiv \frac{\sin x}{x} \tag{7.3.32}$$

in terms of the classical si-sampling function, which, as usual, interpolates the function $f(t)$ at the sampling points $\{\frac{k}{W}\}$.

In order to obtain a better convergent sampling sum than that of (7.3.32),
he introduced known kernels of approximation theory $x(t) \in L(R)$, which
are normalized, i.e.,

$$\frac{1}{\sqrt{2\pi}} \int_{-\infty}^{\infty} x(u)du = 1. \tag{7.3.33}$$

He then established the following more general sampling sum, where $X(w)$
is the Fourier transform of $x(t)$. Let $x \in L(R) \cap C(R)$ have the prop-
erty (7.3.33) with $|x(t)| \leq \text{const.}|t|^{-\gamma}$ for some $\gamma > 1$ and $X(w) = 0$ for all
$|w| > \Omega, \Omega > 0$. Moreover, let $X(w)$ be absolutely continuous on $(-\Omega, \Omega)$.
Then for any $f \in C(R) \cap L(R)$

$$f(t) = \lim_{W \to \infty} \frac{\pi}{\sqrt{2\pi\Omega}} \sum_{k=-\infty}^{\infty} f\left(\frac{k}{W}\right) x\left[\frac{\pi}{\Omega}(Wt - k)\right] \tag{7.3.34}$$

uniformly for all $t \in R$.

As we have already stated, this approximation sum demonstrated better
convergence [845] than that of (7.3.32), especially with such kernels as the
de la Valée Poussin means,

$$x(u) = \theta(u) = \frac{3}{3\sqrt{2\pi}} \text{si}\left(\frac{3}{2}u\right) \text{si}\left(\frac{u}{2}\right) \tag{7.3.35}$$

whose Fourier transform is

$$\theta(w) = \begin{cases} 1, & |w| \leq 1 \\ 2 - |w|, & 1 < w \leq 2 \\ 0, & |w| > 2. \end{cases} \tag{7.3.36}$$

Unfortunately, such a better approximation (7.3.34) with (7.3.35) lacks the
important property for approximating signal functions $f(t)$, namely, the
interpolation at the sampling points $\{\frac{k}{W}\}$.

Splettstösser raised the question as to whether such approximations be
modified in such a way that they interpolate, and that the resulting order
of approximation is at least as good. He demonstrated this point for the
case of the kernel of Fejér,

$$\theta(u) = \frac{1}{\sqrt{2\pi}} \left[\text{si}\left(\frac{u}{2}\right)\right]^2 \tag{7.3.37}$$

with its Fourier transform

$$\Sigma(w) = \begin{cases} 1 - |w|, & |w| \leq 1 \\ 0, & |w| \geq 1 \end{cases} \tag{7.3.38}$$

where the sampling sum (7.3.34) becomes

$$f(t) = \lim_{W \to \infty} \sum_{k=-\infty}^{\infty} f\left(\frac{k}{W}\right) [\text{si}(Wt - k)]^2 \tag{7.3.39}$$

which obviously does not interpolate at $t = \frac{k}{W}$. He then gave a counterexample with the differentiable function

$$f(t) = \begin{cases} 1 + \sin t, & t \in (-\frac{\pi}{2}, \frac{3\pi}{2}) \\ 0, & \text{elsewhere} \end{cases} \tag{7.3.40}$$

and offered a modified version of (7.3.39) that would interpolate. as a Fejér-Hermite type interpolation,

$$f(t) = \lim_{W \to \infty} \sum_{k=-\infty}^{\infty} \left[ f\left(\frac{2k}{W}\right) + \left(t - \frac{2k}{W}\right) f'\left(\frac{2k}{W}\right) \right] \left[ \sin \frac{\pi}{2}(Wt - 2k) \right]^2. \tag{7.3.41}$$

We note that the infinite series in (7.3.41) is also known as the sampling expansion for the band-limited function $f_{\pi W}(t)$ in terms of the samples of the function and its first derivative. This is known [440, Sect. IVB], even with higher derivatives, on the basis of the classical sampling series (7.3.31); hence the involvement of powers of the same si-function as in (7.3.39). We also note that for this particular example (7.3.40) the modified series (7.3.41) would have worse convergence than that of the approximation sum (7.3.39), since the sequence $f'(\frac{2k}{W}) = \cos\frac{2k}{W}$ would behave like $f(\frac{2k}{W}) = 1 + \sin\frac{2k}{W}$ and that the second term $(t - \frac{2k}{W})\cos\frac{2k}{W}$ in (7.3.41) would be slower than the first term $1 + \sin\frac{2k}{W}$.

In the next section we shall develop a method that uses complex contour integration which will modify all approximation sums considered in [845] and extends to a more general analysis of this type associated with integral transforms, other than the usual Fourier transform [448].

**Interpolation for the Sampling Sum of Approximation Theory**

Consider the approximation sum in (7.3.34),

$$\frac{\pi}{\sqrt{2\pi\Omega}} \sum_{k=-\infty}^{\infty} f_{\pi W}\left(\frac{k}{W}\right) x \left[\frac{\pi}{\Omega}(Wt - k)\right]. \tag{7.3.42}$$

for the band-limited function $f_{\pi W}(t)$, i.e., $F(w) = 0. |w| > \pi W$. Let $g(u)$ be the transcendental factor in $x(u)$ of (7.3.42). Then the contour integral along the path $C_R : |z| = R$,

$$\int_{C_R} \frac{f(z)}{(z - t)g(z)}, \tag{7.3.43}$$

as $R \to \infty$. would give the modified version that is needed for the approximate (7.3.34) to have the interpolation property. The proof is straightforward as $f_{\pi W}(t)$ and the main desired part of the modified series would result, respectively. from the contribution of the residue at $z = t$ and the residues at the zeros of $g(z)$. What remains is to show that the contour integral vanishes as $R \to \infty$. This would put a condition on the exponential order $o(e^{\pi W |y|})$ of $f_{\pi W}(z)$ versus that of $g(z)$, which determines the sampling spacing. For example, in the case of the Fejér kernel expansion, the transcendental part is $g(z) = [\sin \frac{\pi}{2} W'z]^2 = o(e^{\pi W'|y|})$, which makes the contour integral vanish for $W' > W$, in other words, as long as the sampling spacing $\frac{1}{W'}$ is less than $\frac{1}{W}$. As a specific example, we consider the approximation sum associated with the the de la Valée Poussin kernel (7.3.35)

$$f(t) = \lim_{W \to \infty} 3 \sum_{k=-\infty}^{\infty} f\left(\frac{k}{W}\right) \operatorname{si}\left[\frac{3\pi}{4}(Wt - 2k)\right] \operatorname{si}\left[\frac{\pi}{4}(Wt - 2k)\right] \quad (7.3.44)$$

where $g(z) = [\sin \frac{3\pi}{4} Wz][\operatorname{si} \frac{\pi}{4} Wz] = o(e^{W|y|})$, which has double zeros at $z_k = \frac{4k}{W}$ and simple zeros at $z_k = \frac{4k}{3W}, k \neq 3j$. If we perform the contour integration (7.3.43) for the band-limited function $f_{\pi W}(z)$, we obtain

$$
\begin{aligned}
f_{\pi W}(t) = & \sum_{k=-\infty}^{\infty} \left[f\left(\frac{4k}{W}\right) + \left(t - \frac{4k}{W}\right)\right] f'\left(\frac{4k}{W}\right) \\
& \times \operatorname{si}\left[\frac{3\pi}{4}(Wt - 4k)\right] \operatorname{si}\left[\frac{\pi}{4}(Wt - k)\right] \\
& + \sum_{k \neq 3j} f\left(\frac{4k}{3W}\right) \operatorname{si}\left[\frac{3\pi}{4}(Wt - \frac{4k}{3})\right] \left[\frac{\sin \frac{\pi}{4} Wt}{\sin \frac{\pi k}{3}}\right] \quad (7.3.45)
\end{aligned}
$$

where the second infinite sum does not involve the samples at $\frac{4k}{W}$. This approximate expansion possesses the interpolation property, where the first series and second series in (7.3.45) give the samples $f(\frac{4k}{W})$ and $f(\frac{4k}{3W}), k \neq j$, respectively. This should answer the open question raised in [845], besides all its other examples and general cases.

## The Extension to the Generalized Sampling Theorem

In order to extend the method to the case of the generalized sampling theorem (7.1.20)-(7.1.21), we note that in this chapter we have already presented all tools necessary for this analysis. This includes the generalized convolution product (7.1.36) and its discretization (7.1.47) as the parallel of Splettstösser's work [845] (only for the Fourier transform of the Shannon sampling expansion). With such tools, one can extend the above results to approximation theory with the retainment of the important interpolation property for a generalized sampling series like the Bessel sampling series. The importance of this development will, of course, depend on how much the general orthogonal expansion is used in approximation theory (aside from the familiar trigonometric Fourier series).

The generalization of Splettstösser's result (7.3.31) to other integral transforms is obtained from our result (7.1.49) when limited to only two band-limited functions $f(t)$ and $g(t)$,

$$\int \rho(\tau) f(\tau) g(t\theta\tau) d\tau = \sum_n \frac{f_I(t_n) g_I(t\theta t_n)}{\|K(.,t_n)\|_2^2}. \qquad (7.3.46)$$

We note again that here the two functions $f(t)$ and $g(t)$ must be band-limited since, for example, $\{(K(x,t_n)\} = \{J_0(\frac{xj_{0,n}}{a})\}$ are orthogonal on $(0,a)$, where $j_{0,n}$ is the $n$th zero of $J_0(x)$. However, these results may allow *non-band-limited* functions $f(t)$, $g(t)$ in (7.3.46), but only for the transforms (7.1.32)-(7.1.33) whose $\{K(x,t_n)\}$ is orthogonal on the infinite interval involved in the definition (7.1.33). An example would be the Laguerre transform with $K(x,t) = L_t(x)$ on $(0,\infty)$, $\rho(x) = e^{-x}$ and $K(x,t_n) = L_n(x)$, the Laguerre polynomials, which are orthogonal on the same infinite interval $(0,\infty)$ [439]. Another example is that of the parabolic cylinder (Hermite-Weber) function $K(x,t) = D_t(x)$ on $(0,\infty)$ with $\rho(x) = 1$, $K(x,t_n) = D_n(X)$ as orthogonal polynomials on $(0,\infty)$ [445]. This statement is supported when (7.3.46) is derived via the orthogonal expansion for $f(t)$, in the integral of the left side of (7.3.46). then integrating term by term.

### Modified Iteration for Non-Linear Problems—The Generalized Convolution for Cylindrical Geometry

The origin of the *modified iteration method* for solving *non-linear problems*, started with the following boundary-value problem for the non-linear chemical concentration $y(x)$ in a *planar* catalyst pellet [263, 449],

$$\frac{d^2y}{dx^2} = \varphi^2 y^2, \qquad 0 < x < a, \qquad (7.3.47)$$

$$y'(0) = 0, \qquad (7.3.48)$$

$$y'(a) = Sh(1 - y(a)), \qquad (7.3.49)$$

where $\varphi$ and Sh are called the Theile and Sherwood numbers, respectively. Do and Weiland [263] used a finite Fourier cosine transform to algebrize the linear derivative term in (7.3.47) and utilized the boundary conditions (7.3.48)-(7.3.49). But for the Fourier transform of the non-linear term $y^2(x)$ in (7.3.47), they used *an approximation* which, in the language of the Fourier transform. amounts to the following:

$$\mathcal{F}\{y^2(x)\} = \frac{Y^2(\lambda)}{\mathcal{F}\{1\}}. \qquad (7.3.50)$$

They followed this by an *iterative method* that added an error term and reported very good results [263]. Such an approximation and its success captured our attention, as we recognized that it must be a very special solution to allow such a violation of (7.3.50) to the rule of the Fourier transform of $y^2(x)$, which should be the following *self-convolution* of $Y(\lambda)$, the Fourier transform of $y(x)$,

$$(Y * Y)(\lambda) = \frac{1}{2\pi} \int_{-\infty}^{\infty} Y(\lambda - \mu)Y(\mu)d\mu, \quad Y(\lambda) = \mathcal{F}\{y(x)\}. \quad (7.3.51)$$

Clearly, their relation (7.3.50) on the interval $(0, a)$ is exact for $y(x) = p_a(x)$, the gate function; a very special case indeed. With our analysis, it turned out that the usual *direct iteration* [451] of this non-linear problem in $Y(\lambda)$ is valid for $\varphi < 1.35$ which corresponds to a solution that is slowly varying between 0.7 and 1.0 on the interval $(0,1)$, which is very close to $p_1(x)$.

### The Modified Iterative Method—Its Origin

In the self-convolution (7.3.51) we could not exactly take $Y(\lambda - \mu)$ as $Y(\lambda)$ out of the integral. Instead we started with a first approximation $Y^{(1)}(\lambda - \mu)$ to divide $Y(\lambda - \mu)$. Then we divided and multiplied inside the integral of (7.3.51) by $Y^{(1)}(\lambda - \mu)$:

$$\int_{-\infty}^{\infty} Y(\lambda - \mu)Y(\mu)d\mu = \int_{-\infty}^{\infty} \frac{Y(\lambda - \mu)}{Y^{(1)}(\lambda - \mu)} Y^{(1)}(\lambda - \mu)Y(\mu)d\mu.$$

Now if $Y^{(1)}(\lambda)$ is close to $Y(\lambda)$, then a Taylor series expansion of $\frac{Y(\lambda - \mu)}{Y^{(1)}(\lambda - \mu)}$ about $\mu = 0$ may well be approximated by its first term as $\frac{Y(\lambda)}{Y^{(1)}(\lambda)}$ and our *approximation* to the above integral becomes

$$\int_{-\infty}^{\infty} Y(\lambda - \mu)Y(\mu)d\mu \cong \frac{Y(\lambda)}{Y^{(1)}(\lambda)} \int_{-\infty}^{\infty} Y^{(1)}(\lambda - \mu)Y(\mu)d\mu, \quad (7.3.52)$$

where the first approximation $Y^{(1)}(\lambda - \mu)$ acts now as a kernel. If we use this approximation for the Fourier transform of $y^2(x)$, and $-\lambda^2 Y(\lambda) + B(\lambda)$ for the Fourier transform of $d^2y/dx^2$ on $(-\infty, \infty)$ in (7.3.47), we have

$$-\lambda^2 Y(\lambda) + B(\lambda) = \frac{\varphi^2}{2\pi} \frac{Y(\lambda)}{Y^{(1)}(\lambda)} \int_{-\infty}^{\infty} Y^{(1)}(\lambda - \mu)Y(\mu)d\mu. \quad (7.3.53)$$

The $B(\lambda)$ term here is symbolic for covering the boundary conditions, as in most practical cases we may use the cosine, sine or finite Fourier transforms. We considered (7.3.53) as an *integral equation* [451, 449] in $Y(\lambda)$ with kernel $Y^{(1)}(\lambda - \mu)$ as the first approximation to its solution.

Moreover, we took the $Y(\lambda)$ outside the integral as (second approximation or output) $Y^{(2)}(\lambda)$, inputted for the first time by the $Y(\mu)$ inside the integral as $Y^{(1)}(\mu)$,

$$-\lambda^2 Y^{(2)}(\lambda) + B(\lambda) = \frac{\varphi^2 Y^{(2)}(\lambda)}{2\pi Y^{(1)}(\lambda)} \int Y^{(1)}(\lambda - \mu)Y^{(1)}(\mu)d\mu. \quad (7.3.54)$$

If we solve for $Y^{(2)}(\lambda)$, we have

$$Y^{(2)}(\lambda) = \frac{\frac{1}{\lambda^2}B(\lambda)}{1 + \frac{\varphi^2}{2\pi}\frac{1}{\lambda^2}\frac{\int Y^{(1)}(\lambda-\mu)Y^{(1)}(\mu)d\mu}{Y^{(1)}(\lambda)}}, \qquad (7.3.55)$$

An iterative process with a *new form* whose general form for the $(m + 1)$ iterate is

$$Y^{(m+1)}(\lambda) = \frac{\frac{1}{\lambda^2}B(\lambda)}{1 + \frac{\varphi^2}{2\pi}\frac{1}{\lambda^2}\frac{1}{Y^{(m)}(\lambda)}\int Y^{(m)}(\lambda-\mu)Y^{(m)}(\mu)d\mu}. \qquad (7.3.56)$$

This new form proved to have much better convergence [449. 460], to the point that the same problem of Do's and Weiland's problem [263] was convergent for $\varphi$ up to 100. Such convergence is supported by the *Banach fixed point theorem* [451], where it is shown that the form of the modified iteration (7.3.56) stays contractive for $\varphi$ up to 100. This is to be compared with a convergence for $\varphi < 1.35$ had we stayed with the following original form, i.e., without the modification that resulted in (7.3.55), (7.3.56),

$$-\lambda^2 Y^{(m+1)}(\lambda) + B(\lambda) = \int_{-\infty}^{\infty} Y^{(m)}(\lambda - \mu)Y^{(m)}(\mu)d\mu. \qquad (7.3.57)$$

This method was applied with success to a variety of non-linear problems [455, see the references therein] which included the above pellet problem in planar as well as cylindrical and spherical coordinates, non-linear waves, and Poison-Boltzmann equation [455]. Also, the present modified iteration was applied to other representations of such nonlinear problems besides the above self-convolution type in (7.3.51) and (7.3.52). which is limited (basically) to quadratic non-linearity. The other representations include Green's function integral representation and finite difference (see [460] and references 9, 10 in [455]). The latter representations are highly suitable for general non-linearity.

We must attribute the development of this method to the familiarity with the simple, but powerful, properties of band-limited functions.

In the case of the above planar coordinates. the role of the Shannon sampling theorem appeared due to the need for interpolating values of the finite cosine transform, which we shall discuss next.

## The Planar Pellet and the Use of Shannon Sampling Theorem

Here we will show where and how the Shannon sampling theorem was used in the analysis of the planar pellet problem (7.3.47)-(7.3.49). For the actual pellet problem, a change of variable $u(x) = 1 - y(x)$ (with $a = 1$) was used [263] to have it in the form

$$\frac{d^2u}{dx^2} = -\varphi^2(1 - u)^2, \qquad 0 < x < 1, \qquad (7.3.58)$$

$$u'(0) = 0, \qquad (7.3.59)$$
$$u'(1) = -\text{Sh} \cdot u(1). \qquad (7.3.60)$$

The compatible transform [456] for this problem is the *finite cosine transform*

$$U_c(\lambda_n) = \int_0^1 u(x) \, \cos(\lambda_n x) \, dx \equiv \mathcal{F}_c\{u\}, \qquad (7.3.61)$$

where $\lambda_n$, due to (7.3.60), are the zeros of

$$\tan \lambda_n = \frac{\text{Sh}}{\lambda_n}, \qquad n = 0, 1, 2, \ldots, \qquad (7.3.62)$$

and with $\text{Sh} \to \infty$ we can see that $\lambda_n = (n + \frac{1}{2})\pi$. Also

$$u(1) = 0.$$

The inverse transform to $U_c(\lambda_n)$ in (7.3.61) is the Fourier-cosine series

$$u(x) = 2 \, \text{Sh} \sum_{n=0}^{\infty} \frac{U_c(\lambda_n) \, \cos(\lambda_n x)}{\text{Sh} + \sin^2 \lambda_n} \qquad (7.3.63)$$

which, as $\text{Sh} \to \infty$, becomes

$$u(x) = 2 \sum_{n=0}^{\infty} U_c(\lambda_n) \cos(\lambda_n x), \qquad \lambda_n = (n + \frac{1}{2})\pi, \qquad n = 0, 1, \ldots \, .$$

$$(7.3.64)$$

If we apply this transform on the boundary-value problem (7.3.58)-(7.3.59), (7.3.63) with the identity

$$\mathcal{F}_c\{u''(x)\} = -\lambda_n^2 U_c(\lambda_n) - u'(0) + [u'(1) + \text{Sh} \cdot u(1)] \, \cos \lambda_n, \qquad (7.3.65)$$

we have

$$\lambda_n^2 U_c(\lambda_n) = \varphi^2 \mathcal{F}_c\{(1 - u)^2\} \qquad (7.3.66)$$

and in terms of $Y(\lambda_n)$,

$$\lambda_n^2 \left[ \frac{\sin \lambda_n}{\lambda_n} - Y_c(\lambda_n) \right] = \varphi^2 \mathcal{F}_c\{y^2\} \qquad (7.3.67)$$

where $\mathcal{F}_c\{1\} = (\sin \lambda_n / \lambda_n)$.

Now we write $\mathcal{F}_c\{y^2\}$ as a convolution product $(Y_c * Y_c)(\lambda_n)$ for this finite cosine transform,

$$(Y_c * Y_c)(\lambda_n) = \sum_{k=0}^{\infty} Y_c(\lambda_n)[Y_c(\lambda_k + \lambda_n) + Y_c(|\lambda_k - \lambda_n|)]. \qquad (7.3.68)$$

If we use this in (7.3.67), we have a non-linear "summation" type equation (as compared with integral equation), in $Y_c(\lambda_n)$,

$$Y_c(\lambda_n) = \frac{\sin \lambda_n}{\lambda_n} - \frac{\varphi^2}{\lambda_n^2} \sum_{k=0}^{\infty} Y_c(\lambda_k)[Y_c(\lambda_k + \lambda_n) + Y_c(|\lambda_k - \lambda_n|)]. \quad (7.3.69)$$

The role of the *Shannon sampling expansion* appears in computing the convolution sum in (7.3.69) where we must note that the arguments $\lambda_k + \lambda_n$ and $\lambda_k - \lambda_n$ would be integer multiples of $\pi$, where the function $Y_c(\lambda_n)$ as in (7.3.64) is defined only for $\lambda_n = (n + \frac{1}{2}\pi)$. For this we can interpolate for $Y_c(n\pi)$ in terms of $Y_c((n + \frac{1}{2})\pi)$. and use the following Shannon sampling expansion

$$Y_c(\lambda) = \sum_{n=-\infty}^{\infty} Y_c(\lambda_n) \frac{\sin(\lambda - \lambda_n)}{(\lambda - \lambda_n)}. \quad (7.3.70)$$

### The Cylindrical Pellet and Bessel Sampling Series

In a manner parallel to the development of the non-linear problem for the planar pellet (7.3.58)-(7.3.60), we find the need for the tools of the generalized convolution theorem (7.1.36) and the generalized (Bessel) sampling series (7.1.26) for the *cylindrical pellet*. Here we consider the same type problem of the non-linear chemical concentration $u(r)$ for a *cylindrical catalyst pellet* with a quadratic non-linear term $\varphi^2 u^2$:

$$\frac{1}{r}\frac{d}{dr}\left[r\frac{du}{dr}\right] = \varphi^2 u^2. \qquad 0 < r < 1. \qquad (7.3.71)$$

$$u_r(0) = 0, \qquad u_r \equiv \frac{du}{dr}, \qquad (7.3.72)$$

$$u(1) = 1. \qquad (7.3.73)$$

We will use the finite $J_0$-Hankel transform to algebraize the differential operator on the left side of (7.3.71),

$$U(j_{0,n}) = \int_0^1 r u(r) J_0(j_{0,n}r)dr \equiv \mathcal{H}_0\{u\}, \qquad (7.3.74)$$

where $\{j_{0,n}\}$ are the zeros of the $J_0$-Bessel function, i.e., $J_0(j_{0,n}) = 0$, $n = 1, 2, \ldots$ . The inverse Hankel transform is the Fourier-Bessel series of $u(r)$ on $(0, 1)$,

$$u(r) = 2 \sum_{n=1}^{\infty} \frac{U(j_{0,n}) J_0(j_{0,n}r)}{J_1^2(j_{0,n})}. \qquad (7.3.75)$$

Clearly, $U(j_{0,n})$ in (7.3.75) is a transform limited function and the role of the *Bessel sampling series* is evident if we are to obtain values of $U(\lambda)$ at $\lambda$ different from $j_{0,n}$. More important is the use of the generalized convolution theorem (7.1.36) for us to be satisfied with a general transform result in parallel to that of the modified iteration (7.3.56) in the Fourier transform space. Due to the lack of space, we can only mention that the *modified iterative method* was successful in solving this problem for high values of $\varphi$ [460]. For more details the reader may consult the tutorial article on the subject [455], or [460]. The basic subject of the convergence is discussed briefly in [451] (see also Ref. 5 in [455]). We also mention here that for efficient computations of the above finite Hankel transform (7.3.74) and its inverse (7.3.75), we employed an "approximate" discrete Hankel transform [442] with good accuracy. This discrete Hankel transform was developed with the help of the Bessel sampling series. The same modified iterative method, with parallel analysis, was used for spherical pellet [460, 455]. These pellet problems were also tried with success [455, 460] for general non-linearity $f(y)$ with some conditions on $f(y)$.

# Acknowledgments

The author would like to thank Professor Robert J. Marks II, the editor of the volume, for inviting me to write this chapter on my own contributions to the generalized sampling theorem. Thanks to my wife Suad and daughter Huda for their patience and continued support. Thanks are also due to Mrs. S. A. Khan for typing the original manuscript with care and patience.

# Appendix A

## A.1 Analysis of Gibbs' Phenomena

As we indicated in Subsection 7.3.2, most of the basic analysis of the well-known Gibbs' phenomenon are relegated to this Appendix. In the following we present a basic treatment of the Gibbs' phenomenon associated with integrals approximation of discontinuous functions. This will be followed by a parallel treatment of the Gibbs' phenomenon for the Fourier series approximation of discontinuous periodic functions.

The simplest example of a function with jump discontinuity in the interior of its domain is the *signum-function* sgn(t) (Fig. A.I):

$$\operatorname{sgn}(t) = \frac{t}{|t|} = \begin{cases} 1, & t > 0 \\ -1, & t < 0. \end{cases} \tag{A.1}$$

The very important *gate function* $p_a(t)$,

$$p_a(t) = \begin{cases} 1, & |t| < a \\ 0, & |t| > a \end{cases} \tag{A.2}$$

with its jump discontinuities at $t = \mp a$ (see Fig. 7.2), can also be expressed in terms of the signum function as

$$p_a(t) = \frac{1}{2}[\operatorname{sgn}(t + a) - \operatorname{sgn}(t - a)]. \tag{A.3}$$

In the following discussion we will concentrate our effort on this basic signum function and the Gibbs' phenomenon associated with its truncated Fourier (integral) representation. For the Gibbs' phenomenon of the truncated series we will use the square wave function of (7.3.11), (A.16) as seen in Fig. 7.5 with its jump discontinuities at $x = 0, x = \mp \pi$. We shall concentrate here on the analysis of the Gibbs phenomenon for both the Fourier integral and the Fourier series representation. However, due to lack of space, we shall in both cases refer to only the illustration of the latter in Figs. 7.6-7.8 and 7.14. More details including this treatment are in [456] (see also [380]).

Let us recall the Fourier integral representation of the gate function $p_a(t)$, of Fig. 7.2,

$$p_a(t) = \int_{-\infty}^{\infty} \frac{\sin 2\pi a f}{\pi f} e^{j2\pi f t} df. \tag{A.4}$$

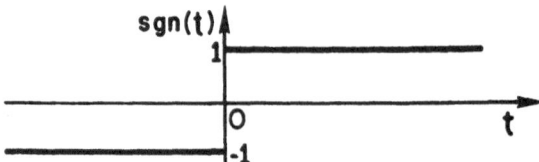

FIGURE A.I: The signum function sgn($t$) (Most of the figures and analysis here are from Jerri, *Integral and Discrete Transforms with Applications and Error Analysis*, 1992, courtesy of Marcel Dekker Inc.).


This gives us a means of computing for $p_a(t)$ via its Fourier transform $\frac{\sin 2\pi a f}{\pi f}$. However, in the actual numerical computations of this infinite integral representation (A.4), the moment we truncate its infinite limits to $-A, A$, with a gate function $p_A(f)$, we expect a Gibbs' phenomenon for the resulting approximation of $p_a(t)$ near its jump discontinuities at $t = \mp a$. This is a windowing effect on $p_a(t)$ with a window $p_A(f)$, which results in convolving $p_a(t)$ with $\frac{\sin 2\pi A t}{\pi t}$, the transform of $p_A(f)$ to give $p_{a,A}(t)$,

$$
\begin{aligned}
p_{a,A}(t) &= p_a(t) * \frac{\sin 2\pi A t}{\pi t} \\
&= \int_{-\infty}^{\infty} \frac{\sin 2\pi A(t-\tau)}{\pi(t-\tau)} p_a(\tau) d\tau.
\end{aligned}
\tag{A.5}
$$

This result represents an exposure to convolving a continuous function $\frac{\sin 2\pi A t}{\pi t}$ with a function $p_a(t)$ that has jump discontinuities. It is such *jump discontinuities* that will contribute an extra term $\gamma(t)$ as the source of the Gibbs' phenomenon. Such a source of error $\gamma(t)$ can be well understood when isolated in the very basic representation of the signum function sgn($t$): once isolated, it is a simple matter to translate our understanding to the case of $p_a(t)$.

We will use the following integral:

$$
\int_0^{\infty} \frac{\sin x}{x} dx = \frac{\pi}{2},
\tag{A.6}
$$

which can be computed by direct complex integration and which makes the basis for the Fourier representation of sgn(t).

If we let $x = 2\pi f t$ in the above integral, we have

$$
\frac{\pi}{2} \int_0^{\infty} \frac{\sin 2\pi f t}{f} df = \begin{cases} 1, & t > 0 \\ -1, & t < 0 \end{cases} \equiv \text{sgn}(t)
\tag{A.7}
$$

as the Fourier (sine) integral representation of sgn($t$). The two different branches of sgn($t$) in (A.7) are easily obtained from (A.6), where the integrand is positive or negative according to $t > 0$ or $t < 0$. If we approximate the integral (A.7) by truncating to $B$ in the upper limit of integration, we obtain the following band-limited function (band-limited to $B$), which we denote by $\text{sgn}_B(t)$, as an approximation to $\text{sgn}(t)$:

$$
\begin{aligned}
\text{sgn}_B(t) &= \frac{2}{\pi} \int_0^B \frac{\sin 2\pi ft}{f} \, df \\
&= \frac{2}{\pi} \int_0^{2\pi Bt} \frac{\sin x}{x} \, dx = \frac{2}{\pi} \, \text{Si}(2\pi Bt) \qquad \text{(A.8)}
\end{aligned}
$$

after letting $x = 2\pi ft$ and recognizing the last truncated integral as the known and well-tabulated *sine integral* $\text{Si}(2\pi Bt)$ (as shown in Fig. A.II for $\text{Si}(t)$).

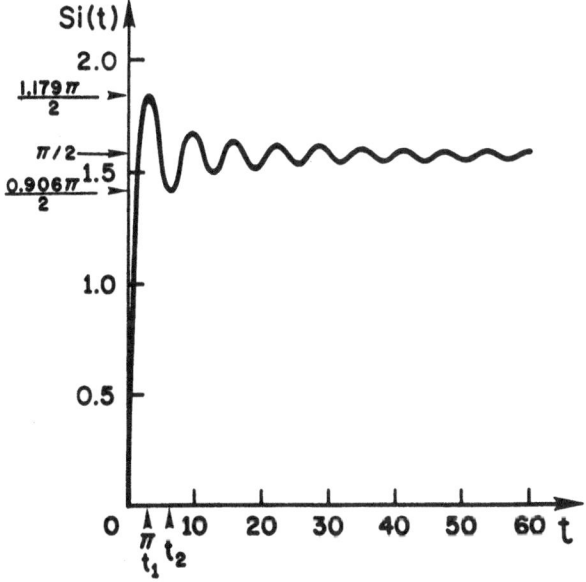

FIGURE A.II: The sine integral $\text{Si}(t) = \int_0^t \sin x/x \, dx$ and the essence of Gibbs' phenomenon near $t = 0$. Note that $\text{sgn}_B(t) = 2/\pi\text{Si}(2\pi Bt)$ in (A.8). See [380].

It is the study of the behavior of this $\text{Si}(2\pi Bt)$, as the truncated approximation of $\text{sgn}(t)$, that would give us the first glimpse at the actual reason behind the Gibbs' phenomenon of the truncated Fourier approximation of functions with jump discontinuities, especially those in the interior of their domain. The first sine integral in (A.8) can be looked at as a representation for the output of passing (first example of a discontinuous function) $\text{sgn}(t)$

through an ideal low-pass filter with a $p_B(t)$ window,

$$\text{sgn}_B(t) = \frac{2}{\pi} \text{Si}(2\pi Bt) = \frac{1}{\pi} \int_{-\infty}^{\infty} \frac{\sin 2\pi B(t-\tau)}{\pi(t-\tau)} \text{sgn}(\tau)d\tau \qquad (A.9)$$

with input $G(t) = \text{sgn}(t)$. Usually $G(t)$ is assumed to be continuous which allows $\lim_{B\to\infty} G_B(t) = G(t)$. In other words, $G_B(t)$ can be brought as close as we desire to the original continuous $G(t)$ by increasing $B$, the width of the truncating gate window. In contrast, as seen in Fig. A.II, the sine integral representation in $\text{sgn}_B(t)$ of (A.8) cannot be brought closer to $\text{sgn}(t)$ by increasing $B$. Indeed, and as we shall illustrate shortly, it turns out that by increasing $B$ we will only change the time scale in Fig. A.II, which will bring the peaks together without changing their relative magnitude. In fact, the size of the first maximum above $\pi/2$ of the Gibbs' phenomenon in Fig. A.II is about 9% of the jump size $J = \pi$ at $t = 0$ regardless of how large we take $B$ to be. To derive and illustrate this phenomenon, we will find the locations and magnitudes of the first maximum and minimum of $\text{Si}(2\pi Bt)$. If we take the first derivative of $\text{Si}(2\pi Bt)$ in (A.8) and equate it to zero, we have

$$\frac{d}{d2\pi Bt} \text{Si}(2\pi Bt) = \frac{\sin 2\pi Bt}{2\pi Bt} = 0,$$

which has zeros at $t_n = \frac{n\pi}{2\pi B} = \frac{n}{2B}$, the locations of the possible maxima and minima of $\text{sgn}_B(t)$. The first maximum is at $t_1 = \frac{1}{2B}$ whose magnitude can be found from substituting $t = t_1 = \frac{1}{2B}$ in (A.8) to have $\frac{2}{\pi} \text{Si}(\pi) = \frac{2}{\pi} \cdot \frac{\pi}{2}(1.17898) = 1.17898$ where $\text{Si}(\pi) = \frac{\pi}{2}(1.17898)$ is found from mathematical tables. Since the size of the jump at $t = 0$ is $\pi$, this maximum would represent 0.17898 which is close to 9% of the jump size as an overshoot from $\text{sgn}(\frac{1}{2B}) = 1$. We must note here that with the location of the extrema $t_n = \frac{n}{2B}$ in (A.8), their magnitudes $\frac{2}{\pi} \text{Si}(n\pi)$ for $\text{sgn}_B(t)$ are independent of $B$, the width of the truncation (gate) window. This should make clear that the Gibbs' phenomenon at a jump discontinuity of any kind cannot be remedied by merely increasing $B$ the width of the truncating gate function window used for $\text{sgn}_B(t)$ in (A.8).

The next extrema is a minimum at $t_2 = \frac{2}{2B}$ with a value from (A.8) as

$$\frac{2}{\pi} \int_0^{2\pi} \frac{\sin x}{x} dx = \frac{2}{\pi} \text{Si}(2\pi) = \frac{2}{\pi} \cdot \frac{\pi}{2}(0.906)$$

which represents $(1 - 0.906)/2 = 0.047$ or about 4.7% of the jump discontinuity at $t = 0$, as an undershoot from $\text{sgn}(\frac{1}{B}) = 1$ (see Fig A.II). The analysis of the Gibbs' phenomenon for more general functions, with jump discontinuities in the interior of the domain, will follow the same way. Such functions can be constructed from the signum function, or its combinations, by adding them to a continuous function. Thus in our attempt to find a possible remedy for the Gibbs' phenomenon, it is sufficient to stay with the more basic signum function.

## A POSSIBLE REMEDY FOR THE GIBBS' PHENOMENON

We have already shown that there is no way that the Gibbs' effect disappears for the truncated Fourier representation $\text{sgn}_B(t)$ of the signum function with its jump discontinuity at zero. So the only alternative is to replace $\text{sgn}(t)$ by a *continuous* function which approaches $\text{sgn}(t)$ as $B \to \infty$. A practical justification for such change of the function is that we may look at two different functions (signals) with the same band limit $B$ and may call them equivalent by insisting that their transforms on $(-B, B)$ carry the same energy $\int_{-B}^{B} |G(f)|^2 df$. The first, and obvious, choice is to replace the part of $\text{sgn}(t)$ on $(-t_1, t_1) = (-\frac{1}{2B}, \frac{1}{2B})$ around $t = 0$ in Fig. 7.3 by a straight line as indicated in Fig. A.III.

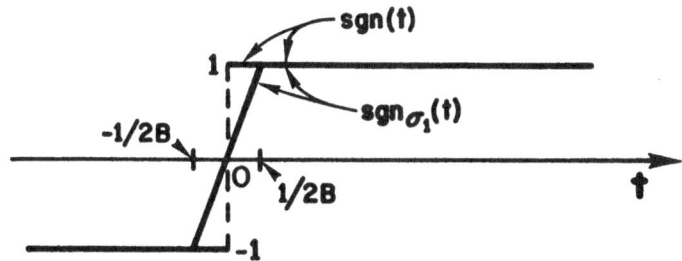

FIGURE A.III: A possible remedy for the Gibbs' phenomenon—approximating $\text{sgn}(t)$ by a continuous function $\text{sgn}_{\sigma_1}(t)$.

The new function $\text{sgn}_{\sigma_1}(t)$ can be shown as the result of convolving $\text{sgn}(t)$ with the gate function $B p_{\frac{1}{2B}}(t)$,

$$
\begin{aligned}
\text{sgn}_{\sigma_1}(t) &= \text{sgn}(t) * B p_{\frac{1}{2B}}(t) \\
&= B \int_{-\infty}^{\infty} p_{\frac{1}{2B}}(t - \tau)\text{sgn}(\tau)d\tau \\
&= \frac{1}{2\frac{1}{2B}} \int_{t-1/2B}^{t+1/2B} \text{sgn}(\tau)d\tau \\
&= \begin{cases} 1, & t > \frac{1}{2B} \\ -1, & t < \frac{1}{2B} \\ 2Bt, & -\frac{1}{2B} < t < \frac{1}{2B} \end{cases}
\end{aligned}
\tag{A.10}
$$

after noting that $p_{1/2B}(t - \tau) = 0$ for $|t - r| > 1/2B$ (or for $\tau < t - 1/2B$ and $\tau > t + 1/2B$), and performing the last integration.

We can see from the last integral in (A.10) that $\text{sgn}_{\sigma_1}(t)$ is also the result of averaging $\text{sgn}(t)$, with constant weight 1, over intervals of width $2 \cdot \frac{1}{2B} =$

$\frac{1}{B}$, where $\frac{1}{2B}$ is the location of the first overshoot (maximum) of the Gibbs' phenomenon near the jump discontinuity at $t = 0$. In the following we will show how such an averaging process as $\mathrm{sgn}_{\sigma_1}(t)$ results in a reduced Gibbs' effect.

The Fourier transform of this averaged function $\mathrm{sgn}_{\sigma_1}(t)$, is the following product of $\frac{1}{j\pi f}$, the transform of $\mathrm{sgn}(t)$, and $B\frac{\sin(\pi/B)f}{\pi f}$ the transform of $Bp_{1/2B}(t)$:

$$F_{\sigma_1}(f) = \frac{1}{j\pi f} B \frac{\sin(\pi/B)f}{\pi f}, \qquad (A.11)$$

$$\mathrm{sgn}_{\sigma_1}(t) = \int_{-\infty}^{\infty} \frac{1}{j\pi f} B \frac{\sin(\pi/B)f}{\pi f} e^{j2\pi ft} dt. \qquad (A.12)$$

where we have used $\mathcal{F}\{\mathrm{sgn}(t)\} = \frac{1}{j\pi f}, j = \sqrt{-1}$.

Now if the truncation for this infinite integral representation of (the continuous) $\mathrm{sgn}_{\sigma_1}(t)$ is done with a gate window $p_A(f)$, we have its band-limited approximation:

$$
\begin{aligned}
\mathrm{sgn}_{\sigma_{1,A}}(t) &= \int_{-A}^{A} \frac{1}{j\pi f} B \frac{\sin(\pi/B)f}{\pi f} e^{j2\pi ft} df \\
&= \int_{-\infty}^{\infty} \frac{\sin 2\pi A(\pi t - \tau)}{\pi(t - \tau)} \mathrm{sgn}_{\sigma_1}(\tau) d\tau. \qquad (A.13)
\end{aligned}
$$

In contrast to $\mathrm{sgn}_B(t)$ in (A.9), where the input to the truncation gate window was a *discontinuous* $\mathrm{sgn}(t)$, we have here its modification (averaged) as a *continuous* input $\mathrm{sgn}_{\sigma_1}(t)$. This would allow $\mathrm{sgn}_{\sigma_{1,A}}(t)$ to get as close as we wish to $\mathrm{sgn}_{\sigma_1}(t)$ of $\mathrm{sgn}(t)$.

It is important to find the relation between $\mathrm{sgn}_B(t)$ of (A.8), with its Gibbs' phenomenon, and $\mathrm{sgn}_{\sigma_{1,A}}(t)$, the approximation to $\mathrm{sgn}_{\sigma_1}(t)$ the continuous replacement of $\mathrm{sgn}(t)$.

From the first integral of (A.13) we have

$$\mathrm{sgn}_{\sigma_{1,A}}(t) = 2 \int_0^A B \frac{\sin(\pi/B)f}{\pi f} \frac{\sin 2\pi ft}{\pi f} df, \qquad (A.14)$$

since the Fourier-transformed function in the first integral in (A.13) is an odd function. If we choose $A = B$, we can compare the above result with that of $\mathrm{sgn}_B(t)$ in (A.8) :

$$\mathrm{sgn}_B(t) = \frac{2}{\pi} \int_0^{2\pi Bt} \frac{\sin x}{x} dx.$$

With $A = B$ and $2\pi ft = x$ in (A.14) we have

$$
\begin{aligned}
\mathrm{sgn}_{\sigma_{1,B}}(t) &= 2 \int_0^B \frac{\sin(\pi/B)f}{(\pi/B)f} \frac{\sin 2\pi ft}{\pi f} df \\
&= \frac{2}{\pi} \int_0^{2\pi Bt} \frac{\sin(1/2Bt)x}{(1/2Bt)x} \frac{\sin x}{x} dx. \qquad (A.15)
\end{aligned}
$$

If we compare (A.8) with (A.15), we note that the above averaging or smoothing process for managing the Gibbs' phenomenon of $\text{sgn}(t)$, as in (A.15), introduces a decaying factor $\frac{\sin(1/2Bt)x}{(1/2Bt)x}$ inside the sine integral. This factor seems to have brought the sine integral to a balance close to $\text{sgn}(t)$. For example, at $t_1 = \frac{1}{2B}$, where (A.8) has a maximum as $\text{sgn}_B(1/2B) = \frac{2}{\pi}\,\text{Si}(\pi)$, we now have from (A.15)

$$\text{sgn}_{\sigma_1,B}\left(\frac{1}{2B}\right) = \frac{2}{\pi}\int_0^\pi \left[\frac{\sin x}{x}\right]^2 dx,$$

where the integrand decays much faster than the $\sin(x)/x$ of $\text{sgn}_B(t)$, hence yielding a lower value than that of $\frac{2}{\pi}\,\text{Si}(\pi)$.

The first integral (A.13) suggests that the averaging process seems to have made the Fourier transform of the original function tend more toward a band-limited function (see the extra $1/f$ factor in the integrand). This is in general due to the convolution by $p_{1/2B}(t)$ in the time space, which enlarges the domain of the resulting function (see (A.10)) and according to the uncertainty principle causes a narrowing in the frequency space domain. This is in the direction of helping the truncation process, which is good as long as it does not change the main character of the original signum function, in particular the fast rise from 0 to 1 near $t = 0$. In the present case the averaging is done with the flat top $p_{1/2B}(t)$ which has affected the original function only to close the discontinuity at $t = 0$, where it replaced the jump discontinuity by a line of slope $2B$.

What remains is to see if, in the light of the above analysis for the Gibbs' phenomenon, we can suggest further improvement to the above averaging remedy. We precede such improvement by stating that the above $\sigma_1$-averaging process moves the location of the extrema further away from the discontinuity as can be shown from differentiating $\text{sgn}_{\sigma_1,B}(t)$ in (A.15) (see (A.8)).

The most direct way for further improvement is to repeat the averaging process, which should decrease the magnitude of the extrema. The $n$ times repeated averaging of $\text{sgn}(t)$ would amount to convolving it $n$ times with $p_{1/2B}(t)$. This is equivalent to convolving $\text{sgn}(t)$ with the hill function $\phi_n(t)$ (B-spline) of order $n$, since the latter is defined as the $n - 1$th self-convolution of the gate function $p_{1/2B}(t)$. We could also look at this process as averaging with a weight which is a spline of order $n$ instead of (the flat top) $B\phi_1(t) = Bp_{1/2B}(t)$ used in the above analysis. The convolution of $\text{sgn}(t)$ with $B\phi_n(t)$ would amount to multiplying inside its integral in (A.15) by $\sigma_n = \left[\frac{\sin(\pi/B)f}{(\pi/B)f}\right]^n$ as the transform of $B^n\phi_n(t) = B^n(p_{1/2B}* \overset{n-1}{\cdots} *p_{1/2B})(t)$ .

There is no doubt that such a self-truncating factor inside the integral would improve its convergence to give smoother results. However, the wider domain of the higher order hill function $\varphi_n(t)$ would cause a slower rise for the approximate function near the jump discontinuity. This is not an extremely difficult problem since it can be remedied by increasing $B$ which, of course, will increase the number of computations. The latter is not much of a price to pay when we are using the fast Fourier transform (FFT). Figure A.III (which is the same as Fig. 7.13 in Subsection 7.3.2) also illustrates such remedy for the most familiar case of the Gibb's phenomenon in the (truncated) Fourier series approximation of the square wave function near its jump discontinuity at $x = 0$, where it was discussed in Subsection 7.3.2.

In the next section, we will treat the Gibbs' phenomenon for the Fourier series approximation of (periodic) functions with jump discontinuities. The analysis will follow much in parallel to that of the integrals, where the $\sigma_n$ factor improvements are illustrated for the (periodic) square wave of Fig. 7.5. In essence, the numerical computations show a larger rise time near $t = 0$, but much smoother approximation with $\sigma_2$ than that of $\sigma_1$ (Fig. 7.8). With $\sigma_6$ the rise time doubled that of $\sigma_3$ but the accuracy is exceedingly better when the same number of terms was used for both cases (see Fig. 7.14).

## THE GIBBS' PHENOMENON IN FOURIER SERIES APPROXIMATION

Now we turn to the usual Gibbs' phenomenon of the truncated Fourier series for periodic function with jump discontinuities. The example we take is that of the square wave of unit amplitude on $(-\pi. \pi)$ as illustrated in Fig. 7.5,

$$
f(t) = \begin{cases} 1, & 0 < t < \pi \\ -1, & -\pi < t < 0, \end{cases} \tag{A.16}
$$

whose Fourier sine series is

$$
f(t) \sim \frac{4}{\pi} \sum_{n=1}^{\infty} \frac{1}{2n-1} \sin(2n-1)t. \tag{A.17}
$$

The $N$th partial sum,

$$
S_N(t) = \frac{4}{\pi} \sum_{n=1}^{N} \frac{1}{2n-1} \sin(2n-1)t. \tag{A.18}
$$

is illustrated for $N = 5$ in Fig. 7.6, and for $N = 10, 40$ in Figs. 7.7 and 7.8.

To find the locations of the extrema of the Gibbs' phenomenon of $S_N(t)$ in (A.18), we will express the $S_N(t)$ as an integral, then follow the last section's analysis used with the Fourier integral approximation of the signum function in (A.8).

Inside the partial sum $S_N(t)$ in (A.18) we can write

$$
\frac{\sin[(2n-1)t]}{(2n-1)} = \int_0^t \cos[(2n-1)x]dx \tag{A.19}
$$

to have

$$S_N(t) = \frac{4}{\pi} \sum_{n=1}^{N} \int_0^t [\cos(2n-1)x] dx;  \tag{A.20}$$

and after interchanging the integral and sum we have

$$S_N(t) = \frac{4}{\pi} \int_0^t \left[ \sum_{n=1}^{N} \cos(2n-1)x \right] dx.  \tag{A.21}$$

But we can also show that the sum inside (A.21) is $\sin 2Nx/2\sin x$ by simply writing

$$
\begin{aligned}
2\sin x \sum_{n=1}^{N} \cos(2n-1)x &= \sum_{n=1}^{N} 2\sin x \cos(2n-1)x \\
&= \sum_{n=1}^{N} [\sin(2n-1+1)x + \sin(1-2n+1)x] \\
&= \sum_{n=1}^{N} [\sin 2nx - \sin(2n-2)x] \\
&= \sin 2x + \sin 4x - \sin 2x + \sin 6x - \\
&\quad \cdots + \sin 2Nx - \sin(2N-2)x \\
&= \sin 2Nx  \tag{A.22}
\end{aligned}
$$

after using simple trigonometric identities that helped in the cancellation of all terms except $\sin 2Nx$. From (A.21) and (A.22), we have

$$S_N(t) = \frac{2}{\pi} \int_0^t \frac{\sin 2Nx}{\sin x} dx.  \tag{A.23}$$

To search for the extrema of the Gibb's phenomenon, as we did for the truncated Fourier integrals in (A.8), we differentiate $S_N(t)$ with respect to $t$ to have

$$\frac{dS_N(t)}{dt} = \frac{2}{\pi} \frac{\sin 2Nt}{\sin t}$$

which has its extrema at $t_k = \frac{k\pi}{2N}$ with the first maximum at $t_1 = \frac{\pi}{2N}$, as may be seen in Fig. 7.7 with $N = 10$.

The magnitude of the first maximum can be evaluated from (A.23) with $t_1 = \frac{\pi}{2N}$ as

$$S_N(\frac{\pi}{2N}) = \frac{2}{\pi} \int_0^{\pi/2N} \frac{\sin 2Nx}{\sin x} dx,  \tag{A.24}$$

which can be computed numerically. If $N$ is large enough, then we are integrating in (A.24) over a small interval $(0, \frac{\pi}{2N})$ where $x$ is small enough to allow the approximation of $\sin x \sim x$, and thus the integral (A.24) can be approximated by the following sine integral:

$$
S_N\left(\frac{\pi}{2N}\right) \sim \frac{2}{\pi}\int_0^{\pi/2N}\frac{\sin 2Nx}{\sin x}dx. \quad N \text{ large}
$$

$$
= \frac{2}{\pi}\int_0^{\pi/2N}\frac{\sin y}{y}dy = \frac{2}{\pi}\,\mathrm{Si}(\pi)
$$

$$
= \frac{2}{\pi}\cdot\frac{\pi}{2}(1.17898) = 1.17898,
$$

as we had found before using $\mathrm{Si}(\pi)$. It is interesting to note that for the Fourier integrals the maximum truncation error was 9% and was independent of the size $2\pi B$ of the truncation window, whereas for the Fourier series here it is a lower bound of 9% attained as $N \to \infty$. For smaller $N$, we see $\sin x < x$ which makes $S_N(\frac{\pi}{2N})$ larger than 1.18, hence an even larger overshoot of the Gibbs' effect. The same type of analysis can be done for the Gibb's phenomenon near the discontinuity at $t = \pi$. Figure 7.7 illustrates this Gibbs' phenomenon with $N = 10$ as well as the following $\sigma_1$ averaging for reducing it.

As to the remedy for removing this Gibbs' phenomenon, it should be of the same nature as the one we used for the Fourier integral representation, namely, to modify by removing the jump discontinuity. This means that we approximate the discontinuous function by a function which is continuous in a small neighborhood around the jump discontinuity at $x = 0$ by a straight line on the interval $(-\frac{\pi}{2N}, \frac{\pi}{2N})$ as we did for the signum function $\mathrm{sgn}(t)$ around $t = 0$ in Fig. A.III. This again would mean averaging the function over intervals of width $\frac{\pi}{N}$ or convolving the function with $p_{\pi/2N}(t)$. The latter modification is also equivalent to multiplying the Fourier coefficients $\frac{4}{\pi}\cdot\frac{1}{2n-1}$ of the original function in (A.18) by $\frac{\sin(2n-1)\frac{\pi}{2N}}{(2n-1)\frac{\pi}{2N}}$ (as Fourier coefficients of the Fourier expansion of $Np_{\pi/2N}(t)$ on $(-\pi, \pi)$). Therefore, the $\sigma_1$-average of the partial sum $S_N(t)$ becomes

$$
S_{\sigma_1,N}(t) = \frac{N}{\pi}\int_{t-\pi/2N}^{t+\pi/2N} S_N(x)dx
$$

$$
= \frac{N}{\pi}\int_{t-\pi/2N}^{t+\pi/2N}\frac{4}{\pi}\sum_{n=1}^{N}\frac{1}{2n-1}\sin(2n-1)x
$$

$$
= \frac{4}{\pi}\frac{N}{\pi}\sum_{n=1}^{N}\frac{1}{2n-1}\int_{t-\pi/2N}^{t+\pi/2N}\sin(2n-1)x\,dx
$$

$$
= \frac{4}{\pi}\frac{N}{\pi}\sum_{n=1}^{N}\frac{1}{2n-1}\left.\frac{-\cos(2n-1)x}{(2n-1)}\right|_{x=t-\frac{\pi}{2N}}^{x=t+\frac{\pi}{2N}}
$$

$$= \frac{4}{\pi} \sum_{n=1}^{N} \frac{1}{2n-1} \left[ \frac{\sin(2n-1)\frac{\pi}{2N}}{(2n-1)\frac{\pi}{2N}} \right] \sin(2n-1)t. \quad (A.25)$$

As we mentioned in Subsection 7.3.2, we observe that the averaging process introduces a $\frac{\sin(2n-1)\frac{\pi}{2N}}{(2n-1)\frac{\pi}{2N}}$ "decaying" factor to the original coefficients of the series (A.18), which is exactly analogous to the development of the Fourier integrals in (A.15). This $\sigma_1$-averaging of the Gibbs' phenomenon is illustrated near the jump discontinuity $t = 0$ in Fig. 7.7, where a ten-term ($N = 10$) series was used.

## A Possible Improvement for Reducing the Gibbs' Phenomenon

As to further improvements in eliminating the Gibbs' phenomenon, we suggest the same repeated averagings that were employed in the previous case of the Fourier integral representation. The $m$ repeated averagings are equivalent to convolving the function $m$ times with $Np_{\pi/2N}(t)$, which in turn will introduce the self-truncating factor

$$\sigma_m = \left[ \frac{\sin(2n-1)\frac{\pi}{2N}}{(2n-1)\frac{\pi}{2N}} \right]^m \quad (A.26)$$

to the Fourier coefficients of $S_N(t)$ in (A.18). Figure 7.8 illustrates the Gibbs' phenomenon and the improvements attained by applying $\sigma_1$ and $\sigma_2 = \sigma_1^2$ averagings with $N = 40$. The important initial case of $m = 1$ is due to Lancos [380]. Figure 7.14 clarifies how smooth the higher averagings ($m = 2, 3$ and $6$) are compared to the $\sigma_1$-averaging.

We note that with the $\sigma_1$ factor, the oscillations of the Gibbs' phenomenon decrease; moreover, with $\sigma_3$, in Fig. 7.14, we observe that they have almost disappeared. The only minor drawback to $\sigma_3$ and its excellent accuracy is that it has about double the rise time as that of $\sigma_1$. However, this can be made up for by doubling $N$ to have the same or better rise time than that of $\sigma_1$ as we illustrate in Fig. 7.14 for $\sigma_1$ with $N = 20$ and $\sigma_3$ with $N = 40$. In Fig. 7.14 we also illustrate the $\sigma_6$ case with $N = 80$ while it still has double the rise time compared to that of $\sigma_3$ with $N = 40$, it also has very high accuracy for approximating the original square wave function where it is continuous.

### THE GIBBS' PHENOMENON FOR GENERAL ORTHOGONAL EXPANSION AND ITS POSSIBLE REMEDY

As discussed in Subsection 7.3.2, the above detailed analysis of the Gibbs phenomenon in this Appendix was done here, primarily, for the purpose of setting the background for discussions and parallel analytical treatment regarding the same Gibbs' type phenomena for the general orthogonal expansion. Such illustrations and discussions were done for the Fourier-Bessel series, as an example, at the end of Subsection 7.3.2 (see Figs. 7.9, 7.12 and the figures in [366]).

In summary, we presented the $\sigma^m$-averaging as a remedy to the Gibbs phenomenon of the (trigonometric) Fourier series (or integrals). We used the transform of the hill function $\phi_{m+1}(x)$ in (7.1.47) (on a very narrow window around the jump discontinuity) as the modifying factor inside the series representing the function with jump discontinuity.

In a very close parallel to this treatment, the remedy for the Gibbs phenomenon of the general orthogonal expansions (or transforms) lies in using the transform of the generalized hill function $\psi_{m+1}(x)$ in (7.1.52) for a modifying (decaying) factor inside the series (or integral) representing the function with jump discontinuity. More recent tentative results in this direction are promising.

# Bibliography

[1] J.B. Abbiss, C. De Mol, and H.S. Dhadwal. Regularized iterative and non-iterative procedures for object restoration from experimental data. *Opt. Acta*, **pages** 107–124, 1983.

[2] J.B. Abbiss, B.J. James, J.S. Bayley, and M.A. Fiddy. Super-resolution and neural computing. *Proc. SPIE, Int. Soc. Opt. Eng.*, pages 100–106, 1988.

[3] M. Abramowitz and I.A. Stegun. *Handbook of Mathematical Functions*. Dover, New York, 9th ed., 1964.

[4] H.C. Agarwal. Heat transfer in laminar flow between parallel plates at small Peclet numbers. *Appl. Sci. Res., Vol. 9*, pages 177–189, 1960.

[5] T.H. Ahberg, E.N. Nilson, and T.L. Walsh. *The Theory of Splines and Their Applications*. Academic Press, New York, 1964.

[6] C.B. Ahn, J.H. Kim, and Z.H. Cho. High-speed spiral echo planar NMR imaging – I. *IEEE Trans. Medical Imaging, Vol. MI-5*, pages 2–7, 1986.

[7] S. T. Alexander. Fast Adaptive Filters: A Geometrical Approach *IEEE ASSP Magazine*, pages 18–28, October 1986.

[8] J.P. Allebach. Analysis of sampling pattern dependence in time-sequential sampling of spatio-temporal signals. *J. Opt. Soc. Am., Vol. 71*, pages 99–105, 1981.

[9] H.C. Andrews and C.L. Patterson. Singular value decompositions and digital image processing. *IEEE Trans. Acoust., Speech, Signal Processing*, pages 26–53, 1976.

[10] K. Araki. Sampling and recovery formula in finite Hankel transform based on the minimum norm principle. *Trans. Inst. Electron. Commun. Eng. Jpn., Vol. J68A*, pages 643–649, 1985.

[11] H. Arsenault. Diffraction theory of Fresnel zone plates. *J. Opt. Soc. Am., Vol. 58*, page 1536, 1968.

[12] H. Arsenault and A. Boivin. An axial form of the sampling theorem and its application to optical diffraction. *J. Appl. Phys., Vol. 38*, pages 3988–3990, 1967.

[13] H. Arsenault and A. Boivin. Optical filter synthesis by holographic methods. *J. Opt. Soc. Am., Vol. 58*, pages 1490–1493, 1968.

[14] H. Arsenault and K. Chalasinska-Macukow. The solution to the phase retrieval problem using the sampling theorem. *Opt. Comm., Vol. 47*, pages 380–386, 1983.

[15] H. Arsenault and B. Genestar. Deconvolution of experimental data. *Can. J. Phys.*, Vol. *49*, pages 1865–1868, 1971.

[16] R.B. Ash. *Information Theory.* Interscience, New York, 1965.

[17] C. Atzeni and L. Masotti. A new sampling procedure for the synthesis of linear transversal filters. *IEEE Trans. Aerospace & Electronic Systems,* Vol. *AES-7*, pages 662–670, 1971.

[18] T. Auba and Y. Funahashi. The structure of all-pass matrices that satisfy directional interpolation requirements. *IEEE Trans. Automatic Control,* Vol. *36*, pages 1485–1489, 1991.

[19] T. Auba and Y. Funahashi. The structure of all-pass matrices that satisfy two-sided interpolation requirements. *IEEE Trans. Automatic Control,* Vol. *36*, pages 1489–1493, 1991.

[20] L. Auslander and R. Tolimieri. Characterizing the radar ambiguity functions. *IEEE Trans. Information Theory,* Vol. *IT-30*, pages 832–836, 1984.

[21] G.S. Azarov. Transmission of analog messages in a delta-modulation system when the sampling is not uniform. *Telecommunications and Radio Engineering (English Translation of Elektrosvyaz and Radiotekhnika),* Vol. *31-32, No. 2*, pages 82–86, February 1977.

[22] A. Bachl and W. Lukosz. Experiment on super-resolution imaging of a reduced object field. *J. Opt. Soc. Am.,* Vol. *57*, pages 163–169, 1967.

[23] A.V. Balakrishnan. A note on the sampling principle for continuous signals. *IRE Trans. Information Theory,* Vol. *IT-3*, pages 143–146, 1957.

[24] A.V. Balakrishnan. On the problem of time jitter in sampling. *IRE Trans. Information Theory,* Vol. *IT-8*, pages 226–236, 1962.

[25] A.V. Balakrishnan. Essentially bandlimited stochastic processes. *IEEE Trans. Information Theory,* Vol. *IT-11*, pages 145–156, 1965.

[26] I. Bar-David. An implicit sampling theorem for bounded bandlimited functions. *Inform. Contr.,* Vol. *24*, pages 36–44, 1974.

[27] I. Bar-David. Sample functions of a Gaussian process cannot be recovered from their zero crossings. *IEEE Trans. Information Theory,* Vol. *IT-21*, pages 86–87, 1975.

[28] I. Bar-David. On the degradation of bandlimiting systems by sampling. *IEEE Trans. Communications,* Vol. *COM-25*, pages 1050–1052, 1977.

[29] R. Barakat. Application of the sampling theorem to optical diffraction theory. *J. Opt. Soc. Am.,* Vol. *54*, pages 920–930, 1964.

[30] R. Barakat. Determination of the optical transfer function directly from the edge spread function. *J. Opt. Soc. Am.,* Vol. *55*, pages 1217–1221, 1965.

[31] R. Barakat. Nonlinear transformation of stochastic processes associated with Fourier transforms of bandlimited positive functions. *Int. J. Contr.,* Vol. *14, No. 6*, pages 1159–1167, 1971.

[32] R. Barakat. Note on a sampling expansion posed by O'Neill and Walther. *Opt. Comm.,* Vol. *23*, pages 207–208, 1977.

[33] R. Barakat. Shannon numbers of diffraction images. *Opt. Comm.*, pages 391–394, 1982.

[34] R. Barakat and E. Blackman. Application of the Tichonov regularization algorithm to object restoration. *Opt. Comm.*, pages 252–256, 1973.

[35] R. Barakat and A. Houston. Line spread and edge spread functions in the presence of off-axis aberrations. *J. Opt. Soc. Am., Vol. 55*, pages 1132–1135, 1965.

[36] R. Barakat and J.E. Cole III. Statistical properties of $n$ random sinusoidal waves in additive Gaussian noise. *J. Sound and Vibration, Vol. 63*, pages 365–377, 1979.

[37] L.A. Baranov. Error estimates of the restoration of continuous random signals when the sampling is irregular. *Telecommunication and Radio Engineering (English Translation of Elektrosvyaz and Radiotekhnika), Vol. 37-38, No. 8*, pages 37–39, August 1983.

[38] V. Bargmann, P. Butera, L. Girardello, and J.R. Klauder. On the completeness of the coherent states, *Rep. Math. Phys., 2*, pages 221–228, 1971.

[39] H.A. Barker. Synchronous sampling theorem for nonlinear systems. *Electron. Lett., Vol. 5*, page 657, 1969.

[40] C.W. Barnes. Object restoration in a diffraction-limited imaging system. *J. Opt. Soc. Am., Vol. 56*, pages 575–578, 1966.

[41] H.O. Bartelt, K.H. Brenner, and A.W. Lohmann. The Wigner distribution function and its optical production. *Opt. Comm., Vol. 32*, pages 32–38, 1980.

[42] H.O. Bartelt and J. Jahns. Interferometry based on the Lau effect. *Opt. Comm.*, pages 268–274, 1979.

[43] M.J. Bastiaans. A frequency-domain treatment of partial coherence. *Opt. Acta, Vol. 24*, pages 261-274, 1977.

[44] M.J. Bastiaans. A generalized sampling theorem with application to computer-generated transparencies. *J. Opt. Soc. Am., Vol. 68*, pages 1658–1665, 1978.

[45] M.J. Bastiaans. The Wigner distribution function applied to optical signals and systems. *Opt. Comm., Vol. 25*, pages 26–30, 1978.

[46] M.J. Bastiaans. An approximate realization of band-limited functions by low-pass filtering of width-modulated pulses, in *Optica Hoy y Mañana*, Proceedings of the 11th Congress of the International Commission for Optics, ed. J. Bescos, A. Hidalgo, L. Plaza, and J. Santamaría, Sociedad Española de Optica, Madrid, 1978, pages 327–330.

[47] M.J. Bastiaans. The Wigner distribution function and Hamilton's characteristics of a geometrical-optical system. *Opt. Comm., Vol. 30*, pages 321–326, 1979.

[48] M.J. Bastiaans. Gabor's expansion of a signal into Gaussian elementary signals. *Proc. IEEE*, pages 538–539, 1980.

[49] M.J. Bastiaans. Sampling theorem for the complex spectrogram, and Gabor's expansion of a signal in Gaussian elementary signals, in *1980 International Optical Computing Conference*, ed. W.T. Rhodes, *Proc. SPIE, Vol. 231*, pages 274–280, 1980 (also published in *Opt. Eng., Vol. 20*, pages 594–598, 1981).

[50] M.J. Bastiaans. The expansion of an optical signal into a discrete set of Gaussian beams. *Optik, Vol. 57*, pages 95–102, 1980.

[51] M.J. Bastiaans. Wigner distribution function display: a supplement to ambiguity function display using a single 1–D input. *Appl. Opt., Vol. 19*, page 192, 1980.

[52] M.J. Bastiaans. A sampling theorem for the complex spectrogram, and Gabor's expansion of a signal in Gaussian elementary signals. *Opt. Eng., Vol. 20*, pages 594–598, 1981.

[53] M.J. Bastiaans. Gabor's signal expansion and degrees of freedom of a signal. *Opt. Acta, Vol. 29*, pages 1223–1229, 1982.

[54] M.J. Bastiaans. Optical generation of Gabor's expansion coefficients for rastered signals. *Opt. Acta, Vol. 29*, pages 1349–1357, 1982.

[55] M.J. Bastiaans. Signal description by means of a local frequency spectrum. *Proc. SPIE, Int. Soc. Opt. Eng., Vol. 373*, pages 49–62, 1981.

[56] M.J. Bastiaans. Gabor's signal expansion and degrees of freedom of a signal. *Opt. Acta, Vol. 29*, pages 1223–1229, 1982.

[57] M.J. Bastiaans. Use of the Wigner distribution function in optical problems. *1984 European Conference on Optics, Optical Systems and Applications, Proc. SPIE, Int. Soc. Opt. Eng., Vol. 492*, pages 251–262, 1984.

[58] M.J. Bastiaans. On the sliding-window representation in digital signal processing. *IEEE Trans. Acoust., Speech, Signal Processing, Vol. ASSP-33*, pages 868–873, 1985.

[59] M.J. Bastiaans. Application of the Wigner distribution function to partially coherent light. *J. Opt. Soc. Am. A, Vol. 3*, pages 1227–1238, 1986.

[60] M.J. Bastiaans. Error reduction in computer-generated half-tone transparencies that are based on one-dimensional pulse-width modulation. *Progress in Holography*, ed. J. Ebbeni, *Proc. SPIE, Vol. 812*, pages 100–107, 1987.

[61] M.J. Bastiaans. Error reduction in two-dimensional pulse-area modulation, with application to computer-generated transparencies. *14th Congress of the International Commission for Optics*, ed. H.H. Arsenault, *Proc. SPIE, Vol. 813*, pages 341–342, 1987.

[62] M.J. Bastiaans. Error reduction in one-dimensional pulse-area modulation, with application to computer-generated transparencies. *J. Opt. Soc. Am. A, Vol. 4*, pages 1879–1886, 1987.

[63] M.J. Bastiaans. Local-frequency descriptions of optical signals and systems. EUT Report 88–E–191 (Faculty of Electrical Engineering, Eindhoven University of Technology, Eindhoven, Netherlands, 1988).

[64] M.J. Bastiaans. Gabor's signal expansion applied to partially coherent light. *Opt. Comm., Vol. 86*, pages 14–18, 1991.

[65] M.J. Bastiaans. On the design of computer-generated transparencies based on area modulation of a regular array of pulses. *Optik, Vol. 88*, pages 126–132, 1991.

[66] M.J. Bastiaans. Second-order moments of the Wigner distribution function in first-order optical systems. *Optik, Vol. 88*, pages 163–168, 1991.

[67] M.J. Bastiaans, M.A. Machado, and L.M. Narducci. The Wigner distribution function and its applications in optics. *Optics in Four Dimensions-1980, AIP Conf. Proc. 65*, eds. (American Institute of Physics, New York, 1980), pages 292-312, 1980.

[68] J.S. Bayley and M.A. Fiddy. On the use of the Hopfield model for optical pattern recognition. *Opt. Comm.*, pages 105–110, 1987.

[69] G.A. Bekey and R. Tomovic. Sensitivity of discrete systems to variation of sampling interval. *IEEE Trans. Automatic Control, Vol. AC-11*, pages 284–287, 1966.

[70] B.W. Bell and C.L. Koliopoulos. Moiré topography, sampling theory, and charged-coupled devices. *Opt. Lett., Vol. 9*, pages 171–173, 1984.

[71] V.K. Belov and S.N. Grishkin. Digital bandpass filters with nonuniform sampling. *Priborostroemie, Vol. 23, No. 3 (in Russian)*, pages 3–7, March 1980.

[72] V.I. Belyayev. Processing and theoretical analysis of oceanographic observations. *Translation from Russian JPRS 6080*, 1973.

[73] M. Bendinelli, A. Consortini, L. Ronchi, and R.B. Frieden. Degrees of freedom and eigenfunctions of the noisy image. *J. Opt. Soc. Am., Vol. 64*, pages 1498–1502, 1974.

[74] M. Bertero. Linear inverse and ill-posed problems. In P.W. Hawkes, editor, *Advances in Electronics and Electron Physics*. Academic Press, New York, 1989.

[75] M. Bertero, P. Boccacci, and E.R. Pike. Resolution in diffraction-limited imaging, a singular value analysis. II. The case of incoherent illumination. *Opt. Acta*, pages 1599–1611, 1982.

[76] M. Bertero, C. de Mol, E.R. Pike, and J.G. Walker. Resolution in diffraction-limited imaging, a singular value analysis. IV. The case of uncertain localization or non-uniform illumination of the object. *Opt. Acta*, pages 923–946, 1984.

[77] M. Bertero, C. De Mol, and E.R. Pike. Analytic inversion formula for confocal scanning microscopy. *J. Opt. Soc. Am. A*, pages 1748–1750, 1987.

[78] M. Bertero, C. De Mol, and G.A. Viano. Restoration of optical objects using regularization. *Opt. Lett.*, pages 51–53, 1978.

[79] M. Bertero, C. De Mol, and G.A. Viano. On the problems of object restoration and image extrapolation in optics. *J. Math. Phys.*, pages 509–521, 1979.

[80] M. Bertero and E.R. Pike. Resolution in diffraction-limited imaging, a singular value analysis. I. The case of coherent illumination. *Opt. Acta*, pages 727–746, 1982.

[81] M. Bertero and E.R. Pike. Exponential sampling method for Laplace and other dilationally invariant transforms: I. Singular system analysis. *Inverse Problems*, pages 1–20, 1991.

[82] M. Bertero and E.R. Pike. Exponential sampling method for Laplace and other dilationally invariant transforms: II. Examples in photon correlation spectroscopy and Fraunhofer diffraction. *Inverse Problems*, pages 21–41, 1991.

[83] A.S. Besicovitch. *Almost Periodic Functions*. Wiley Interscience, New York, 1968.

[84] F.E. Beutler. Sampling theorems and bases in a Hilbert space. *Inform. Contr.*, Vol. 4, pages 97–117, 1961.

[85] F.J. Beutler. Error-free recovery of signals from irregularly spaced samples. *SIAM Review, Vol. 8, No. 3*, pages 328–335, 1966.

[86] F.J. Beutler. Alias-free randomly timed sampling of stochastic processes. *IEEE Trans. Information Theory, Vol. IT-16*, pages 147–152, 1970.

[87] F.J. Beutler. Recovery of randomly sampled signals by simple interpolators. *Inform. Contr., Vol. 26, No. 4*, pages 312–340, 1974.

[88] F.J. Beutler. On the truncation error of the cardinal sampling expansion. *IEEE Trans. Information Theory, Vol. IT-22*, pages 568–573, 1976.

[89] F.J. Beutler and D.A. Leneman. Random sampling of random processes: stationary point processes. *Inform. Contr., Vol. 9*, pages 325–344, 1966.

[90] F.J. Beutler and D.A. Leneman. The theory of stationary point processes. *Acta Math, Vol. 116*, pages 159–197, September 1966.

[91] H.S. Black. *Modulation Theory*. Van Nostrand, New York, 1953.

[92] R.B. Blackman and J.W. Tukey. The measurement of power spectra from the point of view of communications engineering. *Bell Systems Tech. J.*, Vol. 37, page 217, 1958.

[93] W.E. Blash. Establishing proper system parameters for digitally sampling a continuous scanning spectrometer. *Appl. Spectroscopy., Vol. 30*, pages 287–289, 1976.

[94] V. Blažek. Sampling theorem and the number of degrees of freedom of an image. *Opt. Comm, Vol. 11*, pages 144–147, 1974.

[95] E.M. Bliss. Watch out for hidden pitfalls in signal sampling. *Electron. Eng., Vol. 45*, pages 59–61, 1973.

[96] R.P. Boas, Jr. *Entire Functions*. Academic Press, New York, 1954.

[97] R.P. Boas, Jr. Summation formulas and bandlimited signals. *Tohoku Math. J., Vol. 24*, pages 121–125, 1972.

[98] H. Bohr. *Almost Periodic Functions*. Wiley Interscience, New York, 1968.

[99] A. Boivin and C. Deckers. A new sampling theorem for the complex degree of coherence for reconstruction of a luminous isotropic source. *Opt. Comm., Vol. 26*, pages 144–147, 1978.

[100] F.E. Bond and C.R. Cahn. On sampling the zeros of bandwidth limited signals. *IRE Trans. Information Theory, Vol. IT-4*, pages 110–113, 1958.

[101] F.E. Bond, C.R. Cahn, and J.C. Hancock. A relation between zero crossings and Fourier coefficients for bandlimited functions. *IRE Trans. Information Theory, Vol. IT-6*, pages 51–55, 1960.

[102] C. de Boor. *A Practical Guide to Splines*. Springer-Verlag, New York, 1978.

[103] E. Borel. Sur l'interpolation. *C.R. Acad. Sci. Paris, Vol. 124*, pages 673–676, 1877.

[104] E. Borel. Mémoire sur les séries divergentes. *Ann. Ecole Norm. Sup., Vol. 3*, pages 9–131, 1899.

[105] E. Borel. La divergence de la formule de Lagrange a été établie également. In *Leçons sur les fonctions de variables réelles et les développements en séries de polynômes*. Gauthier-Villars, Paris, 1905, pages 74-79.

[106] M. Born and E. Wolf. *Principles of Optics*. Pergamon Press, Oxford, 1980.

[107] G. E. T. Box and G. M. Jenkins, *Time Series Analysis. Forecasting and Control* (revised edition). Holden-Day, San Francisco, 1976.

[108] G.D. Boyd and J.P. Gordon. Confocal multimode resonator for millimeter through optical wavelength masers. *Bell Systems Tech. J.*, pages 489–508, 1961.

[109] G.D. Boyd and H. Kogelnik. Generalized confocal resonator theory. *Bell Systems Tech. J.*, pages 1347–1369, 1962.

[110] G.R. Boyer. Realization d'un filtrage super-resolvant. *Opt. Acta*, pages 807–816, 1983.

[111] R.N. Bracewell. *The Fourier Transform and its Applications*, 2nd edition, Revised. McGraw-Hill, New York, 1986.

[112] R.N. Bracewell. *The Hartley Transform*. Oxford University Press, New York, 1986.

[113] L.M. Bregman. Finding the common point of convex sets by the method of successive projections. *Dokl. Akad. Nauk. USSR, Vol. 162, No. 3*, pages 487–490, 1965.

[114] E.O. Brigham. *The Fast Fourier Transform*. Prentice-Hall, Englewood Cliffs, NJ, 1974.

[115] L. Brillouin. *Science and Information Theory*. Academic Press, New York, 1962.

[116] J.L. Brown, Jr. Anharmonic approximation and bandlimited signals. *Inform. Contr., Vol. 10*, pages 409–418, 1967.

[117] J.L. Brown, Jr. On the error in reconstructing a non-bandlimited function by means of the bandpass sampling theorem. *J. Math. Anal. Appl., Vol. 18*, pages 75–84, 1967; Erratum, *Vol. 21*, page 699, 1968.

[118] J.L. Brown, Jr. A least upper bound for aliasing error. *IEEE Trans. Automatic Control, Vol. AC-13*, pages 754–755, 1968.

[119] J.L. Brown, Jr. Sampling theorem for finite energy signals. *IRE Trans. on Information Theory*, Vol. *IT-14*, pages 818–819, 1968.

[120] J.L. Brown, Jr. Bounds for truncation error in sampling expansions of bandlimited signals. *IRE Trans. Information Theory*, Vol. *IT-15*, pages 669–671, 1969.

[121] J.L. Brown, Jr. Truncation error for bandlimited random processes. *Inform. Sci.*, Vol. *1*, pages 261–272, 1969.

[122] J.L. Brown, Jr. Uniform linear prediction of bandlimited processes from past samples. *IEEE Trans. Information Theory*, Vol. *IT-18*, pages 662–664, 1972.

[123] J.L. Brown, Jr. On mean-square aliasing error in the cardinal series expansion of random processes. *IEEE Trans. Information Theory*, Vol. *IT-24*, pages 254–256, 1978.

[124] J.L. Brown, Jr. Comments on energy processing techniques for stress wave emission signals. *J. Acoust. Soc. Am.*, Vol. *67*, page 717, 1980.

[125] J.L. Brown, Jr. First-order sampling of bandpass signals: a new approach. *IEEE Trans. Information Theory*, Vol. *IT-26*, pages 613–615, 1980.

[126] J.L. Brown, Jr. Sampling bandlimited periodic signals – an application of the DFT. *IEEE Trans. Education*, Vol. *E-23*, pages 205–206, 1980.

[127] J.L. Brown, Jr. A simplified approach to optimum quadrature sampling. *J. Acoust. Soc. Am.*, Vol. *67*, pages 1659–1662, 1980.

[128] J.L. Brown, Jr. Multi-channel sampling of low-pass signals. *IEEE Trans. Circuits & Systems*, Vol. *CAS-28*, pages 101–106, 1981.

[129] J.L. Brown, Jr. Cauchy and polar-sampling theorems. *J. Opt. Soc. Am. A*, Vol. *1*, pages 1054–1056, 1984.

[130] J.L. Brown, Jr. An RKHS analysis of sampling theorems for harmonic-limited signals. *IEEE Trans. Acoust., Speech, Signal Processing*, Vol. *ASSP-33*, pages 437–440, 1985.

[131] J.L. Brown, Jr. Sampling expansions for multiband signals. *IEEE Trans. Acoust., Speech, Signal Processing*, Vol. *ASSP-33*, pages 312–315, 1985.

[132] J.L. Brown, Jr. On the prediction of bandlimited signal from past samples. *Proc. IEEE*, Vol. *74*, pages 1596–1598, 1986.

[133] J.L. Brown, Jr. Sampling reconstruction of $n$-dimensional bandlimited images after multilinear filtering. *IEEE Trans. Circuits & Systems*, Vol. *CAS-36*, pages 1035–1038, 1989.

[134] J.L. Brown, Jr. and O. Morean. Robust prediction of bandlimited signals from past samples. *IEEE Trans. Information Theory*, Vol. *IT-32*, pages 410–412, 1986.

[135] W.M. Brown. Optimum prefiltering of sampled data. *IRE Trans. Information Theory*, Vol. *IT-7*, pages 269–270, 1961.

[136] W.M. Brown. Sampling with random jitter. *J. SIAM*, Vol. *2*, pages 460–473, 1961.

[137] W.M. Brown. Fourier's integral. *Proc. Edinburgh Math. Soc.*, *Vol. 34*, pages 3–10, 1915–1916.

[138] W.M. Brown. On a class of factorial series. *Proc. London Math. Soc.*, *Vol. 2*, pages 149–171, 1924.

[139] N.G. de Bruijn. A theory of generalized functions, with applications to Wigner distribution and Weyl correspondence. *Nieuw Arch. Wiskunde (3)*, *Vol. 21*, 205–280, 1973.

[140] O. Bryngdahl. Image formation using self-imaging techniques. *J. Opt. Soc. Am.*, *Vol. 63*, pages 416–419, 1973.

[141] O. Bryngdahl and F. Wyrowski. Digital holography. Computer generated holograms. In *Progress in Optics*, North-Holland, Amsterdam, 1990, pages 1–86.

[142] G.J. Buck and J.J. Gustincic. Resolution limitations of a finite aperture. *IEEE Trans. Antennas & Propagation*, *Vol. AP-15*, pages 376–381, 1967.

[143] J. Bures, P. Meyer, and G. Fernandez. Reconstitution d'un profil d'eclairement echantillonne par une legne de microphotodiodes. *Opt. Comm.*, pages 39–44, 1979.

[144] P.L. Butzer. A survey of the Whittaker-Shannon sampling theorem and some of its extensions. *J. Math. Res. Exposition*, *Vol. 3*, pages 185–212, 1983.

[145] P.L. Butzer. The Shannon sampling theorem and some of its generalizations. An Overview. in *Constructive Function Theory, Proc. Conf. Varna, Bulgaria, June 1-5, 1981*; Bl. Sendov et al., eds., Publishing House Bulgarian Academy of Science, Sofia, 1983, pages 258–274.

[146] P.L. Butzer. Some recent applications of functional analysis to approximation theory. In *Zum Werk Leonhard Eulers (Vorträge des Euler-Kolloquiums Berlin)*, Knobloch, Louhivaara, Winkler, eds., Birkhäuser Verlag, Basel-Stuttgart-Boston, 1983/1984, pages xii, 238 and 133–155.

[147] P.L. Butzer and W. Engels. Dyadic calculus and sampling theorems for functions with multidimensional domain, I. General theory. *Inform. Contr.*, *Vol. 52*, pages 333–351, 1982.

[148] P.L. Butzer and W. Engels. Dyadic calculus and sampling theorems for functions with multidimensional domain, II. Applications to dyadic sampling representations. *Inform. Contr.*, *Vol. 52*, pages 352–363, 1982.

[149] P.L. Butzer and W. Engels. On the implementation of the Shannon sampling series for bandlimited signals. *IEEE Trans. Information Theory*, *Vol. IT-29*, pages 314–318, 1983.

[150] P.L. Butzer, W. Engels, S. Ries, and R. L. Stens. The Shannon sampling series and the reconstruction of signals in terms of linear, quadratic and cubic splines. *SIAM J. Appl. Math.*, *Vol. 46*, pages 299–323, 1986.

[151] P.L. Butzer, W. Engels, and U. Scheben. Magnitude of the truncation error in sampling expansions of bandlimited signals. *IEEE Trans. Acoust., Speech, Signal Processing. ASSP-30*, pages 906–912, 1982.

[152] P.L. Butzer, A. Fischer, and R.L. Stens. Generalized sampling approximation of multivariable signals; theory and some applications. *Note di Mat., Köthe memorial volume* (to appear).

[153] P.L. Butzer, A. Fischer, and R.L. Stens. Generalized sampling approximation of multivariable signals; general theory. *Proc. 4th Meeting on Real Analysis and Measure Theory, Capri, 1990*, Atti Sem. Mat. Fis. Univ. Modena (to appear).

[154] P.L. Butzer, M. Hauss, and R.L. Stens. The sampling theorem and its unique role in various branches of mathematics. *Mathematical Sciences, Past and Present, 300 years of Mathematische Gesellschaft in Hamburg,* Mitteilungen Math. Ges. Hamburg (to appear).

[155] P.L. Butzer and G. Hinsen. Reconstruction of bounded signals from pseudo-periodic irregularly spaced samples. *Signal Processing, Vol. 17,* pages 1–17, 1988.

[156] P.L. Butzer and G. Hinsen. Two-dimensional nonuniform sampling expansions – an iterative approach I, II. *Appl. Anal., Vol. 32,* pages 53–85, 1989.

[157] P.L. Butzer and C. Markett. The Poisson summation formula for orthogonal systems. In *Anniversary Volume on Approximation Theory and Functional Analysis (Proc. Conf. Math. Res. Inst. Oberwolfach, Black Forest)*, Butzer, Stens, Nagy, eds., *ISNM Vol. 65*, Birkhäuser Verlag, Basel-Stuttgart-Boston, 1984, pages 595–601

[158] P.L. Butzer and R.J. Nessel. Contributions to the theory of saturation for singular integrals in several variables I. General theory. *Nederland Akad. Wetensch. Proc. Ser. A, Vol. 69* and *Indag. Math., Vol. 28*, pages 515–531, 1966.

[159] P.L. Butzer and R.J. Nessel. *Fourier Analysis and Approximation Vol. I: One-Dimensional Theory.* Birkhäuser Verlag, Basel, Academic Press, New York, 1971.

[160] P.L. Butzer and R.J. Nessel. De la Vallée Poussin's work in approximation and its influence. In *Charles Baron de la Vallée, Collected Works*, P.L. Butzer and J. Mawhin, eds., Bibliographisches Institut, Mannheim, Germany, to appear.

[161] P.L. Butzer, S. Ries, and R.L. Stens. The Whittaker-Shannon sampling theorem, related theorems and extensions; a survey. In *Proc. JIEEEC '83 (Proceedings Jordan - International Electrical and Electronic Engineering Conference, Amman, Jordan)*, pages 50–56, 1983.

[162] P.L. Butzer, S. Ries, and R.L. Stens. Shannon's sampling theorem, Cauchy's integral formula and related results. In *Anniversary Volume on Approximation Theory and Functional Analysis (Proceedings Conf. Math. Res. Inst. Oberwolfach, Black Forest)* Butzer, Stens, and Nagy, eds., *ISNM Vol. 65*, Birkhäuser Verlag, Basel-Stuggart-Boston, 1984, pages 363–377.

[163] P.L. Butzer, S. Ries, and R.L. Stens. Approximation of continuous and discontinuous functions by generalized sampling series. *J. Approx. Theory, Vol. 50,* pages 25–39, 1987.

[164] P. L. Butzer, M. Schmidt, E. L. Stark, and L. Vogt, Central factorial numbers; their main properties and some applications. *Num. Funct. Anal. Optim. Vol. 10*, pages 419–488, 1989.

[165] P. L. Butzer and M. Schmidt, Central factorial numbers and their role in finite difference calculus and approximation. *Proc. Conf. on Approximation Theory, Kecskemét, Hungary.* Colloquia Mathematica Societatis János Bolyai, 1990.

[166] P.L. Butzer and W. Splettstösser. *Approximation und Interpolation durch verallgemeinerte Abtastsummen. Forschungsberichte* No. 2515, Landes Nordrhein-Westfalen, Köln-Opladen, 1977.

[167] P.L. Butzer and W. Splettstösser. A sampling theorem for duration-limited functions with error estimates. *Inform. Contr., Vol. 34*, pages 55–65, 1977.

[168] P.L. Butzer and W. Splettstösser. Sampling principle for duration-limited signals and dyadic Walsh analysis. *Inform. Sci., Vol. 14*, pages 93–106, 1978.

[169] P.L. Butzer and W. Splettstösser. On quantization, truncation and jitter errors in the sampling theorem and its generalizations. *Signal Processing, Vol. 2*, pages 101–102, 1980.

[170] P.L. Butzer, W. Splettstösser, and R.L. Stens. The sampling theorem and linear prediction in signal analysis. *Jahresber. Deutsch Math.-Verein. Vol. 90*, pages 1–70, 1988.

[171] P.L. Butzer and R.L. Stens. *Index of Papers on Signal Theory: 1972-1989.* Lehrstuhl A für Mathematik, Aachen University of Technology, Aachen, Germany, 1990.

[172] P.L. Butzer and R.L. Stens. The Euler-Maclaurin summation formula, the sampling theorem, and approximate integration over the real axis. *Linear Algebra Applications, Vol. 52/53*, pages 141–155, 1983.

[173] P.L. Butzer and R.L. Stens. The Poisson summation formula, Whittaker's cardinal series and approximate integration. In *Proceedings Second Edmonton Conference on Approximation Theory*, Z. Ditzian, et al., eds., *American Math. Soc., Providence, Rhode Island, Canadian Math Soc. Proc., Vol. 3*, pages 19–36, 1983.

[174] P.L. Butzer and R.L. Stens. A modification of the Whittaker-Kotelnikov-Shannon sampling series. *Aequationés Math., Vol. 28*, pages 305–311, 1985.

[175] P.L. Butzer and R.L. Stens. Prediction of non-bandlimited signals from past samples in terms of splines of low degree. *Math. Nachr., Vol. 132*, pages 115–130, 1987.

[176] P.L. Butzer and R.L. Stens. Linear prediction in terms of samples from the past; an overview. In *Numerical Methods and Approximation Theory III (Proc. Conf. on Numerical Methods and Approximation Theory)*, Nis 18.–21.8.1987; Ed. G.V. Milovanovic, Faculty of Electronic Engineering, University of Nis, Yugoslavia, 1988, pages 1–22.

[177] P.L. Butzer and R.L. Stens. Sampling theory for not necessarily bandlimited functions; an historical overview. *SIAM Review, Vol. 34*, pages 40-53, March 1992.

[178] P.L. Butzer, R.L. Stens, and M. Wehrens. The continuous Legendre transform, its inverse transform and applications. *Int. J. Math. Sci., Vol. 3*, pages 47-67, 1980.

[179] P. L. Butzer and R. L. Stens. De la Vallée Poussin's paper of 1908 on interpolation and sampling theory, and its influence. *Ch.-J. de la Vallée Poussin, Collected Works*, P. L. Butzer and J. Mawhin, eds., Bibliographisches Institut, Mannheim, to appear.

[180] P.L. Butzer and H.J. Wagner. Walsh-Fourier series and the concept of a derivative. *Applicable Analysis, Vol. 3*, pages 29-46, 1973.

[181] P.L. Butzer and H.J. Wagner. On dyadic analysis based on the pointwise dyadic derivative. *Analysis Math., Vol. 1*, pages 171-196, 1975.

[182] A. Buzo, A.H. Gray Jr., R.M. Gray, and J.D. Markel. Speech coding based upon vector quantization. *IEEE Trans. Acoust., Speech, Signal Processing, Vol. ASSP-28*, pages 562-574, Oct. 1980.

[183] A. Buzo, F. Kuhlmann, and C. Blas. Rate distortion bounds for quotient-based distortions with applications to Itakura-Saito distortion measures. *IEEE Trans. Inform. Theory, Vol. IT-32*, pages 141-147, March 1986.

[184] J. Cadzow. An extrapolation procedure for bandlimited signals. *IEEE Trans. Acoust., Speech, Signal Processing, Vol. ASSP-27*, pages 4-12, 1979.

[185] D. Cahana and H. Stark. Bandlimited image extrapolation with faster convergence. *Appl. Opt.*, pages 2780-2786, 1981.

[186] G. Calvagio and D. Munson, Jr. New results on Yen's approach to interpolation from nonuniformly spaced samples. *Proc. IASSP*, pages 1533-1538, 1990.

[187] S. Cambanis and M.K. Habib. Finite sampling approximation for non-bandlimited signals. *IEEE Trans. Information Theory, Vol. IT-28*, pages 67-73, 1982.

[188] S. Cambanis and E. Masry. Truncation error bounds for the cardinal sampling expansion of bandlimited signals. *IEEE Trans. Information Theory, Vol. IT-28*, pages 605-612, 1982.

[189] L.L. Campbell. A comparison of the sampling theorems of Kramer and Whittaker. *J. SIAM, Vol. 12*, pages 117-130, 1964.

[190] L.L. Campbell. Sampling theorem for the Fourier transform of a distribution with bounded support. *SIAM J. Appl. Math., Vol. 16*, pages 626-636, 1968.

[191] L.L Campbell. Further results on a series representation related to the sampling theorem. *Signal Processing, Vol. 9*, pages 225-231, 1985.

[192] A.B. Carlson. *Communications Systems, An Introduction to Signals and Noise in Electrical Communication*, 2nd Ed. McGraw-Hill, New York, 1975.

[193] W.H. Carter and E.Wolf. Coherence and radiometry with quasihomogeneous sources. *J. Opt. Soc. Am., Vol. 67*, pages 785–796, 1977.

[194] D. Casasent. Optical signal processing. In *Optical Data Processing*, D. Casasent, editor, Springer-Verlag, Berlin, 1978, pages 241–282.

[195] W.T. Cathey, B.R. Frieden, W.T. Rhodes, and C.K. Rushforth. Image gathering and processing for enhanced resolution. *J. Opt. Soc. Am. A*, pages 241–250, 1984.

[196] A.L. Cauchy. Memoire sur diverses formules d'analyser, *Comptes Rendus, Vol. 12*, 1841, pages 283–298.

[197] H.J. Caulfield. Wavefront sampling in holography. *Proc. IEEE, Vol. 57*, pages 2082–2083, 1969.

[198] H.J. Caulfield. Correction of image distortion arising from nonuniform sampling. *Proc. IEEE, Vol. 58*, page 319, 1970.

[199] G. Cesini, P. Di Filippo, G. Lucarini, G. Guattari, and P. Pierpaoli. The use of charge-coupled devices for automatic processing of interferograms. *J. Opt.*, pages 99–103, 1981.

[200] G. Cesini, G. Guattari, G. Lucarini, and C. Palma. An iterative method for restoring noisy images. *Opt. Acta*, pages 501–508, 1978.

[201] J. Cezanne and A. Papoulis. The use of modulated splines for the reconstruction of bandlimited signals. *IEEE Trans. Acoust., Speech, Signal Processing, Vol. ASSP-36, No. 9*, pages 1521–1525, 1988.

[202] K. Chalasinska-Macukow and H.H. Arsenault. Fast iterative solution to exact equations for the two-dimensional phase-retrieval problem. *J. Opt. Soc. Am. A, Vol. 2*, pages 46–50, 1985.

[203] S.S. Chang. Optimum transmission of a continuous signal over a sampled data link. *AIEE Trans., Vol. 79, Pt. II (Applications and Industry)*, pages 538–542, 1961.

[204] J.I. Chargin and V.P. Iakovlev. *Finite Functions in Physics and Technology*. Nauka, Moscow, 1971.

[205] P. Chavel and S. Lowenthal. Noise and coherence in optical image processing. I. The Callicr effect and its influence on image contrast. *J. Opt. Soc. Am., Vol. 68*, pages 559–568, 1978.

[206] P. Chavel and S. Lowenthal. Noise and coherence in optical image processing. II. Noise fluctuations. *J. Opt. Soc. Am., Vol. 68*, pages 721–732, 1978.

[207] D.S. Chen and J.P. Allebach. Analysis of error in reconstruction of two-dimensional signals from irregularly spaced samples. *IEEE Trans. Acoust., Speech, Signal Processing, Vol. ASSP-35*, pages 173–179, 1987.

[208] K.H. Chen and C.C. Yang. On $n$-dimensional sampling theorems. *Appl. Math. Comput., Vol. 7*, pages 247–252, 1980.

[209] D.K. Cheng and D.L. Johnson. Walsh transform of sampled time functions and the sampling principle. *Proc. IEEE, Vol. 61*, pages 674–675, 1973.

[210] K.F. Cheung. The Generalized Sampling Expansion – Its Stability, Posedness and Discrete Equivalent. M.S. thesis, Dept. Electrical Engr., Univ.
of Washington, Seattle, Washington, 1983.

[211] K.F. Cheung. Image Sampling Density Reduction Below That of Nyquist.
Ph.D. thesis, University of Washington, Seattle, 1988.

[212] K.F. Cheung and R.J. Marks II. Ill-posed sampling theorems. *IEEE
Transactions on Circuits and Systems, Vol. CAS-32*, pages 829–835, 1985.

[213] K.F. Cheung, R.J. Marks II and L.E. Atlas. Neural net associative mem-
ories based on convex set projections, *IEEE Proc. 1st International Conference on Neural Networks, San Diego, June 1987.*

[214] K.F. Cheung and R.J. Marks II. Papoulis' generalization of the sampling
theorem in higher dimensions and its application to sample density reduction. In *Proceedings of the International Conference on Circuits and
Systems, Nanjing, China, July 6–8, 1989.*

[215] K.F. Cheung and R.J. Marks II. Image sampling below the Nyquist density without aliasing. *J. Opt. Soc. Am. A, Vol. 7*, pages 92–105, 1990.

[216] K.F. Cheung, R.J. Marks II, and L.E. Atlas. Convergence of Howard's
minimum-negativity-constraint extrapolation algorithm. *J. Opt. Soc. Am.
A, Vol. 5*, pages 2008–2009, 1988.

[217] R. Cheung and Y. Rahmat-Samii. Experimental verification of
nonuniform sampling technique for antenna far-field construction. *Elec-
tromagnetics, Vol. 6, No. 4*, pages 277–300, 1986.

[218] D.G. Childers. Study and experimental investigation on sampling rate and
aliasing in time-division telemetry systems. *IRE Trans. Space Electronics
& and Telemetry*, pages 267–283, 1962.

[219] H. K. Ching. Truncation Effects in the Estimation of Two–Dimensional
Continuous Bandlimited Signals. M.S. thesis, Dept. Electrical Engineering, Univ. of Washington, Seattle, 1985.

[220] J.J. Chou and L.A. Piegl. Data reduction using cubic rational B-splines.
*IEEE Computer Graphics and Applications Magazine, Vol. 12, No. 3*,
pages 60–68, May 1992.

[221] R.V. Churchill. *Operational Mathematics*, McGraw–Hill, New York, 1972.

[222] T.A.C.M. Claasen and W.F.G. Mecklenbräuker. The Wigner distribution - a tool for time-frequency signal analysis. Part I: Continuous-time
signals; Part II: Discrete-time signals; Part III: Relations with other time-
frequency signal transformations. *Philips J. Res., Vol. 35*, pages 217–250,
276–300, 372–389, 1980.

[223] J.J. Clark. Sampling and reconstruction of non-bandlimited signals. In
*Proc. Visual Communication and Image Processing IV, SPIE, Vol. 1199*,
pages 1556–1562, 1989.

[224] J.J. Clark, M.R. Palmer, and P.D. Lawrence. A transformation method
for the reconstruction of functions from nonuniformly spaced samples.
*IEEE Trans. Acoust., Speech, Signal Processing, Vol. ASSP-33*, 1985.

[225] D. Cochran and J. Clark. On the sampling and reconstruction of time-warped bandlimited signals. *Proc. IASSP*, pages 1539–1541, 1990.

[226] L. Cohen. Generalized phase-space distribution functions. *J. Math. Phys.*, Vol. 7, pages 781-786, 1966.

[227] L. Cohen. Time-frequency distributions - A review. *Proc. IEEE, Vol. 77*, pages 941-961, 1989.

[228] T.W. Cole. Reconstruction of the image of a confined source. *Astronomy and Astrophysics*, pages 41–45, 1973.

[229] T.W. Cole. Quasi-optical techniques in radio astronomy, In *Progress in Optics*, North-Holland, Amsterdam, 1977, pages 187-244.

[230] R.J. Collier, C.B. Burckhardt, and L.H. Lin. *Optical Holography*. Academic Press, Orlando, FL., 1971.

[231] A. Consortini and F. Crochetti. An attempt to remove noise from histograms of probability density of irradiance in atmospheric scintillation. *Proc. SPIE*, pages 524–525, 1990.

[232] J.W. Cooley, P.A.W. Lewis, and P.D. Welch. The finite Fourier transform. *IEEE Trans. Audio Electroacoustics, Vol. AE-17*, pages 77–85, 1969.

[233] J.W. Cooley and J.W. Tukey. An algorithm for the machine calculations of complex Fourier series. *Math. Comput., Vol. 19*, pages 297–301, 1965.

[234] I.J. Cox, C.J.R. Sheppard, and T. Wilson. Reappraisal of arrays of concentric annuli as super-resolving filters. *J. Opt. Soc. Am., Vol. 72*, pages 1287–1291, 1982.

[235] R.E. Crochiere and L.R. Rabiner. Interpolation and decimation of digital signals – a tutorial review. *Proc. IEEE, Vol. 69*, pages 300–331, 1981.

[236] R.E. Crochiere and L.R. Rabiner. *Multirate Digital Signal Processing*. Prentice-Hall, Englewood Cliffs, NJ, 1983.

[237] H.Z. Cummins and E.R. Pike, editors. *Photon Correlation Spectroscopy and Velocimetry*. Plenum Press, New York, 1977.

[238] L.J. Cutrona, E.N. Leith, C.J. Palermo, and L.J. Porcello. Optical data processing and filtering systems. *IRE Trans. Information Theory*, pages 386–400, 1960.

[239] B.S. Cybakov and V.P. Jakovlev. On the accuracy of restoring a function with a finite number of terms of a Kotel'nikov series. *Radio Eng. & Electron., Vol. 4*, pages 274–275, 1959.

[240] W. Dahmen and C.A. Micchelli. Translates of multivariate splines. *Linear Algebra Applications, Vol. 52/53*, pages 217–234, 1983.

[241] H. Dammann and K. Görtker. High-efficiency in-line multiple imaging by means of multiple and filtering systems. *Opt. Comm.*, pages 312–315, 1971.

[242] R. Dandliker, E. Marom, and F.M. Mottier. Wavefront sampling in holographic interferometry. *Opt. Comm.*, pages 368–371, 1972.

[243] H. D'Angelo. *Linear Time Varying Systems, Analysis and Synthesis*. Allyn and Bacon, Boston, 1970.

[244] I. Daubechies. The wavelet transform, time-frequency localization and signal analysis. *IEEE Trans. Inform. Theory, Vol. IT-36*, pages 961–1005, 1990.

[245] Q.I. Daudpota, G. Dowrick, and C.A. Grwated. Estimation of moments of a Poisson sampled random process. *J. Phys. A: Math. Gen., Vol. 10*, pages 471–483, 1977.

[246] A.M. Davis. Almost periodic extension of bandlimited functions and its application to nonuniform sampling. *IEEE Trans. Circuits & Systems, Vol. CAS-33, No. 10*, pages 933–938, October 1986.

[247] J.A. Davis, E.A. Merrill, D.M. Cottrell, and R.M. Bunch. Effects of sampling and binarization in the output of the joint Fourier transform correlator. *Opt. Eng., Vol. 29, No. 9*, pages 1094–1100, 1990.

[248] P.J. Davis. *Interpolation and Approximation.* Blaisdell, New York, 1963.

[249] C. de Boor. On calculating with B–splines. *J. Approx. Theory, Vol. 6*, pages 50–62, 1972.

[250] C. de Boor. Package for calculating with B–splines. Tech. Report No. 1333, Math. Research Center, University of Wisconsin, Madison, 1973.

[251] C. de Boor. *A Practical Guide to Splines.* Springer-Verlag, New York, 1978.

[252] G. Defrancis. Directivity, super-gain and information. *IRE Trans. Antennas & Propagation, Vol. AP-4*, pages 473–478, 1956.

[253] P. Delsarte and Y. Gemin. The split Levinson algorithm. *IEEE Trans. Acoust., Speech, Signal Processing, Vol. ASSP-34*, pages 470–478, 1986.

[254] P. Delsarte and Y. Genin. On the role of orthogonal polynomials on the unit circle in digital signal processing applications. In *Orthogonal Polynomials: Theory and Practice* (ed. by P. Nevai and E. H. Ismail), Kluwer Academic Publishers, Dordrecht, 1990, pages 115–133.

[255] P. DeSantis, G. Guattari F. Gori, and C. Palma. Optical systems with feedback. *Opt. Acta*, pages 505–518, 1976.

[256] P. DeSantis and F. Gori. On an iterative method for super-resolution. *Opt. Acta*, pages 691–695, 1975.

[257] P. DeSantis, F. Gori, G. Guattari, and C. Palma. Emulation of super-resolving pupils through image postprocessing. *Opt. Comm.*, pages 13–16, 1986.

[258] P. DeSantis, F. Gori, G. Guattari, and C. Palma. Synthesis of partially coherent fields. *J. Opt. Soc. Am., Vol. 76*, pages 1258–1262, 1986.

[259] P. DeSantis and C. Palma. Degrees of freedom of aberrated images. *Opt. Acta*, pages 743–752, 1976.

[260] A.J. Devaney and R. Chidlaw. On the uniqueness question in the problem of phase retrieval from intensity measurements. *Opt. Acta*, pages 1352–1354, 1978.

[261] P. DeVincenti, G. Doro, and A. Saitto. Reconstruction of bi-dimensional electromagnetic field by uni-dimensional sampling technique. *Electron. Letters, Vol. 17*, pages 324–325, 1981.

[262] V.A. Ditkin and A.P. Prudnikiv. *Integral Transforms and Operation Calculus*, Fitmatgiz, Moscow, 1961: English translation, Pergamon Press, New York, 1965.

[263] D.D. Do and R.H. Weiland. Self–poissoning of catalyst pellets. *AICHE Symp. Series, Vol. 77, No. 202*, pages 36–45, 1981.

[264] M.M. Dodson and A.M. Silva. Fourier analysis and the sampling theorem. *Proc. Royal Irish Academy, Vol. 85 A*, pages 81–108, 1985.

[265] R.C. Dorf, M.C. Farren, and C.A. Philips. Adaptive sampling frequency for sampled data control systems. *IRE Trans. Automatic Control, Vol. AC-7*, pages 38–47, 1962.

[266] F. Dowla and J.H. McClellan. MEM spectral analysis for nonuniformly sampled signals. In *Proceedings of the 2nd International Conference on Computer Aided Seismic Analysis and Discrimination, North Dartmouth, Mass., August 1981*.

[267] E. Dubois. The sampling and reconstruction of time-varying imagery with application in video systems. *Proc. IEEE, Vol. 23*, pages 502–522, 1985.

[268] D.E. Dudgeon and R.M. Mersereau. *Multidimensional Digital Signal Processing*. Prentice-Hall, Englewood Cliffs, NJ, 1984.

[269] P.M. Duffieux. *L'integral de Fourier et ses applications a l'optique*. Rennes-Oberthur, 1946.

[270] R.J. Duffin and A.C. Schaeffer. Some properties of functions of exponential type. *Bull. Amer. Math. Soc., Vol. 44*, pages 236–240, 1938.

[271] R.J. Duffin and A.C. Schaeffer. A class of nonharmonic Fourier series. *Trans. Am. Math. Soc.*, pages 341–366, 1952.

[272] J. Dunlop and V.J. Phillips. Signal recovery from repetitive non-uniform sampling patterns. *The Radio and Electronic Engineer, Vol. 44*, pages 491–503, 1974.

[273] K. Dutta and J.W. Goodman. Reconstruction of images of partially coherent objects from samples of mutual intensity. *J. Opt. Soc. Am., Vol. 67*, pages 796–803, 1977.

[274] J. Duvernoy. Estimation du nombre d'echantillons et du domaine spectral necessaires pour determiner une forme moyenne par une technique de superposition. *Opt. Comm.*, pages 286–288, 1973.

[275] H. Dym and H. P. McKean. *Fourier series and Integrals*. Academic Press, New York, 1972.

[276] H. Dym and H. P. McKean. *Gaussian Processes, Function Theory, and the Inverse Spectral Problem*. Academic Press, New York, 1976.

[277] R.E. Edwards. *Fourier Series, A Modern Introduction, Vol. 1*. Holt, Rinehart and Winston, New York, 1967.

[278] A.G. Emslie and G.W. King. Spectroscopy from the point of view of the communication theory, II. *J. Opt. Soc. Am., Vol. 43*, pages 658–663, 1953.

[279] W. Engels and W. Splettstösser. A note on Maqusi's proofs and truncation error bounds in the dyadic (Walsh) sampling theorem. *IEEE Trans. Acoust, Speech, Signal Processing, Vol. ASSP-30*, pages 334–335, 1982.

[280] A. Ephremides and L.H. Brandenburg. On the reconstruction error of sampled data estimates. *IEEE Trans. Information Theory, Vol. IT-19*, pages 365–367, 1973.

[281] A. Erdelyi et al. *Higher Transcendental Functions, Vol. 1.* McGraw-Hill, New York, 1953.

[282] A. Erdelyi et al. *Tables of Integral Transforms II.* McGraw-Hill, New York, 1954.

[283] T. Ericson. A generalized version of the sampling theorem. *Proc. IEEE, Vol. 60*, pages 1554–1555, 1972.

[284] A. Erteza and K.S. Lin. On the transform property of a bandlimited function and its samples. *Proc. IEEE, Vol. 68*, pages 1149–1150, 1980.

[285] N.H. Farhat and G.P. Shah. Implicit sampling in optical data processing. *Proc. SPIE, Vol. 201*, pages 100–106, 1979.

[286] H.G. Feichtinger. Wiener amalgam spaces and some of their applications. In *Conf. Proc. Function spaces, Edwardsville, IL*, Marcel Dekker, New York, April 1990.

[287] E.B. Felstead and A.U. Tenne-Sens. Optical interpolation with application to array processing. *Appl. Opt., Vol. 14*, pages 363–368, 1975.

[288] W.L. Ferrar. On the cardinal function of interpolation–theory. *Proc. Royal Soc. Edinburgh, Vol. 45*, pages 267–282, 1925.

[289] W.L. Ferrar. On the cardinal function of interpolation–theory. *Proc. Royal Soc. Edinburgh, Vol. 46*, pages 323–333, 1926.

[290] W.L. Ferrar. On the consistency of cardinal function interpolation. *Proc. Royal Soc. Edinburgh, Vol. 47*, pages 230–242, 1927.

[291] Paulo J.S.G. Ferreira. Incomplete sampling series and the recovery of missing samples from oversampled band-limited signals. *IEEE Trans. Signal Processing, Vol. 40*, January 1992.

[292] H.A. Ferweda. The phase reconstruction problem for wave amplitudes and coherence functions. In *Inverse Source Problems in Optics*, H.P. Baltes, editor, Springer-Verlag, Berlin, 1978.

[293] H.A. Ferweda and B.J. Hoenders. On the reconstruction of a weak phase-amplitude object IV. *Optik*, pages 317–326, 1974.

[294] Alfred Fettweis, Josef A. Nossek, and Klaus Meerkötter. Reconstruction of signals after filtering and sampling rate reduction. *IEEE Trans. Acoust., Speech, Signal Processing, Vol. ASSP-33, No. 4*, pages 893–902, 1985.

[295] M.A. Fiddy and T.J. Hall. Nonuniqueness of super-resolution techniques applied to sampled data. *J. Opt. Soc. Am., Vol. 71*, pages 1406–1407, 1981.

[296] A.R. Figueiras-Vidal, J.B. Marino-Acebal, and R.G. Gomez. On generalized sampling expansions for deterministic signals. *IEEE Trans. Circuits & Systems, Vol. CAS-28*, pages 153–154, 1981.

[297] G. Fix and G. Strang. Fourier analysis of the finite element method in Ritz-Galerkin theory. *Studies Appl. Math., Vol. 48*, pages 268–273, 1969.

[298] I.T. Fjallbrandt. Method of data reduction of sampled speech signals by using nonuniform sampling and a time variable digital filter. *Electron. Lett., Vol. 13*, pages 334–335, 1977.

[299] T.T. Fjallbrandt. Interpolation and extrapolation in nonuniform sampling sequences with average sampling rate below the Nyquist rate. *Electron. Lett., Vol. 11*, pages 264–266, 1975.

[300] G.C. Fletcher and D.J. Ramsay. Photon correlation spectroscopy of poly-disperse samples, I. Histogram method with exponential sampling. *Opt. Acta, Vol. 30*, pages 1183–1196, 1983.

[301] L. Fogel. A note on the sampling theorem. *IRE Trans., Vol. 1*, pages 47–48, 1955.

[302] J. Frank. Computer processing of electron micrographs. In *Advanced Techniques in Biological Electron Microscopy*, J.K. Koehler, editor, Springer-Verlag, New York, 1973.

[303] G. Franklin. Linear filtering of sampled data. *IRE Int. Conv. Rec., Vol. 3, Pt. 4*, pages 119–128, 1955.

[304] L.E. Franks. *Signal Theory*. Prentice-Hall, Englewood Cliffs, NJ, 1969.

[305] H. Freeman. *Discrete Time Systems*. John Wiley, New York, 1965.

[306] A.T. Friberg. On the existence of a radiance function for finite planar sources of arbitrary states of coherence. *J. Opt. Soc. Am., Vol. 69*, pages 192–199, 1979.

[307] B.R. Frieden. Image evaluation by the use of sampling theorem. *J. Opt. Soc. Am., Vol. 56*, pages 1355–1362, 1966.

[308] B.R. Frieden. On arbitrarily perfect imagery with a finite aperture. *Opt. Acta*, pages 795–807, 1969.

[309] B.R. Frieden. Evaluation, design and extrapolation methods for optical signals, based on the use of the prolate functions. In *Progress in Optics*, E. Wolf, editor, North-Holland, Amsterdam, 1971.

[310] B.R. Frieden. Image enhancement and restoration. In *Picture Processing and Digital Filtering*, Springer-Verlag, Berlin, 1975.

[311] B.R. Frieden. Band-unlimited reconstruction of optical objects and spectra. *J. Opt. Soc. Am., Vol. 66*, pages 1013–1019, 1976.

[312] D. Gabioud. Effect of irregular sampling, especially in the case of justification jitter. *Mitteilungen AGEN, No. 44*, pages 39–46, December 1986.

[313] D. Gabor. Theory of communication. *J. IEE (London), Vol. 93, Part III*, pages 429–457, 1946.

[314] D. Gabor. A new microscopic principle. *Nature*, pages 777–779, 1948.

[315] D. Gabor. Light and information. In *Progress in Optics*, E. Wolf, editor, North-Holland, Amsterdam, 1961.

[316] N.C. Gallagher, Jr., and G.L. Wise. A representation for band-limited functions. *Proc. IEEE, Vol. 63*, page 1624, 1975.

[317] H. Gamo. Intensity matrix and degrees of coherence. *J. Opt. Soc. Am., Vol. 47*, page 976, 1957.

[318] H. Gamo. Transformation of intensity matrix by the transmission of a pupil. *J. Opt. Soc. Am., Vol. 48*, pages 136–137, 1958.

[319] H. Gamo. Matrix treatment of partial coherence. In *Progress in Optics*, E. Wolf, editor, North-Holland, Amsterdam, 1964.

[320] M. Gamshadzahi. Bandwidth reduction in delta modulation systems using an iterative reconstruction scheme. Master's thesis, Department of ECE, Illinois Institute of Technology, Chicago, December 1989.

[321] William L. Gans. The measurement and deconvolution of time jitter in equivalent-time waveform samplers. *IEEE Trans. Instrumentation & Measurement, Vol. IM-32, No. 1*, pages 126–133, 1983.

[322] William L. Gans. Calibration and error analysis of a picosecond pulse waveform measurement system at NBS. *Proc. IEEE, Vol. 74, No. 1*, pages 86–90, 1986.

[323] N.T. Garder. A note on multidimensional sampling theorem. *Proc. IEEE, Vol. 60*, pages 247–248, 1972.

[324] W.A. Gardner. A sampling theorem for nonstationary random processes. *IEEE Trans. Information Theory, Vol. IT-18*, pages 808–809, 1972.

[325] J.D. Gaskill. *Linear Systems, Fourier Transforms and Optics*. John Wiley & Sons, Inc., New York, 1978.

[326] M. Gaster and J.B. Roberts. Spectral analysis of randomly sampled signals. *J. Inst. Math. Applic., Vol. 15*, pages 195–216, 1975.

[327] W.S. Geisler and D.B. Hamilton. Sampling theory analysis of spatial vision. *J. Opt. Soc. Am. A, Vol. 3*, pages 62–70, 1986.

[328] T. Gerber and P.W. Schmidt. The sampling theorem and small-angle scattering. *J. Appl. Crystallography, Vol. 16*, pages 581–589, 1983.

[329] R.W. Gerchberg. Super-resolution through error energy reduction. *Optica Acta, Vol. 21*, pages 709–720, 1974.

[330] Allen Gersho. Asymptotically optimal block quantization. *IEEE Trans. Information Theory, Vol. IT-25*, pages 373–380, 1979.

[331] Allen Gersho and Robert M. Gray. *Vector quantization and signal compression*. Kluwer Academic Publishers, Boston, 1992.

[332] G. Gheen and F.T.S. Yu. Broadband image subtraction by spectral dependent sampling. *Opt. Comm.*, pages 335–341, 1986.

[333] R.J. Glauber. Coherent and incoherent states of the radiation field. *Phys. Rev.*, pages 2766–2788, 1963.

[334] M.J.E. Golay. Point arrays having compact, nonredundant autocorrelations. *J. Opt. Soc. Am., Vol. 61*, pages 272–273, 1971.

[335] M.H. Goldburg and R.J. Marks II. Signal synthesis in the presence of an inconsistent set of constraints. *IEEE Trans. Circuits & Systems, Vol. CAS-32*, pages 647–663, 1985.

[336] S. Goldman. *Information Theory*. Dover, New York, 1968.

[337] Rafael C. Gonzalez. *Digital Image Processing*. Addison-Wesley Publishing Company, Inc., Reading, MA, 1977.

[338] I.J. Good. The loss of information due to clipping a waveform. *Inform. Contr., Vol. 10*, pages 220–222, 1967.

[339] J.W. Goodman. *Introduction to Fourier Optics*. McGraw-Hill, New York, 1968.

[340] J.W. Goodman. Imaging with low-redundancy arrays, In *Applications of Holography*, Plenum Press, New York, 1971, pages 49–55.

[341] J.W. Goodman. Linear space-variant optical data processing. In *Optical Information Processing*, S.H. Lee, editor, Springer-Verlag, New York, 1981.

[342] J.W. Goodman. *Statistical Optics*. Wiley, New York, 1985.

[343] G.C. Goodwin, M.B. Zarrop, and R.L. Payne. Coupled design of test signals, sampling intervals, and filters for system identification. *IEEE Trans. Automatic Control, Vol. AC-19*, 1974.

[344] W.J. Gordon and R.F. Riesenfeld. B-spline curves and surfaces. In *Computer Aided Geometric Design*, R. Barnbill and R.F. Riesenfeld, editors, Academic Press, New York, 1974.

[345] F. Gori. Integral equations for incoherent imagery. *J. Opt. Soc. Am., Vol. 64*, pages 1237–1243, 1974.

[346] F. Gori. On an iterative method for super-resolution. *Int. Opt. Computing Conf., IEEE Catalog no. 75 CH0941-5C*, pages 137–141, 1975.

[347] F. Gori. Lau effect and coherence theory. *Opt. Comm.*, pages 4–8, 1979.

[348] F. Gori. Fresnel transform and sampling theorem. *Opt. Comm., Vol. 39*, pages 293–298, 1981.

[349] F. Gori and R. Grella. The converging prolate spheroidal functions and their use in Fresnel optics. *Opt. Comm.*, pages 5–10, 1983.

[350] F. Gori and G. Guattari. Effects of coherence on the degrees of freedom of an image. *J. Opt. Soc. Am., Vol. 61*, pages 36–39, 1971.

[351] F. Gori and G. Guattari. Holographic restoration of nonuniformly sampled bandlimited functions. *Opt. Comm., Vol. 3*, pages 147–149, 1971.

[352] F. Gori and G. Guattari. Imaging systems using linear arrays of non-equally spaced elements. *Phys. Lett., Vol 32A*, pages 38–39, 1970.

[353] F. Gori and G. Guattari. Nonuniform sampling in optical processing. *Optica Acta, Vol. 18*, pages 903–911, 1971.

[354] F. Gori and G. Guattari. Optical analog of a non-redundant array. *Phys. Lett. Vol. 32A*, pages 446–447, 1970.

[355] F. Gori and G. Guattari. Use of nonuniform sampling with a single correcting operation. *Opt. Comm., Vol. 3*, pages 404–406, 1971.

[356] F. Gori and G. Guattari. Shannon number and degrees of freedom of an image. *Opt. Comm.*, pages 163–165, 1973.

[357] F. Gori and G. Guattari. Degrees of freedom of images from point-like element pupils. *J. Opt. Soc. Am., Vol. 64*, pages 453–458, 1974.

[358] F. Gori and G. Guattari. Eigenfunction technique for point-like element pupils. *Opt. Acta*, pages 93–101, 1975.

[359] F. Gori and G. Guattari. Signal restoration for linear systems with weighted inputs; singular value analysis for two cases of low-pass filtering. *Inverse Problems*, pages 67–85, 1985.

[360] F. Gori, G. Guattari, C. Palma, and M. Santarsiero. Superresolving postprocessing for incoherent imagery. *Opt. Comm.*, pages 98–102, 1988.

[361] F. Gori and F. Mallamace. Interference between spatially separated holographic recordings. *Opt. Comm. Vol. 8*, pages 351–354, 1973.

[362] F. Gori and C. Palma. On the eigenvalues of the sinc kernel. *J. Phys. A, Math. Gen.*, pages 1709–1719, 1975.

[363] F. Gori, S. Paolucci, and L. Ronchi. Degrees of freedom of an optical image in coherent illumination, in the presence of aberrations. *J. Opt. Soc. Am., Vol. 65*, pages 495–501, 1975.

[364] F. Gori and L. Ronchi. Degrees of freedom for scatterers with circular cross section. *J. Opt. Soc. Am., Vol. 71*, pages 250–258, 1981.

[365] F. Gori and S. Wabnitz. Modifications of the Gerchberg method applied To electromagnetic imaging, In *Inverse Methods in Electromagnetic Imaging*, D. Reidel Publishing Co., Dordrecht, 1985, pages 1189–1203.

[366] D. Gottlieb and S.A. Orszag. *Numerical Analysis of Spectral Methods-Theory and Applications*. SIAM Publ., Philadelphia, 1977.

[367] I.S. Gradshteyn and I.M. Ryzhik. *Tables of Integrals, Products and Series*. Academic Press, New York, 1965.

[368] R.M. Gray, A. Buzo, A.H. Gray, Jr., and Y. Matsuyama. Distortion Measures for speech processing. *IEEE Trans. Acoust., Speech, Signal Processing, Vol. ASSP-28*, pages 367–376, Aug. 1980.

[369] U. Grenader and G. Szegö. *Toeplitz Forms and their Applications*. University of California Press, Berkeley, CA, 1958.

[370] B. Grobler. Practical application of the sampling theorem. *Feingerätetechnik, Vol. 26*, pages 342–345, 1977.

[371] C.R. Guarino. A new method of spectral analysis and sample rate reduction for bandlimited signals. *Proc. IEEE, Vol. 69*, pages 61–63, 1981.

[372] L.G. Gubin, B.T. Polyak, and E.V. Ralik. The method of projections for finding the common point of convex sets. *USSR Comput. Math. Phys., Vol. 7, No. 6*, pages 1–24, 1967.

[373] S.C. Gupta. Increasing the sampling efficiency for a control system. *IEEE Trans. Automatic Control, Vol. AC-8*, pages 263–264, 1963.

[374] M.K. Habib and S. Cambanis. Dyadic sampling approximations for non-sequency-limited signals. *Inform. Contr., Vol. 49*, pages 199–211, 1981.

[375] N.K. Habib and S. Cambanis. Sampling approximation for non-bandlimited harmonizable random signals. *Inform. Sci., Vol. 23*, pages 143–152, 1981.

[376] A.H. Haddad and J. B. Thomas. Integral representation for non-uniform sampling expansions. In *Proc. 4th Allerton Conference of Circuit System Theory*, pages 322–333, 1966.

[377] A.H. Haddad, K. Yao, and J.B. Thomas. General methods for the derivation of sampling theorems. *IEEE Trans. Information Theory, Vol. IT-13*, 1967.

[378] M.O. Hagler, R.J. Marks II, E.L. Kral, J.F. Walkup, and T.F. Krile. Scanning technique for coherent processors. *Appl. Opt., Vol. 19*, pages 1670–1672, 1980.

[379] H. Hahn. Über das Interpolationproblem. *Math. Z., Vol. 1*, pages 115–142, 1918.

[380] R. W. Hamming. *Numerical Methods for Scientists and Engineers.* McGraw-Hill, New York, 1962.

[381] J.L. Harris. Diffraction and resolving power. *J. Opt. Soc. Am., Vol. 54*, pages 931–936, 1964.

[382] C.J. Hartley. Resolution of frequency aliases in ultrasonic pulsed doppler velocimeters. *IEEE Trans. Sonics & Ultrasonics, Vol. SU-28*, pages 69–75, 1981.

[383] S. Haykin. *Introduction to Adaptive Filters.* MacMillan, New York, 1976.

[384] Z.S. Hegedus. Annular pupil arrays; application to confocal scanning. *Opt. Acta*, pages 815–826, 1985.

[385] H.D. Helms and J.B. Thomas. Truncation error of sampling theorem expansion. *Proc. IRE, Vol. 50*, pages 179–184, 1962.

[386] C.W. Helstrom. An expansion of a signal in Gaussian elementary signals. *IEEE Trans. Information Theory, Vol. IT-12*, pages 81–82, 1966.

[387] G.T. Herman, ed., *Image Reconstruction from Projections* Springer-Verlag, Berlin, 1979.

[388] G.T. Herman, J. Zheng and C.A. Bucholtz. Shape-based interpolation. *IEEE Computer Graphics and Applications Magazine, Vol. 12, No. 3*, pages 69–79, May 1992.

[389] R.S. Hershel, P.W. Scott, and T.E. Shrode. Image sampling using phase gratings. *J. Opt. Soc. Am., Vol. 66*, pages 861–863, 1976.

[390] J.R. Higgins. An interpolation series associated with the Bessel-Hankel transform. *J. London Math. Soc., Vol. 5*, pages 707–714, 1972.

[391] J.R. Higgins. A sampling theorem for irregularly spaced sample points. *IEEE Trans. Information Theory, Vol. IT-22*, pages 621–622, 1976.

[392] J.R. Higgins. *Completeness and Basis Properties of Sets of Special Functions.* Cambridge University Press, Cambridge, 1977.

[393] J.R. Higgins. Five short stories about the cardinal series. *Bull. Am. Math. Soc., Vol. 12*, pages 45–89, 1985.

[394] J.R. Higgins. Sampling theorems and the contour integral method. *Appl. Anal., Vol. 41*, pages 155–169, 1991.

[395] N.R. Hill and G.E. Ioup. Convergence of the Van Cittert iterative method of deconvolution. *J. Opt. Soc. Am., Vol. 66*, pages 487–489, 1976.

[396] G. Hinsen and D. Klösters. The sampling series as a limiting case of Lagrange interpolation. *Appl. Anal.* (to appear 1991).

[397] K. Hirano, M. Sakane and M.Z. Mulk. Design of 3–D Recursive Digital Filters. *IEEE Trans. Circuits and Systems, Vol. 31*, pages 550–561, June 1984.

[398] E. Holzler. Some remarks on special cases of the sampling theorem. *Frequenz, Vol. 31*, pages 265–269, 1977.

[399] I. Honda. Approximation for a class of signal functions by sampling series representation. *Keio Eng. Rep., Vol. 31*, pages 21–26, 1978.

[400] W. Hoppe. The finiteness postulate and the sampling theorem of the three dimensional electron microscopical analysis of aperiodic structures. *Optik, Vol. 29*, pages 619–623, 1969.

[401] H. Horiuchi. Sampling principle for continuous signals with time-varying bands. *Inform. Contr., Vol. 13*, pages 53–61, 1968.

[402] L. Hormander. *An Introduction to Complex Analysis in Several Variables.* Van Nostrand, Princeton, NJ, 1966.

[403] R.F. Hoskins and J.S. Pinto. Generalized sampling expansion in the sense of Papoulis. *SIAM J. Appl. Math., Vol. 44*, pages 611–617, 1984.

[404] R.F. Hoskins and J.S. Pinto. Sampling expansions and generalized translation invariance. *J. Franklin Inst., Vol. 317*, pages 323–332, 1984.

[405] G.H. Hostetter. Generalized interpolation of continuous-time signal samples. *IEEE Trans. Systems, Man, Cybernetics, Vol. SMC-11*, pages 433–439, 1981.

[406] S.J. Howard. Continuation of discrete Fourier spectra using minimum-negativity constraint. *J. Opt. Soc. Am., Vol. 71*, pages 819–824, 1981.

[407] S.J. Howard. Method for continuing Fourier spectra given by the fast Fourier transform. *J. Opt. Soc. Am., Vol. 71*, pages 95–98, 1981.

[408] S.J. Howard. Fast algorithm for implementing the minimum-negativity constraint for Fourier spectrum extrapolation. *Appl. Opt., Vol. 25*, pages 1670–1675, 1986.

[409] T.C. Hsiah. Comparisons of adaptive sampling control laws. *IEEE Trans. Automatic Control, Vol. AC-17*, pages 830–831, 1972.

[410] T.C. Hsiah. Analytic design of adaptive sampling control law in sampled data systems. *IEEE Trans. Automatic Control, Vol. 19*, pages 39–41, 1974.

[411] F.O. Huck, N. Halyo, and S.K. Park. Aliasing and blurring in 2-D sampled imagery. *Appl. Opt., Vol. 19*, pages 2174–2181, 1980.

[412] F.O. Huck and S.K. Park. Optical-mechanical line-scan imaging process: Its information capacity and efficiency. *Appl. Opt., Vol. 14*, pages 2508–2520, 1975.

[413] R. Hulthen. Restoring causal signals by analytic continuation: A generalized sampling theorem for causal signals. *IEEE Trans. Acoust., Speech, Signal Processing, Vol. ASSP-31*, pages 1294–1298, 1983.

[414] Y. Ichioka and N. Nakajima. Iterative image restoration consideration visibility. *J. Opt. Soc. Am., Vol. 71*, pages 983–988, 1981.

[415] O. Ikeda and T. Sato. Super-resolution imaging system using waves with a limited frequency bandwidth. *J. Acoust. Soc. Am.*, Vol. 6, pages 75–81, 1979.

[416] G. Indebetow. Propagation of spatially periodic wavefields. *Opt. Acta*, pages 531–539, 1984.

[417] G. Indebetow. Polychromatic self-imaging. *J. Mod. Opt.*, pages 243–252, 1988.

[418] D.L. Jagerman. Bounds for truncation error of the sampling expansion. *SIAM J. Appl. Math.*, Vol. 14, pages 714–723, 1966.

[419] D.L. Jagerman. Information theory and approximation of bandlimited functions. *Bell Systems Tech. J.*, Vol. 49, pages 1911–1941, 1970.

[420] D.L. Jagerman and L. Fogel. Some general aspects of the sampling theorem. *IRE Trans. Information Theory*, Vol. IT-2, pages 139–146, 1956.

[421] J. Jahns and A.W. Lohmann. The Lau effect (a diffraction experiment with incoherent illumination). *Opt. Comm.*, pages 263–267, 1979.

[422] A.K. Jain. *Fundamentals of Digital Image Processing*. Prentice-Hall, Englewood Cliffs, NJ, 1989.

[423] A.K. Jain and S. Ranganath. Extrapolation algorithms for discrete signals with application to spectral estimation. *IEEE Trans. Acoust., Speech, Signal Processing*, Vol. ASSP-29, pages 830–845, 1981.

[424] T. Jannson. Shannon number of an image and structural information capacity in volume holography. *Opt. Acta*, pages 1335–1344, 1980.

[425] A.J.E.M. Janssen. Weighted Wigner distributions vanishing on lattices. *J. Math. Anal. Appl.*, Vol. 80, pages 156–167, 1981.

[426] A.J.E.M. Janssen. Gabor representation of generalized functions. *J. Math. Anal. Appl.*, Vol. 83, pages 377–396, 1981.

[427] A.J.E.M. Janssen. Positivity of weighted Wigner distributions. *SIAM J. Math. Anal.*, Vol. 12, pages 752–758, 1981.

[428] A.J.E.M. Janssen. Bargmann transform, Zak transform and coherent states. *J. Math. Phys.*, Vol. 23, pages 720–731, 1982.

[429] A.J.E.M. Janssen. The Zak transform: a signal transform for sampled time-continuous signals. *Philips J. Res.*, Vol. 43, pages 23–69, 1988.

[430] A.J.E. Janssen, R.N. Veldhuis, and L.B. Vries. Adaptive interpolation of discrete time signals that can be modeled as autoregressive processes. *IEEE Trans. Acoust., Speech, Signal Processing*, Vol. ASSP-34, pages 317–330, 1986.

[431] P.A. Jansson, editor. *Deconvolution. With applications in Spectroscopy.* Academic Press, Orlando, FL, 1984.

[432] P.A. Jansson, R.H. Hunt, and E.K.Plyer. Resolution enhancement of spectra. *J. Opt. Soc. Am.*, Vol. 60, pages 596–599, 1970.

[433] Y.C. Jenq. Digital spectra of nonuniformly sampled signals: Digital look up tunable sinusoidal oscillators. *IEEE Trans. Instrumentation & Measurement*, Vol. 37, No. 3, pages 358–362, 1988.

[434] A.J. Jerri. On the application of some interpolating functions in physics. *J. Res. National Bureau of Standards -B. Math. Sciences, Vol. 73B, No. 3*, pages 241–245, 1969.

[435] A.J. Jerri. On the equivalence of Kramer's and Shannon's generalized sampling theorems. *IEEE Trans. Information Theory, Vol. IT-15*, pages 469–499, 1969.

[436] A.J. Jerri. Some applications for Kramer's generalized sampling theorem. *J. Eng. Math., Vol. 3, No. 2*, April 1969.

[437] A.J. Jerri. Application of the sampling theorem to time-varying systems. *J. Franklin Inst., Vol. 293*, pages 53–58, 1972.

[438] A.J. Jerri. Sampling for not necessarily finite energy signals. *Int. J. System Sci., Vol. 4, No. 2*, pages 255–260, 1973.

[439] A.J. Jerri. Sampling expansion for the $L_\nu^2$ -Laguerre integral transform. *J. Res. National Bureau of Standards -B. Math. Sciences, Vol. 80B*, pages 415–418, 1976.

[440] A.J. Jerri. The Shannon sampling theorem - its various extensions and applications: A tutorial review. *Proc. IEEE, Vol. 65, No. 11*, pages 1565–1596, 1977.

[441] A.J. Jerri. Computation of the hill functions of higher order. *Math. Comp., Vol. 31*, pages 481–484, 1977.

[442] A.J. Jerri. Towards a discrete Hankel transform and its applications. *J. of Appl. Anal., Vol. 7*, pages 97–109, 1978.

[443] A.J. Jerri. The application of general discrete transforms to computing orthogonal series and solving boundary value problems. *Bull. Calcutta Math. Soc., Vol. 71*, pages 177–187, 1979.

[444] A.J. Jerri. General transforms hill functions. *J. Applicable Analysis, Vol. 14*, pages 11–25, 1982.

[445] A.J. Jerri. A note on sampling expansion for a transform with parabolic cylinder kernel. *Inform. Sci., Vol. 26*, pages 1–4, 1982.

[446] A.J. Jerri. On a recent application of the Shannon sampling theorem. In *IEEE International Symposium on Information Theory, Quebec, Canada*, 1983.

[447] A.J. Jerri. A definite integral. *SIAM Rev., Vol. 25, No.1*, page 101, Jan. 1983.

[448] A.J. Jerri. Interpolation for the generalized sampling sum of approximation theory. *The AMS Summer Meeting, Albany, NY*, Aug. 1983.

[449] A.J. Jerri. Application of the transform–iterative method to nonlinear concentration boundary value problem. *Chem. Eng. Commun., Vol. 23*, pages 101–113, 1983.

[450] A.J. Jerri. The generalized sampling theorem for transforms of not necessarily square integrable functions. *J. Math., Math. Sci., Vol. 8*, pages 355–358, 1985.

[451] A.J. Jerri. *Introduction to Integral Equations with Applications*. Marcel Dekker, New York, 1985.

[452] A.J. Jerri. Part II: The Sampling Expansion – A Detailed Bibliography. 1986.

[453] A.J. Jerri. Equalities and Inequalities for Fourier Analysis and the Sampling Expansions with Detailed Bibliography, 1986.

[454] A.J. Jerri. An extended Poisson type sum formula for general integral transforms and aliasing error bound for the generalized sampling theorem. *J. Appl. Analysis, Vol. 26*, pages 199–221, 1988.

[455] A.J. Jerri. A recent modification of iterative methods for solving nonlinear problems. *The Math. Heritage of C.F. Gauss*, G.M. Rassias, editor, World Scientific Publ. Co., Singapore, 1991, pages 379–404.

[456] A.J. Jerri. *Integral and Discrete Transforms with Applications and Error Analysis*. Marcel Dekker Inc., New York, 1992.

[457] A.J. Jerri and E.J. Davis. Application of the sampling theorem to boundary value problems. *J. Eng. Math., Vol. 8*, pages 1–8, 1974.

[458] A.J. Jerri and D.W. Kreisler. Sampling expansions with derivatives for finite Hankel and other transforms. *SIAM J. Math. Anal., Vol. 6*, pages 262–267, 1975.

[459] A.J. Jerri. Truncation error for the generalized Bessel type sampling series. *J. Franklin Inst. (USA) Vol. 314, No. 5*, pages 323–328, Nov. 1982.

[460] A.J. Jerri, R.L. Herman and R.H. Weiland. A modified iterative method for nonlinear chemical concentration in cylindrical and spherical pellets. *Chem. Eng. Commun., Vol. 52*, pages 173–193, 1987.

[461] H. Johnen. Inequalities connected with moduli of smoothness. *Mat. Vesnik, Vol. 9 (24)*, 289–303, 1972.

[462] G. Johnson and J.H.S. Hutchinson. The limitations of NMR recalled–echo imaging techniques. *J. Magnetic Resonance, Vol. 63*, pages 14–30, 1985.

[463] M.C. Jones. The discrete Gerchberg algorithm. *IEEE Trans. Acoust., Speech, Signal Processing, Vol. ASSP-34*, pages 624–626, 1986.

[464] M. Kac, W.L. Murdock, and G. Szegö. On the eigenvalues of certain Hermitian forms. *J. Rat. Mech., Anal.*, pages 767–800, 1953.

[465] M.I. Kadec. The exact value of the Paley-Wiener constant *Soviet Math. Dokl., Vol. 5*, pages 559–561, 1964.

[466] R.E. Kahn and B. Liu. Sampling representation and optimum reconstruction of signals. *IEEE Trans. Information Theory, Vol. IT-11*, pages 339–347, 1965.

[467] T. Kailath. Channel characterization: time-variant dispersive channels. In *Lectures on Communications Systems Theory*, E.J. Baghdady, editor, McGraw-Hill, New York, 1960.

[468] S.C. Kak. Sampling theorem in Walsh-Fourier analysis. *Electron. Lett., Vol. 6*, pages 447–448, 1970.

[469] G. Kakoza and D. Munson. A frequency-domain characterization of interpolation from nonuniformly spaced data. In *Proceedings of the International Symposium on Circuits and Systems, Portland, Oregon*, pages 288–291, May 9-11, 1989.

[470] N.S. Kambo and F.C. Mehta. Truncation error for bandlimited non-stationary processes. *Inform. Sci., Vol. 20*, pages 35–39, 1980.

[471] N.S. Kambo and F.C. Mehta. An upper bound on the mean square error in the sampling expansion for non-bandlimited processes. *Inform. Sci., Vol. 21*, pages 69–73, 1980.

[472] L.M. Kani and J.C. Dainty. Super-resolution using the Gerchberg algorithm. *Opt. Comm.*, pages 11–17, 1988.

[473] D. Kaplan and R.J. Marks II. Noise sensitivity of interpolation and extrapolation matrices. *Appl. Opt., Vol. 21*, pages 4489–4492, 1982.

[474] R. Kasturi, T.T. Krile, and J.F. Walkup. Space-variant 2-D processing using a sampled input/sampled transfer function approach. *SPIE Inter. Optical Comp. Conf., Vol. 232*, pages 182–190, 1980.

[475] M. Kawamura and S. Tanaka. Proof of sampling theorem in sequency analysis using extended Walsh functions. *Systems-Comput. Controls, Vol. 9*, pages 10–15, 1980.

[476] Steven M. Kay. The effect of sampling rate on autocorrelation estimation. *IEEE Trans. Acoust., Speech, Signal Processing, Vol. ASSP-29, No. 4*, page 859–867, 1981.

[477] R.B. Kerr. Polynomial interpolation errors for bandlimited random signals. *IEEE Trans. Systems, Man, Cybernetics*, pages 773–774, 1976.

[478] R.B. Kerr. Truncated sampling expansions for bandlimited random signals. *IEEE Trans. Systems, Man, Cybernetics, Vol. SMC-9*, pages 362–364, 1979.

[479] A.A. Kharkevich. Kotel'nikov's theorem - a review of some new publications. *Radiotekhnika, Vol. 13*, pages 3–10, 1958.

[480] Y.I. Khurgin and V.P. Yakovlev. Progress in the Soviet Union on the theory and applications of bandlimited functions. *Proc. IEEE, Vol. 65*, pages 1005–1029, 1976.

[481] T. Kida. Generating functions for sampling theorems. *Electronics and Communications in Japan, (English Translation), Vol. 65, No. 1*, pages 9–18, January 1982.

[482] T. Kida. Extended formulas of sampling theorem on bandlimited waves and their application to the design of digital filters for data transmission. *Electron Communication Japan, Part 1, Vol. 68, No. 11*, pages 20–29, November 1985.

[483] G.W. King and A.G. Emslie. Spectroscopy from the point of view of the communication theory I. *J. Opt. Soc. Am., Vol. 41*, pages 405–409, 1951.

[484] G.W. King and A.G. Emslie. Spectroscopy from the point of view of the communication theory III. *J. Opt. Soc. Am., Vol. 43*, pages 664–668, 1953.

[485] S.J. Kishner and T.W. Barnard. Detection and location of point images in sampled optical systems. *J. Opt. Soc. Am., Vol. 62*, pages 17–20, 1972.

[486] I. Kluvanek. Sampling theorem in abstract harmonic analysis. *Mat. Casopis Sloven. Akad. Vied, Vol. 15*, pages 43–48, 1965.

[487] J.J. Knab. System error bounds for Lagrange estimation of bandlimited functions. *IEEE Trans. Information Theory, Vol. IT-21*, pages 474–476, 1975.

[488] J.J. Knab. Interpolation of bandlimited functions using the approximate prolate series. *IEEE Trans. Information Theory, Vol. IT-25*, pages 717–720, 1979.

[489] J.J. Knab. Simulated cardinal series errors versus truncation error bounds. *Proc. IEEE, Vol. 68*, pages 1020–1060, 1980.

[490] J.J. Knab. Noncentral interpolation of bandlimited signals. *IEEE Trans. Aerospace & Electronic Systems, Vol. AES-17*, pages 586–591, 1981.

[491] J.J. Knab and M.I. Schwartz. A system error bound for self truncating reconstruction filter class. *IEEE Trans. Information Theory, Vol. IT-21*, pages 341–342, 1975.

[492] Y. Kobayashi and M. Inoue. Combination of two discrete Fourier transforms of data sampled separately at different rates. *J. Inf. Process, Vol. 3*, pages 203–207, 1980.

[493] T. Kodama. Sampling theorem application to SAW diffraction simulation. *Electron. Lett., Vol. 16*, pages 460–462, 1980.

[494] H. Kogelnik and T. Li. Laser beams and resonators. *Proc. IEEE*, pages 1312–1329, 1966.

[495] A. Kohlenberg. Exact interpolation of bandlimited functions. *J. Appl. Phys., Vol. 24*, pages 1432–1436, 1953.

[496] D. Kohler and L. Mandel. Source reconstruction from the modulus of the correlation function: a practical approach to the phase problem of optical coherence theory. *J. Opt. Soc. Am., Vol. 63*, pages 126–134, 1973.

[497] A.N. Kolmogorov. Interpolation und Extrapolation von stationären zufälligen Folgen (Russian). *Bull Acad. Sci. USSR, Ser. Math, Vol. 5*, pages 3–14, 1941.

[498] B.H. Kolner and D.M. Bloom. Direct electric-optic sampling of transmission-line signals propagating on a GaAs substrate. *Electron. Lett., Vol. 20*, pages 818–819, 1984.

[499] A. Kolodziejczyk. Lensless multiple image formation by using a sampling filter. *Opt. Comm.*, pages 97–102, 1986.

[500] R. Konoun. PPM reconstruction using iterative techniques, project report. Technical report, Dept. of ECE, Illinois Institute of Technology, Chicago, September, 1989.

[501] A. Korpel. Gabor: frequency, time and memory. *Appl. Opt.*, pages 3624–3632, 1982.

[502] V.A. Kotel'nikov. On the carrying capacity of "ether" and wire in electrocommunications. In *Idz. Red. Upr. Svyazi RKKA (Moscow)*, Material for the first all-union conference on the questions of communications, 1933.

[503] H.P. Kramer. A generalized sampling theorem. *J. Math. Phys., Vol. 38*, pages 68–72, 1959.

[504] H.P. Kramer. The digital form of operators on bandlimited functions. *J. Math. Anal. Appl.*, Vol. 44, No. 2, pages 275–287, 1973.

[505] G.M. Kranc. Comparison of an error-sampled system by a multirate controller. *Trans. Am. Inst. Elec. Eng.*, Vol. 76, pages 149–159, 1957.

[506] M. G. Krein. On a fundamental approximation problem in the theory of extrapolation and filtration of stationary random processes (Russian). *Dokl. Akad. Nauk SSSR*, Vol. 94, pages 13–16, 1954. [English translation: *Amer. Math. Soc. Selected Transl. Math. Statist. Prob.*, Vol. 4, pages 127–131], 1964.

[507] R. Kress. On the general Hermite cardinal interpolation. *Math. Comput.*, Vol. 26, pages 925–933, 1972.

[508] H.N. Kritikos and P.T. Farnum. Approximate evaluation of Gabor expansions. *IEEE Trans. Syst., Man, Cybern.*, Vol. SMC-17, pages 978–981, 1987.

[509] A. Kumar and O.P. Malik. Discrete analysis of load-frequency control problem. *Proc. IEEE*, Vol. 131, pages 144–145, 1984.

[510] G.M. Kurajian and T.Y. Na. Elastic beams on nonlinear continuous foundations. *ASME Winter Annual Meeting*, pages 1–7, Nov. 15–20, 1981.

[511] H.J. Kushner and L. Tobias. On the stability of randomly sampled systems. *IEEE Trans. Automatic Control*, Vol. AC-14, pages 319–324, 1969.

[512] B. Lacaze. A generalization of $n$-th sampling formula. *Ann. Telecomm.*, Vol. 30, pages 208–210, 1975.

[513] P. Lancaster and M. Tiemenesky. *The Theory of Matrices*. Academic Press, Orlando, FL, 1985.

[514] H.J. Landau. On the recovery of a bandlimited signal after instantaneous companding and subsequent band limiting. *Bell Systems Tech. J.*, Vol. 39, pages 351–364, 1960.

[515] H.J. Landau. Necessary density conditions for sampling and interpolation of certain entire functions. *Acta Math.*, pages 37–52, 1967.

[516] H.J. Landau. Sampling, data transmission and the Nyquist rate. *Proc. IEEE*, Vol. 55, pages 1701–1706, 1967.

[517] H.J. Landau. On Szegö's eigenvalue distribution theorem and non-Hermitian kernels. *J. Analyse Math.*, pages 335–357, 1975.

[518] H.J. Landau and W.L. Miranker. The recovery of distorted bandlimited signals. *J. Math. Anal. & Appl.*, Vol. 2, pages 97–104, 1961.

[519] H.J. Landau and H.O. Pollak. Prolate spheroidal wave functions Fourier analysis and uncertainty II. *Bell Systems Tech. J.*, Vol. 40, pages 65–84, 1961.

[520] H.J. Landau and H.O. Pollak. Prolate spheroidal wave functions Fourier analysis and uncertainty III: The dimension of the space of essentially time- and bandlimited signals. *Bell Systems Tech. J.*, Vol. 41, pages 1295–1336, 1962.

[521] H.J. Landau and H. Widom. Eigenvalue distribution of time and frequency limiting. *J. Math. Anal., Appl.*, Vol. 77, pages 469–481, 1980.

[522] A. Landé. Optik und Thermodynamik, In *Handbuch der Physik*, Springer-Verlag, Berlin, 1928, pages 453–479.

[523] L. Landweber. An iteration formula for Fredholm integral equations of the first kind. *Amer. J. Math.*, pages 615–624, 1951.

[524] B.P. Lathi. *Communication Systems*. John Wiley & Sons, Inc., New York, 1965.

[525] A.J. Lee. On bandlimited stochastic processes. *SIAM J. Appl. Math.*, Vol. *30*, pages 269–277, 1976.

[526] A.J. Lee. Approximate interpolation and the sampling theorem. *SIAM J. Appl. Math.*, Vol. *32*, pages 731–744, 1977.

[527] A.J. Lee. Sampling theorems for nonstationary random processes. *Trans. Amer. Math. Soc.*, Vol. *242*, pages 225–241, 1978.

[528] A.J. Lee. A note on the Campbell sampling theorem. *SIAM J. Appl. Math.*, Vol. *41*, pages 553–557, 1981.

[529] H. Lee. On orthogonal transformations. *IEEE Trans. Circuits & Systems*, Vol. *CAS-32*, pages 1169–1177, 1985.

[530] Y. W. Lee. *Statistical Theory of Communication*. John Wiley & Sons, New York, 1960.

[531] W. H. Lee. Computer-generated holograms: techniques and applications, In *Progress in Optics*, North-Holland, Amsterdam, 1978, pages 119–232.

[532] E.N. Leith and J. Upatnieks. Reconstructed wavefronts and communication theory. *J. Opt. Soc. Am.*, pages 1123–1130, 1962.

[533] O.A.Z. Leneman. Random sampling of random processes: impulse processes. *Inform. Contr.*, Vol. *9*, pages 347–363, 1966.

[534] O.A.Z. Leneman. On error bounds for jittered sampling. *IEEE Trans. Automatic Control*, Vol. *AC-11*, page 150, 1966.

[535] O.A.Z. Leneman. Random sampling of random processes: optimum linear interpolation. *J. Franklin Inst.*, Vol. *281*, pages 302–314, 1966.

[536] O.A.Z. Leneman and J. Lewis. A note on reconstruction for randomly sampled data. *IEEE Trans. Automatic Control*, Vol. *AC-10*, page 626, 1965.

[537] O.A.Z. Leneman and J. Lewis. Random sampling of random processes: Mean square comparison of various interpolators. *IEEE Trans. Automatic Control*, Vol. *AC-10*, pages 396–403, 1965.

[538] O.A.Z. Leneman and J. Lewis. On mean-square reconstruction error. *IEEE Trans. Automatic Control*, Vol. *AC-11*, pages 324–325, 1966.

[539] A. Lent and H. Tuy. An iterative method for the extrapolation of bandlimited functions. *J. Math. Anal., Appl.*, pages 554–565, 1981.

[540] L. Levi. Fitting a bandlimited signal to given points. *IEEE Trans. Information Theory*, Vol. *IT-11*, pages 372–376, 1965.

[541] N. Levinson. Gap and density theorems. In *Colloq. Pub. 26*. Amer. Math. Soc., New York, 1940.

[542] G.B. Lichtenberger. A note on perfect predictability and analytic processes. *IEEE Trans. Information Theory, Vol. IT-20*, pages 101–102, 1974.

[543] Y. Linde, A. Buzo and R.M. Gray. An algorithm for vector quantizer design. *IEEE Trans. Commun., Vol. COM-28*, pages 89–95, Jan. 1980.

[544] D.A. Linden. A discussion of sampling theorems. *Proc. IRE, Vol. 47*, pages 1219–1226, 1959.

[545] D.A. Linden and N.M. Abramson. A generalization of the sampling theorem. *Inform. Contr., Vol. 3*, pages 26–31, 1960.

[546] E.H. Linfoot. Information theory and optical images. *J. Opt. Soc. Am.*, pages 808–819, 1955.

[547] E.H. Linfoot. Quality evaluation of optical systems. *Opt. Acta*, pages 1–14, 1958.

[548] C. L. Liu and Jane W. S. Liu. *Linear Systems Analysis*. McGraw-Hill, New York, 1975.

[549] S.P. Lloyd. A sampling theorem for stationary (wide sense) stochastic processes. *Trans. Am. Math. Soc., Vol. 92*, pages 1–12, 1959.

[550] Robert P. Loce and Ronald E. Jodoin. Sampling theorem for geometric moment determination and its application to a laser beam position detector. *Appl. Opt., Vol. 29, No. 26*, pages 3835–3843, 1990.

[551] P.P. Loesberg. Low-frequency A-D input systems. *Instrum. Control Syst., Vol. 43*, pages 124–126, 1970.

[552] A.W. Lohmann and D.P. Paris. Binary Fraunhofer holograms, generated by computer. *Appl. Opt.*, pages 1739–1748, 1967.

[553] A.W. Lohmann. The space-bandwidth product applied to spatial filtering and to holography. *IBM Research Paper, RJ-438*, 1967.

[554] A.W. Lohmann and D.P. Paris. Computer generated spatial filters for coherent optical data processing. *Appl. Opt.*, pages 651–655, 1968.

[555] A.W. Lohmann. An interferometer based on the Talbot effect. *Opt. Comm.*, pages 413–415, 1971.

[556] A.W. Lohmann. Three-dimensional properties of wave-fields. *Optik*, pages 105–117, 1978.

[557] D.G. Luenberger. *Optimization by Vector Space Methods*. Wiley, New York, 1969.

[558] H.D. Luke. Zur Entsehung des Abtasttheorems. *Nachr. Techn. Z., Vol. 31*, pages 271–274, 1978.

[559] W. Lukosz. Optical systems with resolving powers exceeding the classical limit. *J. Opt. Soc. Am., Vol. 56*, pages 1463–1472, 1966.

[560] W. Lukosz. Optical systems with resolving powers exceeding the classical limit, II. *J. Opt. Soc. Am., Vol. 57*, pages 932–941, 1967.

[561] A. Luthra. Extension of Parseval's relation to nonuniform sampling. *IEEE Trans. Acoust., Speech, Signal Processing, Vol. ASSP-36, No. 12*, pages 1909–1911, 1988.

[562] D.M. Mackay. Quantal aspects of scientific information. *Phil. Mag.*, pages 289–311, 1950.

[563] D.M. Mackay. The structural information-capacity of optical instruments. *Inform. Contr.*, pages 148–152, 1958.

[564] J. Maeda and K. Murata. Digital restoration of incoherent bandlimited images. *Appl. Opt.*, pages 2199–2204, 1982.

[565] H. Maitre. Iterative super-resolution. Some new fast methods. *Opt. Acta*, pages 973–980, 1981.

[566] J. Makhoul. Linear prediction: A tutorial review. *Proc. IEEE, Vol. 63*, pages 561–580, 1975.

[567] N.J. Malloy. Nonuniform sampling for high resolution spectrum analysis. In *Proc. ICASSP '84*, 1984.

[568] L. Mandel and E. Wolf. Coherence properties of optical fields. *Rev. Mod. Phys.*, pages 231–287, 1965.

[569] L. Mandel and E. Wolf. Spectral coherence and the concept of cross-spectral purity. *J. Opt. Soc. Am., Vol. 66*, pages 529–535, 1976.

[570] M. Maqusi. A sampling theorem for dyadic stationary processes. *IEEE Trans. Acoust. Speech, Signal Processing, Vol. ASSP-26*, pages 265–267, 1978.

[571] M. Maqusi. Truncation error bounds for sampling expansions of sequency band limited signals. *IEEE Trans. Acoust., Speech, Signal Processing, Vol. ASSP-26*, pages 372–374, 1978.

[572] M. Maqusi. Sampling representation of sequency bandlimited nonstationary random processes. *IEEE Trans. Acoust., Speech, Signal Processing, Vol. ASSP-28*, pages 249–251, 1980.

[573] M. Maqusi. Correlation and spectral analysis of nonlinear transformation of sequency bandlimited signals. *IEEE Trans. Acoust., Speech, Signal Processing, Vol. ASSP-30*, pages 513–516, 1982.

[574] E.W. Marchand and E. Wolf. Radiometry with sources of any state of coherence. *J. Opt. Soc. Am., Vol. 64*, pages 1219–1226, 1974.

[575] A. Maréchal and M. Françon. *Diffraction, structure des images*. Masson, Paris, 1970.

[576] J.W. Mark and T.D. Todd. A nonuniform sampling approach to data compression. *IEEE Trans. Communications, Vol. COM-29*, pages 24–32, 1981.

[577] W.D. Mark. Spectral analysis of the convolution and filtering of non-stationary stochastic processes. *J. Sound Vib., Vol. 11*, pages 19–63, 1970.

[578] J. D. Markel and H. H. Gray, Jr. *Linear Prediction of Speech*. Springer-Verlag, New York, 1982.

[579] R.J. Marks II. Two-dimensional coherent space-variant processing using temporal holography. *Appl. Opt., Vol. 18*, pages 3670–3674, 1979.

[580] R.J. Marks II. Coherent optical extrapolation of two-dimensional signals: processor theory. *Appl. Opt., Vol. 19*, pages 1670–1672, 1980.

[581] R.J. Marks II. Gerchberg's extrapolation algorithm in two dimensions. *Appl. Opt., Vol. 20*, pages 1815–1820, 1981.

[582] R.J. Marks II. Sampling theory for linear integral transforms. *Opt. Lett., Vol. 6*, pages 7–9, 1981.

[583] R.J. Marks II. Posedness of a bandlimited image extension problem in tomography. *Opt. Lett., Vol. 7*, pages 376–377, 1982.

[584] R.J. Marks II. Restoration of continuously sampled bandlimited signals from aliased data. *IEEE Trans. Acoust., Speech, Signal Processing, Vol. ASSP-30*, pages 937–942, 1982.

[585] R.J. Marks II. Noise sensitivity of bandlimited signal derivative interpolation. *IEEE Trans. Acoust., Speech, Signal Processing, Vol. ASSP-31*, pages 1029–1032, 1983.

[586] R.J. Marks II. Restoring lost samples from an oversampled bandlimited signal. *IEEE Trans. Acoust., Speech, Signal Processing, Vol. ASSP-31*, pages 752–755, 1983.

[587] R.J. Marks II. Linear coherent optical removal of multiplicative periodic degradations: processor theory. *Opt. Eng., Vol. 23*, pages 745–747, 1984.

[588] R.J. Marks II. Multidimensional signal sample dependency at Nyquist densities. *J. Opt. Soc. Am. A, Vol. 3*, pages 268–273, 1986.

[589] R.J. Marks II. Class of continuous level associative memory neural nets. *Appl. Opt.*, pages 2005–2010, 1987.

[590] R.J. Marks II. *Introduction to Shannon Sampling and Interpolation Theory*. Springer-Verlag, New York, 1991.

[591] R.J. Marks II and M.W. Hall. Differintegral interpolation from a bandlimited signal's samples. *IEEE Trans. Acoust., Speech, Signal Processing, Vol. ASSP-29*, pages 872–877, 1981.

[592] R.J. Marks II and D. Kaplan. Stability of an algorithm to restore continuously sampled bandlimited images from aliased data. *J. Opt. Soc. Am., Vol. 73*, pages 1518–1522, 1983.

[593] R.J. Marks II and T.F. Krile. Holographic representations of space-variant systems: System theory. *Appl. Opt., Vol. 15*, pages 2241–2245, 1976.

[594] R.J. Marks II, J.F. Walkup, and T.F. Krile. Ambiguity function display: An improved coherent processor. *Appl. Opt., Vol. 16*, pages 746–750, 1977; addendum *Vol. 16*, page 1777, 1977.

[595] R.J. Marks II and M.W. Hall. Ambiguity function display using a single 1-D input. *Appl. Opt., Vol. 18*, pages 2539–2540, 1979.

[596] R.J. Marks II and D. Radbel. Error in linear estimation of lost samples in an oversampled bandlimited signal. *IEEE Trans. Acoust., Speech, Signal Processing, Vol. ASSP-32*, pages 648–654, 1984.

[597] R.J. Marks II and D.K. Smith. Gerchberg-type linear deconvolution and extrapolation algorithms. In *Transformations in Optical Signal Processing*, W.T. Rhodes, J.R. Fienup, and B.E.A. Saleh, editors, *SPIE, Vol. 373*, pages 161–178, 1984.

[598] R.J. Marks II and M.J. Smith. Closed form object restoration from limited spatial and spectral information. *Opt. Lett., Vol. 6*, pages 522–524, 1981.

[599] R.J. Marks II and S.M. Tseng. Effect of sampling on closed form bandlimited signal interval interpolation. *Appl. Opt., Vol. 24*, pages 763–765, Erratum, *Vol. 24*, page 2490, 1985.

[600] R.J. Marks II, J.F. Walkup, and M.O. Hagler. Volume hologram representation of space-variant systems. In *Applications of Holography and Optical Data Processing*, E. Marom, A.A. Friesem, and E. Wiener-Aunear, editors, Pergamon Press, Oxford, 1977, pages 105–113.

[601] R.J. Marks II, J.F. Walkup, and M.O. Hagler. Sampling theorems for linear shift-variant systems. *IEEE Trans. Circuits & Systems, Vol. CAS-25*, pages 228–233, 1978.

[602] R.J. Marks II, J.F. Walkup, and M.O. Hagler. Methods of linear system characterization through response cataloging. *Appl. Opt., Vol. 18*, pages 655–659, 1979.

[603] R.J. Marks II, J.F. Walkup, and M.O. Hagler. Line spread function notation. *Appl. Opt., Vol. 15*, pages 2289–2290, 1976.

[604] R.J. Marks II, J.F. Walkup, and M.O. Hagler. A sampling theorem for space-variant systems. *J. Opt. Soc. Am., Vol. 66*, pages 918–921, 1976.

[605] A.G. Marshall and F.R. Verdun. *Fourier transforms in NMR, Optical and Mass Spectrometry*. Elsevier, Amsterdam, 1990.

[606] H.G. Martinez and T.W. Parks. A class of infinite-duration impulse response digital filters for sampling rate reduction. *IEEE Trans. Acoust., Speech, Signal Processing, Vol. ASSP-27*, pages 154–162, 1979.

[607] F. Marvasti. The extension of Poisson sum formula to nonuniform samples. *Proc. 5th Aachener Kolloquium, Aachen, Germany*, 1984.

[608] F. Marvasti. A note on modulation methods related to sine wave crossings. *IEEE Trans. Communications*, pages 177–178, 1985.

[609] F. Marvasti. Reconstruction of a signal from the zero-crossings of an FM signal. *Trans. IECE of Japan*, page 650, 1985.

[610] F. Marvasti. Signal recovery from nonuniform samples and spectral analysis of random samples. In *IEEE Proceedings on ICASSP, Tokyo*, pages 1649–1652, April 1986.

[611] F. Marvasti. Spectral analysis of random sampling and error free recovery by an iterative method. *IECE Trans. of Inst. Electron. Commun. Engs., Japan (Section E), Vol. E 69, No. 2*, Feb. 1986.

[612] F. Marvasti. A unified approach to zero-crossings, nonuniform and random sampling of signals and systems. In *Proceedings of the International Symposium on Signal Processing and Its Applications, Brisbane, Australia*, pages 93–97, 1987.

[613] F. Marvasti. Minisymposium on zero-crossings and nonuniform sampling, In *Abstract of 4 presentations in the Proceedings of SIAM Annual Meeting, Minneapolis, Minnesota*, July 1988.

[614] F. Marvasti. Analysis and recovery of sample-and-hold signals with irregular samples. In *Annual Proceedings of the Allerton Conference on
Communication, Control and Computing*, October 1989.

[615] F. Marvasti. Minisymposium on nonuniform sampling for 2-D signals. In
*Abstract of 4 presentations in the Proceedings of SIAM Annual Meeting,
San Diego, California*, 1989.

[616] F. Marvasti. Rebuttal on the comment on the properties of two-
dimensional bandlimited signals. *J. Opt. Soc. Am. A, Vol. 6, No. 9*, page
1310, 1989.

[617] F. Marvasti. Reconstruction of 2-D signals from nonuniform samples or
partial information. *J. Opt. Soc. Am. A, Vol. 6*, pages 52–55, 1989.

[618] F. Marvasti. Relationship between discrete spectrum of frequency modulated (FM) signals and almost periodic modulating signals. *Trans. IECE
Japan*, pages 92–94, 1989.

[619] F. Marvasti. Spectral analysis of nonuniform samples of irregular samples
of multidimensional signals. In *6th Workshop on Multidimensional Signal
Processing, California*, September 1989.

[620] F. Marvasti. Extension of Lagrange interpolation to 2-D signals in polar
coordinates. *IEEE Trans. Circuits & Systems*, March 1990.

[621] F. Marvasti. The interpolation of 2-D signals from their isolated zeros. *J.
Multidimensional Systems and Signal Processing*, March 1990.

[622] F. Marvasti and M. Analoui. Recovery of signals from nonuniform samples
using iterative methods. In *IEEE Proceedings of International Conference
on Circuits and Systems, Oregon, July*, 1989.

[623] F. Marvasti, M. Analoui, and M. Gamshadzahi. Recovery of signals
from nonuniform samples using iterative methods. *IEEE Trans. Acoust.,
Speech, Signal Processing*, April 1991.

[624] F. Marvasti, Peter Clarkson, Dobcik Miraslov, and Chuande Liu. Speech
recovery from missing samples. In *Proc. IASTED Conference on Control
and Modeling, Tehran, Iran*, June 1990.

[625] F. Marvasti and Reda Siereg. Digital signal processing using FM zero-
crossing. In *Proceedings of IEEE International Conference on Systems
Engineering, Wright State University, August*, 1989.

[626] F.A. Marvasti. Spectrum of nonuniform samples. *Electron. Lett., Vol. 20,
No. 2*, 1984.

[627] F.A. Marvasti. Comments on a note on the predictability of bandlimited
processes. *Proc. IEEE, Vol. 74*, page 1596, 1986.

[628] F.A. Marvasti. *A Unified Approach to Zero-Crossing and Nonuniform
Sampling of Single and Multidimensional Signals and Systems.*
Nonuniform Publications, Oak Park, IL, 1987.

[629] F.A. Marvasti and A.K. Jain. Zero crossings, bandwidth compression and
restoration of nonlinearly distorted bandlimited signals. *J. Opt. Soc. Am.
A, Vol. 3*, pages 651–654, 1986.

[630] F.A. Marvasti. Extension of Lagrange interpolation to 2-D nonuniform samples in polar coordinates. *IEEE Trans. Circuits & Systems, Vol. 37, No. 4*, pages 567–568, 1990.

[631] F.A. Marvasti and Liu Chuande. Parseval relationship of nonuniform samples of one and two-dimensional signals. *IEEE Trans. Acoust., Speech, Signal Processing, Vol. 38, No. 6*, pages 1061–1063, 1990.

[632] F.A. Marvasti. Oversampling as an alternative to error correction codes. *Minisymposium on Sampling Theory and Practice, SIAM Annual Meeting, Chicago, IL*, July 1990. Also in ICC'92, Chicago, IL, June 1992.

[633] F.A. Marvasti and T.J. Lee. Analysis and recovery of sample-and -hold and linearly interpolated signals with irregular samples. *IEEE Trans. Signal Processing, Vol. 40, No. 8*, pages 1884–1891, 1992.

[634] E. Masry. Random sampling and reconstruction of spectra. *Inf. Contr., Vol. 19*, pages 275–288, 1971.

[635] E. Masry. Alias-free sampling: an alternative conceptualization and its applications. *IEEE Trans. Information Theory, Vol. IT-24*, 1978.

[636] E. Masry. Poisson sampling and spectral estimation of continuous time processes. *IEEE Trans. Information Theory, Vol. IT-24*, 1978.

[637] E. Masry. The reconstruction of analog signals from the sign of their noisy samples. *IEEE Trans. Information Theory, Vol. IT-27*, pages 735–745, 1981.

[638] E. Masry. The approximation of random reference sequences to the reconstruction of clipped differentiable signals. *IEEE Trans. Acoust., Speech, Signal Processing, Vol. ASSP-30*, pages 953–963, 1982.

[639] E. Masry and S. Cambanis. Bandlimited processes and certain nonlinear transformations. *J. Math. Anal., Vol. 53*, pages 59–77, 1976.

[640] E. Masry and S. Cambanis. Consistent estimation of continuous time signals from nonlinear transformations of noisy samples. *IEEE Trans. Information Theory, Vol. IT-27*, pages 84–96, 1981.

[641] E. Masry, D. Kramer, and C. Mirabile. Spectral estimation of continuous time process: Performance comparison between periodic and Poisson sampling schemes. *IEEE Trans. Automatic Control, Vol. AC-23*, pages 679–685, 1978.

[642] J. Mathews and R.L. Walker. *Mathematical Methods of Physics*, 2nd ed. W.A. Benjamin, Menlo Park, CA, 1970.

[643] M.J. McDonnell. A sampling function appropriate for deconnotation. *IEEE Trans. Information Theory, Vol. IT-22*, pages 617–621, 1976.

[644] K.C. McGill and L.J. Dorfman. High-resolution alignment of sampled waveforms. *IEEE Trans. Biomedical Engineering, Vol. BME-31*, pages 462–468, 1984.

[645] J. McNamee, F. Stenger, and E.L. Whitney. Whittaker's cardinal function in retrospect. *Math. Comp., Vol. 25*, pages 141–154, 1971.

[646] F.C. Mehta. A general sampling expansion. *Inform. Sci., Vol. 16*, pages 41–44, 1978.

[647] P.M. Mejías and R. Martínez Herrero. Diffraction by one-dimensional Ronchi grids: on the validity of the Talbot effect. *J. Opt. Soc. Am. A, Vol. 8*, pages 266–269, 1991.

[648] R.M. Mersereau. The processing of hexagonally sampled two dimensional signals. *Proc. IEEE, Vol. 67*, pages 930–949, 1979.

[649] R.M. Mersereau and T.C. Speake. The processing of periodically sampled multidimensional signals. *IEEE Trans. Acoust., Speech, Signal Processing, Vol. ASSP-31, No. 31*, pages 188-194, 1983.

[650] D. Middleton. *An Introduction to Statistical Communication Theory.* McGraw-Hill, New York, 1960.

[651] D. Middleton and D.P. Peterson. A note on optimum presampling filters. *IEEE Trans. Circuit Theory, Vol. CT-10*, pages 108–109, 1963.

[652] R. Mintzer and B. Liu. Aliasing error in the design of multirate filters. *IEEE Trans. Acoust., Speech, Signal Processing, Vol. ASSP-26*, pages 76–88, 1978.

[653] J.R. Mitchell and W.L. McDaniel Jr. Adaptive sampling technique. *IEEE Trans. Automatic Control, Vol. AC-14*, pages 200–201, 1969.

[654] J.R. Mitchell and W.L. McDaniel Jr. Calculation of upper bounds for errors of an approximate sampled frequency response. *IEEE Trans. Automatic Control, Vol. AC-19*, pages 155–156, 1974.

[655] H. Miyakawa. Sampling theorem of stationary stochastic variables in multi-dimensional space (Japanese). *J. Inst. Electron. Commun. Eng. Japan, Vol. 42*, pages 421–427, 1959.

[656] K. Miyamoto. On Gabor's expansion theorem. *J. Opt. Soc. Am., Vol. 50*, pages 856–858, 1960.

[657] K. Miyamoto. Note on the proof of Gabor's expansion theorem. *J. Opt. Soc. Am., Vol. 51*, pages 910–911, 1961.

[658] W.D. Montgomery. The gradient in the sampling of $n$-dimensional bandlimited functions. *J. Electron. Contr., Vol. 17*, pages 437–447, 1964.

[659] W.D. Montgomery. Algebraic formulation of diffraction applied to self-imaging. *J. Opt. Soc. Am., Vol. 58*, pages 1112–1124, 1968.

[660] D.R. Mook, G.V. Fisk, and A.V. Oppenheim, A hybrid numerical-analytical technique for the computation of wave fields in stratified media based on the Hankel transform. *J. Acoustics. Soc. Am., Vol. 76, No. 1*, pages 222–243, 1984.

[661] N. Moray, G. Synnock, and S. Richards. Tracking a static display. *IEEE Trans. Systems, Man, Cybernetics, Vol. SMC-3*, pages 518–521, 1973.

[662] F. Mori and J. De Steffano III. Optimal nonuniform sampling interval and test-input design for identification of physiological systems from very limited data. *IEEE Trans. Automatic Control, Vol. AC-24, No. 6*, pages 893–900, December 1979.

[663] H. Mori, I. Oppenheim, and J. Ross. *Some Topics in Quantum Statistics: The Wigner Function and Transport Theory.* Studies in Statistical Mechanics, J. de Boer and G.E. Uhlenbeck, eds., North-Holland, Amsterdam, 1962, Vol. 1, pages 213–298.

[664] D.H. Mugler and W. Splettstösser. Difference methods and round-off error bounds for the prediction of bandlimited-functions from past samples. *Frequenz, Vol. 39*, pages 182–187, 1985.

[665] D.H. Mugler and W. Splettstösser. Some new difference schemes for the prediction of bandlimited signals from past samples. In *Proceedings of the Conference "Mathematics in Signal Processing"* Bath, September 17–19, 1985.

[666] D.H. Mugler and W. Splettstösser. Difference methods for the prediction of bandlimited signals. *SIAM Journal Appl. Math., Vol. 46*, pages 930–941, 1986.

[667] H.D. Mugler and W. Splettstösser. Linear prediction from samples of a function and its derivatives. *IEEE Trans. Information Theory, Vol. IT-33*, pages 360–366, 1987.

[668] D.C. Munson. Minimum sampling rates for linear shift-variant discrete-time systems. *IEEE Trans. Acoust., Speech, Signal Processing, Vol. ASSP-33*, pages 1556–1561, 1985.

[669] P.K. Murphy and N.C. Gallagher. A new approach to two-dimensional phase retrieval. *J. Opt. Soc. Am., Vol. 72*, pages 929–937, 1982.

[670] T. Nagai. Dyadic stationary processes and their spectral representation. *Bull. Math. Stat., Vol. 17*, pages 65–73, 1976/77.

[671] N. Nakajima and T. Asakura. A new approach to two-dimensional phase retrieval. *Opt. Acta, Vol. 32*, pages 647–658, 1985.

[672] H. Nassenstein. Super-resolution by diffraction of subwaves. *Opt. Comm.*, pages 231–234, 1970.

[673] A. Nathan. On sampling a function and its derivatives. *Inform. Contr., Vol. 22*, pages 172–182, 1973.

[674] A. Nathan. Plain and covariant multivariate Fourier transforms. *Inform. Contr., Vol. 39*, pages 73–81, 1978.

[675] A.W. Naylor and G.R. Sell. *Linear Operator Theory in Engineering and Science*. Springer-Verlag, New York, 1982.

[676] J. Ojeda-Castañeda, J. Ibarra, and J.C. Barreiro. Noncoherent Talbot effect: Coherence theory and applications. *Opt. Comm., Vol. 71*, pages 151–155, 1989.

[677] J. Ojeda-Castañeda and E.E. Sicre. Quasi ray-optical approach to longitudinal periodicities of free and bounded wavefields. *Opt. Acta*, pages 17–26, 1985.

[678] J. von Neumann. *Mathematical Foundations of Quantum Mechanics*. Princeton University Press, Princeton, NJ, 1955, Chap. 5, Sect. 4.

[679] B. Nicoletti and L. Mariani. Optimization of nonuniformly sampled discrete systems. *Automatica, Vol. 7*, pages 747–753, 1971.

[680] K. Niederdrenk. *The Finite Fourier and Walsh Transformation with an Introduction to Image Processing*. Vieweg & Sohn, Braunschweig, 1982.

[681] K.A. Nugent. Three-dimensional optical microscopy: a sampling theorem. *Opt. Comm.*, pages 231–234, 1988.

[682] H. Nyquist. Certain topics in telegraph transmission theory. *AIEE Trans.*, Vol. *47*, pages 617–644, 1928.

[683] K. Ogura. On a certain transcendental integral function in the theory of interpolation. *Tôhoku Math. J.*, Vol. *17*, pages 64–72, 1920.

[684] N. Ohyama, M. Yamaguchi, J. Tsujiuchi, T. Honda, and S. Hiratsuka. Suppression of Moiré fringes due to sampling of halftone screened images. *Opt. Comm.*, pages 364–368, 1986.

[685] K.B. Oldham and J. Spanier. *The Fractional Calculus.* Academic Press, New York, 1974.

[686] E.L. O'Neill. Spatial filtering in optics. *IRE Trans. Information Theory*, pages 56–62, 1956.

[687] E.L. O'Neill. *Introduction to Statistical Optics.* Addison-Wesley, Reading, MA, 1963.

[688] E.L. O'Neill and A. Walther. The question of phase in image formation. *Opt. Acta*, pages 33–40, 1962.

[689] Z. Opial. Weak convergence of the sequence of successive approximation for nonexpansive mappings. *Bull. Amer. Math. Soc.*, Vol. *73*, pages 591–597, 1967.

[690] A. V. Oppenheim and R.W. Shafer. *Digital Signal Processing.* Prentice-Hall, Englewood Cliffs, NJ, 1975.

[691] A.V. Oppenheim. Digital processing of speech. In *Applications of Digital Signal Processing.* Prentice-Hall, Englewood Cliffs, NJ, 1978.

[692] A. V. Oppenheim, Alan S. Willsky, and Ian T. Young. *Signals and Systems.* Prentice-Hall, Englewood Cliffs, NJ, 1983.

[693] J.F. Ormsky. Generalized interpolation sampling expressions and their relationship. *J. Franklin Inst.*, Vol. *310*, pages 247–257, 1980.

[694] T.J. Osler. A further extension of the Leibnitz rule for fractional derivatives and its relation to Parseval's formula. *SIAM J. Math. Anal.*, Vol. *4*, No. *4*, 1973.

[695] L.E. Ostrander. The Fourier transform of spline function approximations to continuous data. *IEEE Trans. Audio Electroacoustics.*, Vol. *AU-19*, pages 103–104, 1971.

[696] J.S. Ostrem. A sampling series representation of the gain and refractive index formulas for a combined homogeneously and inhomogeneously broadened laser line. *Proc. IEEE*, Vol. *66*, pages 583–589, 1978.

[697] N. Ostrowsky, D. Sornette, P. Parker, and E.R. Pike. Exponential sampling method for light scattering polydispersity analysis. *Opt. Acta*, Vol. *28*, pages 1059–1070, 1981.

[698] R.E.A. Paley and N. Wiener. *Fourier Transform in Complex Domain.* Colloq. Publications, Vol. 19, American Math. Soc., New York, 1934.

[699] F. Palmieri. Sampling theorem for polynomial interpolation. *IEEE Trans. Acoust., Speech, Signal Processing*, Vol. *ASSP-34*, pages 846–857, 1986.

[700] A. Papoulis. *The Fourier Integral and Its Applications.* McGraw-Hill, New York, 1962.

[701] A. Papoulis. Error analysis in sampling theory. *Proc. IEEE, Vol. 54,* pages 947–955, 1966.

[702] A. Papoulis. Limits on bandlimited signals. *Proc. IEEE, Vol. 55,* pages 1677–1681, 1967.

[703] A. Papoulis. Truncated sampling expansions. *IEEE Trans. Automatic Control,* pages 604–605, 1967.

[704] A. Papoulis. *Systems and Transforms with Applications in Optics.* McGraw-Hill, New York, 1968.

[705] A. Papoulis. A new algorithm in spectral analysis and bandlimited signal extrapolation. *IEEE Trans. Circuits & Systems, Vol. CAS-22,* pages 735–742, 1975.

[706] A. Papoulis. Generalized sampling expansion. *IEEE Trans. Circuits & Systems, Vol. CAS-24,* pages 652–654, 1977.

[707] A. Papoulis. *Signal Analysis.* McGraw-Hill, New York, 1977.

[708] A. Papoulis. *Circuits and Systems, A Modern Approach.* Holt, Rinehart and Winston, Inc., New York, 1980.

[709] A. Papoulis. A note on the predictability of bandlimited processes. *Proc. IEEE, Vol. 73,* pages 1332–1333, 1985.

[710] A. Papoulis. *Probability, Random Variables, and Stochastic Processes,* 3rd ed. McGraw-Hill, New York, 1991.

[711] A. Papoulis and M.S. Bertran. Digital filtering and prolate functions. *IEEE Trans. Circuits & Systems, Vol. CT-19,* pages 674–681, 1972.

[712] T.W. Parks and R.G. Meier. Reconstructions of signals of a known class from a given set of linear measurements. *IEEE Trans. Information Theory, Vol. IT-17,* pages 37–44, 1971.

[713] E. Parzen. A simple proof and some extensions of sampling theorems. Technical report, Stanford University, Stanford, California, 1956.

[714] E. Parzen. *Stochastic Processes.* Holden-Day, San Francisco, 1962.

[715] C. Pask. Simple optical theory of optical super-resolution. *J. Opt. Soc. Am., Vol. 66,* pages 68–69a, 1976.

[716] K. Patorski. The self-imaging phenomenon and its applications. In *Progress in Optics,* North-Holland, Amsterdam, 1989.

[717] L.E. Pellon. A double Nyquist digital product detector for quadrature sampling. *IEEE Trans. Signal Processing, Vol. 40,* pages 1670–1681, 1992.

[718] R.J. Peterka, D.P. Oleary, and A.C. Sanderson. Practical considerations in the implementation of the French-Holden algorithm for sampling of neuronal spike trains. *IEEE Trans. Biomedical Engineering, Vol. BME-24,* pages 192–195, 1978.

[719] D.P. Petersen. Sampling of space-time stochastic processes with application to information and decision systems, 1963.

[720] D.P. Petersen. Discrete and fast Fourier transformations on $n$-dimensional lattices. *Proc. IEEE, Vol. 58,* 1970.

[721] D.P. Petersen and D. Middleton. Sampling and reconstruction of wave number-limited function in $n$-dimensional Euclidean spaces. *Inform., Contr., Vol. 5,* pages 279–323, 1962.

[722] D.P. Petersen and D. Middleton. Reconstruction of multidimensional stochastic fields from discrete measurements of amplitude and gradient. *Inform. Contr., Vol. 1,* pages 445–476, 1964.

[723] D.P. Petersen and D. Middleton. Linear interpolation, extrapolation and prediction of random space-time fields with limited domain of measurement. *IEEE Trans. Information Theory, Vol. IT-11,* 1965.

[724] J. Peřina. Holographic method of deconvolution and analytic continuation. *Czechoslovak J. Phys., Vol. B21,* pages 731–748, 1971.

[725] J. Peřina and J. Kvapil. A note on holographic method of deconvolution. *Optik,* pages 575–577, 1968.

[726] J. Peřina, V. Perinova, and Z. Braunerova. Super-resolution in linear systems with noise. *Optica Applicata,* pages 79–83, 1977.

[727] E.P. Pfaffelhuber. Sampling series for bandlimited generalized functions. *IEEE Trans. Information Theory, Vol. IT-17,* pages 650–654, 1971.

[728] F. Pichler. Synthese linearer periodisch zeitvariabler Filter mit vorge-schriebenem Sequenzverhalten. *Arch. Elek. Übertr. (AEÜ), Vol. 22,* pages 150–161, 1968.

[729] H.S. Piper, Jr. Best asymptotic bounds for truncation error in sampling expansions of bandlimited signals. *IEEE Trans. Information Theory, Vol. IT-21,* pages 687–690, 1975.

[730] H.S. Piper, Jr. Bounds for truncation error in sampling expansions of finite energy band limited signals. *IEEE Trans. Information Theory, Vol. IT-21,* pages 482–485, 1975.

[731] K. Piwerneta. *A posteriori* compensation for rigid body motion in holographic interferometry by means of a Moiré technique. *Opt. Acta, Vol. 24,* pages 201–209, 1977.

[732] E. Plotkin, L Roytman, and M.N.S. Swamy. Non-uniform sampling of bandlimited modulated signals. *Signal Processing, Vol. 4,* pages 295–303, 1982.

[733] E. Plotkin, L. Roytman, and M.N.S. Swamy. Reconstruction of nonuniformly sampled bandlimited signals and jitter error reduction. *Signal Processing, Vol. 7,* pages 151–160, 1984.

[734] T. Pogány. On a very tight truncation error bound for stationary stochastic processes. *IEEE Trans. Signal Processing, Vol. 39, No. 8,* pages 1918–1919, 1991.

[735] R.J. Polge, R.D. Hays, and L. Callas. A direct and exact method for computing the transfer function of the optimum smoothing filter. *IEEE Trans. Automatic Control, Vol. AC-18,* pages 555–556, 1973.

[736] H.O. Pollak. Energy distribution of bandlimited functions whose samples on a half line vanish. *J. Math. Anal., Appl., Vol. 2*, pages 299–332, 1961.

[737] B. Porat. ARMA spectral estimation of time series with missing observations. *IEEE Trans. Information Theory, Vol. IT-30, No. 6*, pages 823–831, November 1984.

[738] B. Porter and J.J. D'Azzo. Algorithm for closed-loop eigenstructure assignment by state feedback in multivariable linear systems. *Int. J. Control, Vol. 27*, pages 943–947, 1978.

[739] R.P. Porter and A.J. Devaney. Holography and the inverse source problem. *J. Opt. Soc. Am., Vol. 72*, pages 327–330, 1982.

[740] M. Pourahmadi. A sampling theorem for multivariate stationary processes. *J. Multivariate Anal., Vol. 13*, pages 177–186, 1983.

[741] F.D. Powell. Periodic sampling of broad-band sparse spectra. *IEEE Trans. Acoustics, Speech, Signal Processing, Vol. ASSP-31*, pages 1317–1319, 1983.

[742] M. Pracentini and C. Cafforio. Algorithms for image reconstruction after nonuniform sampling. *IEEE Trans. Acoust., Speech, Signal Processing, Vol. ASSP-35, No. 8*, pages 1185–1189, August 1987.

[743] P.M. Prenter. *Splines and Variational Methods*. John Wiley, New York, 1975.

[744] R.T. Prosser. A multi-dimensional sampling theorem. *J. Math. Analysis Appl., Vol. 16*, pages 574–584, 1966.

[745] R. Prost and R. Goutte. Deconvolution when the convolution kernel has no inverse. *IEEE Trans. Acoust., Speech, Signal Processing*, pages 542–549, 1977.

[746] L.R. Rabiner and R.W. Schafer. *Digital Processing of Speech Signals*. Prentice-Hall, Englewood Cliffs, NJ, 1978.

[747] D. Radbel and R.J. Marks II. An FIR estimation filter based on the sampling theorem. *IEEE Trans. Acoust., Speech, Signal Processing, Vol. ASSP-33*, pages 455–460, 1985.

[748] R. Radzyner and P.T. Bason. An error bound for Lagrange interpolation of low pass functions. *IEEE Trans. Information Theory, Vol. IT-18*, pages 669–671, 1972.

[749] J.R. Ragazzini and G.F. Franklin. *Sampled Data Control Systems*. McGraw-Hill, New York, 1958.

[750] Y. Rahmat-Samii and R. Cheung. Nonuniform sampling techniques for antenna applications. *IEEE Trans. Antennas & Propagation, Vol. AP-35, No. 3*, pages 268–279, 1987.

[751] P.K. Rajan. A study on the properties of multidimensional Fourier transforms. *IEEE Trans. Circuits & Systems, Vol. CAS-31*, pages 748–750, 1984.

[752] A. Ralston and P. Rabinowitz. *A First Course in Numerical Analysis*, 2nd ed. McGraw-Hill, New York, 1978.

[753] M.D. Rawn. Generalized sampling theorems for Bessel type transforms of bandlimited functions and distributions. *SIAM J. Appl. Math., Vol. 49, No. 2*, pages 638–649, 1989.

[754] M.D. Rawn. On nonuniform sampling expansions using entire interpolating functions and on the stability of Bessel-type sampling expansions. *IEEE Trans. Information Theory, Vol. 35, No. 3*, 1989.

[755] M.D. Rawn. A stable nonuniform sampling expansion involving derivatives. *IEEE Trans. Information Theory, Vol. 35, No. 6*, 1989.

[756] M. Reed and B. Simon. *Methods and Modern Mathematical Physics. IV. Analysis of Operators.* Academic Press, New York, 1978.

[757] F.M. Reza. *An Introduction to Information Theory.* McGraw-Hill, New York, 1961.

[758] A. Requicha. The zeros of entire functions: Theory and engineering applications. *Proceedings of the IEEE, Vol. 68, No. 3*, pages 308-328, March, 1980.

[759] D.R. Rhodes. The optimum lines source for the best mean-square approximation to a given radiation pattern. *IEEE Trans. Antennas & Propagation*, pages 440–446, 1963.

[760] D.R. Rhodes. A general theory of sampling synthesis. *IEEE Trans. Antennas & Propagation*, pages 176–181, 1973.

[761] W.T. Rhodes. Acousto-optical signal processing: Convolution and correlation. *Proc. IEEE, Vol. 69*, pages 65–79, 1981.

[762] W.T. Rhodes. Space-variant optical systems and processing. In *Applications of Optical Fourier Transforms*, H. Stark, editor, Academic Press, New York, 1982.

[763] S.O. Rice. Mathematical analysis of random noise. *Bell System Technical J., Vol. 23*, pages 282-332, 1944.

[764] S.O. Rice. Mathematical analysis of random noise. *Bell System Technical J., Vol. 24*, pages 46–156, 1945.

[765] S. Ries and W. Splettstösser. On the approximation by truncated sampling series expansions. *Signal Processing, Vol. 7*, pages 191–197, 1984.

[766] S. Ries and R.L. Stens. A localization principle for the approximation by sampling series. In *Proceedings of the International Conference on the Theory of Approximation of Functions (Kiev, USSR), May 30–June 6, 1983*, Nauka, Moscow, pages 507–509.

[767] S. Ries and R.L. Stens. Approximation by generalized sampling series. In *Constructive Theory of Functions (Proceedings of Conference at Varna, Bulgaria, May 27 - June 2, 1984)*, Bl. Sendov et al., editors, Sofia, Publishing House Bulgarian Acad. Sci., 1984, pages 746-756.

[768] A.W. Rihaczek. Signal energy distribution in time and frequency. *IEEE Trans. Inform. Theory, IT-14*, pages 369-374, 1968.

[769] C.L. Rino. Bandlimited image restoration by linear mean-square estimation. *J. Opt. Soc. Am., Vol. 59*, pages 547-553, 1969.

[770] C.L. Rino. The application of prolate spheroidal wave functions to the detection and estimation of bandlimited signals. *Proc. IEEE, Vol. 58*, pages 248–249, 1970.

[771] J. Riordan. *Combinatorial Identities*. Wiley & Sons, New York, 1968.

[772] O. Robaux. Application of the sampling theorem to the reconstitution of partially known objects. III. Decomposition of spectra. *Opt. Acta, Vol. 18*, pages 523–530, 1971.

[773] J.B. Roberts and D.B.S. Ajmani. Spectral analysis of randomly sampled signals using a correlation-based slotting technique. *Proc. IEEE, Vol. 133*, Pt. F, pages 153–162, 1986.

[774] J.B. Roberts and M. Gaster. Rapid estimation of spectra from irregularly sampled records. *Proc. IEEE, Vol. 125*, pages 92–96, 1978.

[775] J.B. Roberts and M. Gaster. On the estimation of spectra from randomly sampled signals – a method of reducing variability. *Proceedings of the Royal Soc. London, Vol. A371*, pages 235–258, 1980.

[776] A. Röseler. Zur Berechnung der optischen abbildung bei teilkohärenter beleuchtung mit hilfe des sampling–theorems. *Opt. Acta, Vol. 16*, pages 641–651, 1969.

[777] G. Ross. Iterative methods in information processing for object restoration. *Opt. Acta*, pages 1523–1542, 1982.

[778] Y.A. Rozanov. To the extrapolation of generalized random stationary processes. *Theor. Probability Appl., Vol. 4*, page 426, 1959.

[779] W. Rozwoda, C.W. Therrien, and J.S. Lim. Novel method for nonuniform frequency-sampling design of 2-D FIR filters. In *Proc. ICASSP '88*, 1988.

[780] D.S. Ruchkin. Linear reconstruction of quantized and sampled random signals. *IRE Trans. Communication Systems, Vol. CS-9*, 1961.

[781] C.K. Rushforth and R.L. Frost. Comparison of some algorithms for reconstructing space-limited images. *J. Opt. Soc. Am., Vol. 70*, pages 1539–1544, 1980.

[782] C.K. Rushforth and R.W. Harris. Restoration, resolution, and noise. *J. Opt. Soc. Am., Vol. 58*, pages 539–545, 1968.

[783] F.D. Russell and J.W. Goodman. Nonredundant arrays and postdetection processing for aberration compensation in incoherent imaging. *J. Opt. Soc. Am., Vol. 61*, pages 182–191, 1971.

[784] M.S. Sabri and W. Steenaart. An approach to bandlimited signal extrapolation: the extrapolation matrix. *IEEE Trans. Circuits & Systems, Vol. CAS-25*, pages 74–78, 1978.

[785] R.A.K. Said and D.C. Cooper. Crosspath real-time optical correlator and ambiguity function processor. *Proc. Inst. Elec. Eng., Vol. 120*, pages 423–428, 1973.

[786] B.E.A. Saleh. *A priori* information and the degrees of freedom of noisy images. *J. Opt. Soc. Am., Vol. 67*, pages 71–76, 1977.

[787] B.E.A. Saleh and M.O. Freeman. Optical transformations. In *Optical Signal Processing*, J.L. Horner, editor, Academic Press, New York, 1987.

[788] M. Salerno, G. Orlandi, G. Martinelli, and P. Burrascano. Synthesis of acoustic well-logging waveforms on an irregular grid. *Geophysical Prospection, Vol. 34, No. 8*, pages 1145–1153, 1986.

[789] I.W. Sandberg. On the properties of some systems that distort signals – I. *Bell Syst. Tech. J.*, pages 2033–2046, 1963.

[790] J.L.C. Sanz. On the reconstruction of bandlimited multidimensional signals from algebraic sampling contours. *Proc. IEEE, Vol. 73*, pages 1334–1336, 1985.

[791] T.K. Sarkar, D.D. Weiner, and V.K. Jain. Some mathematical considerations in dealing with the inverse problem. *IEEE Trans. Antennas & Propagation, Vol. AP-29*, pages 373–379, 1981.

[792] K. Sasakawa. *Application of Miyakawa's Multidimensional Sampling Theorem (Japanese)*. Prof. Group on Inform. Theory, Inst. Electron. Commun. Eng. Japan, I: no. I, 1960; II: no. 9, 1960; III: no. 2, 1961; IV: no. 6, 1961; V: no. 1, 1962.

[793] P. Sathyanarayana, P.S. Reddy, and M.N.S. Swamy. Interpolation of 2-D signals. *IEEE Trans. Circuits & Systems, Vol. 37, No. 5*, pages 623–631, 1990.

[794] K.D. Sauer and J.P. Allebach. Iterative reconstruction of bandlimited images from non-uniformly spaced samples. *IEEE Trans. Circuits & Systems, Vol. 34*, pages 1497–1505, 1987.

[795] R.W. Schafer, R.M. Mersereau, and M.A. Richards. Constrained iterative restoration algorithms. *Proc. IEEE*, pages 432–450, 1981.

[796] R.W. Schafer and L.R. Rabiner. A digital signal processing approach to interpolation. *Proc. IEEE, Vol. 61*, pages 692–702, 1973.

[797] S.A. Schelkunoff. Theory of antennas of arbitrary size and shape. *Proc. IRE*, pages 493–521, 1941.

[798] W. Schempp. Radar ambiguity functions, the Heisenberg group, and holomorphic theta series. *Proc. Am. Math. Soc., Vol. 92*, pages 103–110, 1984.

[799] W. Schempp. Gruppentheoretische aspekte der signal übertragung und der kardinalen interpolationssplines I. *Math. Meth. Appl. Sci., Vol. 5*, pages 195–215, 1983.

[800] I.J. Schoenberg. Contributions to the problem of approximation of equidistant data by analytic functions. Part a: On the problem of smoothing or graduation. A first class of analytic approximation formulae. *Quart. Appl. Math., Vol. 4*, pages 45–99, 1946.

[801] I. J. Schoenberg. Cardinal interpolation and spline functions. *J. Approx. Theory 2*, pages 167–206, 1969.

[802] I. J. Schoenberg. *Cardinal Spline Interpolation*. Regional Conference Series in Applied Mathematics, SIAM, Philadelphia, 1973.

[803] I.J. Schoenberg. *Cardinal Spline Interpolation*. Regional Conference Monograph No. 12, 1973.

[804] A. Schönhage. *Approximationstheorie*. de Gruyter, Berlin - New York, 1971.

[805] L.L. Schumaker. *Spline Functions: Basic Theory.* Wiley, New York, 1981.

[806] G. Schwierz, W. Härer, and K. Wiesent. Sampling and discretization problems in X-ray CT. In *Mathematical Aspects of Computerized Tomography*, Springer-Verlag, Berlin, 1981.

[807] S.C. Scoular and W.J. Fitzgeralds. Periodic nonuniform sampling. *Signal Processing*(to appear).

[808] H.J. Scudder. Introduction to computer aided tomography. *Proc. IEEE*, Vol. *66*, pages 628–637, 1978.

[809] J. Segethova. Numerical construction of the Hill function *SIAM J. Numer. Anal.*, Vol. *9*, pages 199–204, 1972.

[810] K. Seip. An irregular sampling theorem for functions bandlimited in a generalized sense. *SIAM J. Appl. Math.*, Vol. *47*, pages 1112–1116, 1987.

[811] K. Seip. A note on sampling bandlimited stochastic processes. *IEEE Trans. Information Theory*, Vol. *36*, No. *5*, page 1186, 1990.

[812] A. Sekey. A computer simulation study of real-zero interpolation. *IEEE Trans. Audio Electroacoustics*, Vol. *18*, page 43, 1970.

[813] M. De La Sen. Nonperiodic sampling and model matching. *Electron. Lett.*, Vol. *18*, pages 311–313, 1982.

[814] M.De La Sen and S. Dormido. Nonperiodic sampling and identifiability. *Electron. Lett.*, Vol. *17*, pages 922–925, 1981.

[815] M.De La Sen and M.B. Pay. On the use of adaptive sampling in hybrid adaptive error models. *Proc. IEEE*, Vol. *72*, pages 986–989, 1984.

[816] M. Ibrahim Sezan and Henry Stark. Tomographic image reconstruction from incomplete view data by convex projections and direct Fourier inversion. *IEEE Trans. Medical Imaging*, Vol. *MI-3*, No. *2*, pages 91–98, 1984.

[817] C.E. Shannon. A mathematical theory of communication. *Bell Systems Tech. J.*, Vol. *27*, pages 379 and 623, 1948.

[818] C.E. Shannon. Communications in the presence of noise. *Proc. IRE*, Vol. *37*, pages 10–21, 1949.

[819] C.E. Shannon and W. Weaver. *The Mathematical Theory of Communication.* University of Illinois Press, Urbana, IL, 1949.

[820] H.S. Shapiro. *Smoothing and Approximation of Functions.* Van Nostrand Reinhold, New York, 1969.

[821] H.S. Shapiro. Topics in Approximation Theory. *Lecture Notes in Mathematics*, Vol. *187*, Springer-Verlag, Berlin - Heidelberg - New York, 1971.

[822] H.S. Shapiro and R.A. Silverman. Alias free sampling of random noise. *J. SIAM*, Vol. *8*, pages 225–236, 1960.

[823] L. Shaw. Spectral estimates from nonuniform samples. *IEEE Trans. Audio Electroacoustics*, Vol. *AU-19*, 1970.

[824] R.G. Shenoy and T.W. Parks. An optimal recovery approach to interpolation. *IEEE Trans. Signal Processing*, Vol. *40*, No. *6*, pages 1987–1988, 1992.

[825] A. Sherstinsky and C.G. Sodini. A programmable demodulator for over-sampled analog-to-digital modulators. *IEEE Trans. Circuits & Systems, Vol. CAS-37, No. 9*, pages 1092–1103, 1990.

[826] N.K. Sinha and G.J. Lastman. Identification of continuous-time multivariable systems from sampled data. *Int. J. Control, Vol. 35*, pages 117–126, 1982.

[827] D. Slepian. Prolate spheroidal wave functions Fourier analysis and uncertainty IV: Extensions to many dimensions; generalized prolate spheroidal wave functions. *Bell Systems Tech. J., Vol. 43*, pages 3009–3057, 1964.

[828] D. Slepian. Analytic solution of two apodization problems. *J. Opt. Soc. Am., Vol. 55*, pages 1110–1115, 1965.

[829] D. Slepian. Some asymptotic expansions for prolate spheroidal wave functions. *J. Math., Phys. Vol. 44, No. 2*, pages 99–140, June 1965.

[830] D. Slepian. On bandwidth. *Proc. IEEE, Vol. 64*, pages 292–300, 1976.

[831] D. Slepian. Prolate spheroidal wave functions, Fourier analysis and uncertainty – V: The discrete case. *Bell Systems Tech. J., Vol. 57*, pages 1371–1430, 1978.

[832] D. Slepian and H.O. Pollak. Prolate spheroidal wave functions Fourier analysis and uncertainty I. *Bell Systems Tech. J., Vol. 40*, pages 43–63, 1961.

[833] D. Slepian and E. Sonnenblick. Eigenvalues associated with prolate spheroidal wave functions of zero order. *Bell Systems Tech. J., Vol. 44*, pages 1745–1758, 1965.

[834] D.K. Smith and R.J. Marks II. Closed form bandlimited image extrapolation. *Appl. Opt., Vol. 20*, pages 2476–2483, 1981.

[835] M.J. Smith, Jr. An evaluation of adaptive sampling. *IEEE Trans. Automatic Control, Vol. AC-16*, pages 282–284, 1971.

[836] T. Smith, M.R. Smith, and S.T. Nichols. Efficient sinc function interpolation technique for center padded data. *IEEE Trans. Acoust., Speech, Signal Processing, Vol. ASSP-38, No. 9*, pages 1512–1517, 1990.

[837] I.N. Sneddon. *The Use of Integral Transforms.* McGraw-Hill, New York, 1972.

[838] Guy R. Sohie and George N. Maracas. Orthogonality of exponential transients. *Proc. IEEE, Vol. 76, No. 12*, pages 1616–1618, 1988.

[839] S.C. Som. Simultaneous multiple reproduction of space-limited functions by sampling of spatial frequencies. *J. Opt. Soc. Am., Vol. 60*, pages 1628–1634, 1970.

[840] I. Someya. *Waveform Transmission.* Shukyo, Ltd., Tokyo, 1949.

[841] M. Soumekh. Bandlimited interpolation from unevenly spaced sampled data. *IEEE Trans. Acoust., Speech, Signal Processing, Vol. ASSP-36, No. 1*, pages 110–122, 1988.

[842] M. Soumekh. Reconstruction and sampling constraints for spiral data. *IEEE Trans. Acoust., Speech, Signal Processing, Vol. ASSP-37*, pages 882–891, 1989.

[843] J.J. Spilker. Theoretical bounds on the performances of sampled data communications systems. *IRE Trans. Circuit Theory, Vol. CT-7*, pages 335–341, 1960.

[844] W. Splettstösser. Über die Approximation stetiger Funktionen durch die klassischen und durch neue Abtastsummen mit Fehlerabschätzungen. Doctoral dissertation, RWTH Aachen, Aachen, Germany, 1977.

[845] W. Splettstösser. On generalized sampling sums based on convolution integrals. *AEU, Vol.32, No. 7*, pages 267–275, 1978.

[846] W. Splettstösser. Error estimates for sampling approximation of non-bandlimited functions. *Math. Meth. Appl. Sci., Vol. 1*, pages 127–137, 1979.

[847] W. Splettstösser. Error analysis in the Walsh sampling theorem. In *1980 IEEE International Symposium on Electromagnetic Compatibility, Baltimore*, pages 366–370. Vol. 409, 1980.

[848] W. Splettstößer. Bandbegrenzte und effektiv bandbegrenzte funktionen und ihre praediktion aus abtastwerten. Habilitationsschrift, *RWTH Aachen*, 1981.

[849] W. Splettstösser. On the approximation of random processes by convolution processes. *Z. Angew. Math. Mech., Vol. 61*, pages 235–241, 1981.

[850] W. Splettstösser. Sampling series approximation of continuous weak sense stationary processes. *Inform. Contr., Vol. 50*, pages 228–241, 1981.

[851] W. Splettstösser. On the prediction of bandlimited processes from past samples. *Inform. Sci., Vol. 28*, pages 115–130, 1982.

[852] W. Splettstösser. Sampling approximation of continuous functions with multi-dimensional domain. *IEEE Trans. Information Theory, Vol. IT-28*, pages 809–814, 1982.

[853] W. Splettstösser. 75 years aliasing error in the sampling theorem. In *EUSIPCOM - 83, Signal Processing: Theories and Applications (2. European Signal Processing Conference, Erlangen)*, H.W. Schüssler, editor. North-Holland Publishing Company, Amsterdam - New York - Oxford, 1983, pages 1–4.

[854] W. Splettstößer. Lineare praediktion von nicht bandbegrenzten funktionen. *Z. Angew. Math. Mech., Vol. 64*, T393–T395, 1984.

[855] W. Splettstösser, R.L. Stens, and G. Wilmes. On approximation by the interpolating series of G. Valiron. *Funct. Approx. Comment. Math., Vol. 11*, pages 39–56, 1981.

[856] W. Splettstösser and W. Ziegler. The generalized Haar functions, the Haar transform on r⁺, and the Haar sampling theorem. In *Proceedings of the "Alfred Haar Memorial Conference," Budapest*, pages 873–896, 1985.

[857] C.J. Standish. Two remarks on the reconstruction of sampled non-bandlimited functions. *IBM J. Res. Develop., Vol. 11*, pages 648–649, 1967.

[858] H. Stark. Sampling theorems in polar coordinates. *J. Opt. Soc. Am., Vol. 69*, pages 1519–1525, 1979.

[859] H. Stark. Polar sampling theorems of use in optics. *Proc. SPIE, Int. Soc. Opt. Eng.*, *Vol. 358*, pages 24–30, 1982.

[860] H. Stark, editor. *Image Recovery: Theory and Application*. Academic Press, Orlando, FL, 1987.

[861] H. Stark, D. Cahana, and H. Webb. Restoration of arbitrary finite-energy optical objects from limited spatial and spectral information. *J. Opt. Soc. Am.*, *Vol. 71*, pages 635–642, 1981.

[862] H. Stark and C.S. Sarna. Bounds on errors in reconstructing from under-sampled images. *J. Opt. Soc. Am.*, *Vol. 69*, pages 1042–1043, 1979.

[863] H. Stark and C.S. Sarna. Image reconstruction using polar sampling theorems of use in reconstructing images from their samples and in computer aided tomography. *Appl. Opt.*, *Vol. 18*, pages 2086–2088, 1979.

[864] H. Stark and M. Wengrovitz. Comments and corrections on the use of polar sampling theorems in CT. *IEEE Trans. Acoust., Speech, Signal Processing*, *Vol. ASSP-31*, pages 1329–1331, 1983.

[865] H. Stark and John W. Woods. *Probability, Random Processes, and Estimation Theory for Engineers*. Prentice-Hall, Englewood Cliffs, NJ, 1986.

[866] H. Stark, J.W. Woods, I. Paul, and R. Hingorani. Direct Fourier reconstruction in computer tomography. *IEEE Trans. Acoust., Speech, Signal Processing*, *Vol. ASSP-29*, pages 237–245, 1981.

[867] H. Stark, J.W. Woods, I. Paul, and R. Hingorani. An investigation of computerized tomography by direct Fourier inversion and optimum interpolation. *IEEE Trans. Biomedical Engineering*, *Vol. BME-28*, pages 496–505, 1981.

[868] H. Stark. Sampling theorems in polar coordinates. *J. Opt. Soc. Am.*, *Vol. 69, No. 11*, pages 1519–1525, 1979.

[869] O.N. Stavroudis. *The Optics of Rays, Wavefronts, and Caustics*. Academic Press, New York, 1972.

[870] R. Steele and F. Benjamin. Sample reduction and subsequent adaptive interpolation of speech signals. *Bell Systems Tech. J.*, *Vol. 62*, pages 1365–1398, 1983.

[871] F. Stenger. Approximations via Whittaker's cardinal function. *J. Approx. Theory*, *Vol. 17*, pages 222–240, 1976.

[872] F. Stenger. Numerical methods based on Whittaker cardinal, or sinc functions. *SIAM Rev.*, *Vol. 23*, pages 165–224, 1981.

[873] R.L. Stens. Approximation of duration-limited functions by sampling sums. *Signal Processing*, *Vol. 2*, pages 173–176, 1980.

[874] R.L. Stens. A unified approach to sampling theorems for derivatives and Hilbert transforms. *Signal Processing*, *Vol. 5*, pages 139–151, 1983.

[875] R.L. Stens. Error estimates for sampling sums based on convolution integrals. *Inform. Contr.*, *Vol. 45*, pages 37–47, 1980.

[876] R.L. Stens. Approximation of functions by Whittaker's cardinal series. in General Inequalities 4 (*Proc. Conf. Math. Res. Inst. Oberwolfach*, W. Walter, ed.) Birkhäuser Verlag, Basel, 1984, pp. 137–149.

[877] R.L. Stens and A. Fischer. Generalized sampling approximation of multivariable signals; inverse approximation theorems. *Proc. Conf. on Approximation Theory*, Kecskemét, Hungary, August 5–11, 1990; Colloquia Mathematica Societatis János Bolyai (to appear).

[878] R.M. Stewart. Statistical design and evaluation of filters for the restoration of sampled data. *Proc. IRE, Vol. 44*, pages 253–257, 1956.

[879] D.C. Stickler. An upper bound on aliasing error. *Proc. IEEE, Vol. 55*, pages 418–419, 1967.

[880] G. Strang. *Linear Algebra and Its Applications.* Academic Press, New York, 1980.

[881] R.N. Strickland, A.P. Anderson, and J.C. Bennett. Resolution of subwavelength-spaced scatterers by superdirective data processing simulating evanescent wave illumination. *Microwaves, Opt. & Acoust., Vol. 3*, pages 37–42, 1979.

[882] J.A. Stuller. Reconstruction of finite duration signals. *IEEE Trans. Information Theory, Vol. IT-18*, pages 667–669, 1972.

[883] R. Sudhakar, R.C. Agarwal, and S.C. Dutta Roy. Fast computation of Fourier transform at arbitrary frequencies. *IEEE Trans. Circuits & Systems, Vol. CAS-28*, pages 972–980, 1982.

[884] R. Sudol and B.J. Thompson. An explanation of the Lau effect based on coherence theory. *Opt. Comm.*, pages 105–110, 1979.

[885] R. Sudol and B.J. Thompson. Lau effect: theory and experiment. *Appl. Opt.*, pages 1107–1116, 1981.

[886] K. Swaminathan. Signal restoration from data aliased in time. *IEEE Trans. Acoust., Speech, Signal Processing, Vol. ASSP-33*, pages 151–159, 1985.

[887] G.J. Swanson and E.N. Leith. Lau effect and grating imaging. *J. Opt. Soc. Am., Vol. 72*, pages 552–555, 1982.

[888] S. Szapiel. Sampling theorem for rotationally symmetric systems based on dini expansion. *Opt. Lett., Vol. 6*, pages 7–9, 1981.

[889] H.H. Szu and J.A. Blodgett. *Wigner distribution and ambiguity function.* Optics in Four Dimensions-1980, AIP Conf. Proc. 65, M.A. Machado and L.M. Narducci, eds., American Institute of Physics, New York, pages 355–381, 1980.

[890] B. Tatian. Asymptotic expansions for correcting truncation error in transfer function calculations. *J. Opt. Soc. Am., Vol. 61*, pages 1214–1224, 1971.

[891] G.C. Temes, V. Barcilon, and F.C. Marshall III. The optimization of bandlimited systems. *Proc. IEEE, Vol. 61*, pages 196–234, 1973.

[892] M. Theis. Über eine Interpolationsformel von de la Vallée-Poussin. *Math. Z., Vol. 3*, pages 93–113, 1919.

[893] G. Thomas. A modified version of Van Cittert's iterative deconvolution procedure. *IEEE Trans. Acoust., Speech, Signal Processing, Vol. ASSP-29*, pages 938–939, 1981.

[894] J.B. Thomas. *An Introduction to Statistical Communication Theory.* John Wiley & Sons, Inc., New York, 1969.

[895] J.B. Thomas and B. Liu. Error problems in sampling representations, Part 5. *IEEE Int. Conv. Rec., Vol. 12*, pages 269–277, 1964.

[896] B.J. Thompson. Image formation with partially coherent light. In *Progress in Optics*, E. Wolf, editor, North Holland, Amsterdam, 1969, pages 169–230.

[897] B.J. Thompson. Multiple imaging by diffraction techniques. *Appl. Opt.*, page 312, 1976.

[898] J.E. Thompson and L.H. Luessen, editors. *Fast Electrical and Optical Measurements, Volume 1.* Martinus Nijhoff Publishers, 1986.

[899] A.N. Tikhonov and V.Y. Arsenine. *Solutions of Ill-Posed Problems.* Wiston/Wiley, Washington, DC, 1977.

[900] A.F. Timan. *Theory of Approximation of Functions of a Real Variable.* Pergamon Press, Oxford, 1963.

[901] E.C. Titchmarsh. The zeros of certain integral functions. *Proc. London Math. Society, Vol. 25*, pages 283–302, 1926.

[902] E.C. Titchmarsh. *Introduction to the Fourier Integral.* Oxford University Press, Oxford, 1937.

[903] E.C. Titchmarsh. *The Theory of Functions*, 2nd ed. University Press, London, 1939.

[904] E.C. Titchmarsh. *Introduction to the Theory of Fourier Integrals (2nd Edition).* Clarendon Press, Oxford, 1948.

[905] D.E. Todd. Sampled data reconstruction of deterministic bandlimited signals. *IEEE Trans. Information Theory, Vol. IT-19*, pages 809–811, 1973.

[906] G. Toraldo diFrancia. Super-gain antennas and optical resolving power. *Suppl. Nuovo Cimento*, pages 426–438, 1952.

[907] G. Toraldo diFrancia. Resolving power and information. *J. Opt. Soc. Am., Vol. 45*, pages 497–501, 1955.

[908] G. Toraldo diFrancia. Directivity, super-gain and information. *IRE Trans. Antennas & Propagation, Vol. AP-4*, pages 473–478, 1956.

[909] G. Toraldo diFrancia. *La diffrazione della luce.* Einaudi Bornghieri, Torino, 1958.

[910] G. Toraldo diFrancia. Degrees of freedom of an image. *J. Opt. Soc. Am., Vol. 59*, pages 799–804, 1969.

[911] H.J. Trussel, L.L. Arnder, P.R. Morand, and R.C. Williams. Corrections for nonuniform sampling distortions in magnetic resonance imagery. *IEEE Trans. Medical Imaging, Vol. MI-7, No. 1*, pages 32–44, 1988.

[912] Shiao-Min Tseng. *Noise Level Analysis of Linear Restoration Algorithms for Bandlimited Signals.* Ph.D. thesis, University of Washington, 1984.

[913] J. Tsujiuchi. Correction of optical images by compensation of aberrations and by spatial frequency filtering. In *Progress in Optics*, E. Wolf, editor. North-Holland, Amsterdam, 1963, pages 131–180.

[914] Y. Tsunoda and Y. Takeda. High density image-storage holograms by a random phase sampling method. *Appl. Opt., Vol. 13*, pages 2046–2051, 1974.

[915] J. Turunen, A. Vasara, and A.T. Friberg. Propagation invariance and self-imaging in variable coherence optics. *J. Opt. Soc. Am. A, Vol. 8*, pages 282–289, 1991.

[916] J. Turunen, A. Vasara, J. Werterholm, and A. Salin. Strip-geometry two-dimensional Dammann gratings. *Opt. Comm., Vol. 74*, pages 245–252, 1989.

[917] Kazunori Uchida. Spectral domain method and generalized sampling theorems. *Transactions of IEICE, Vol. E73, No. 3*, pages 357–359, 1990.

[918] M. Unser, A. Aldroubi, and M. Eden Polynomial spline signal approximations: Filter design and asymptotic equivalence with Shannon's sampling theorem. *IEEE Trans. Information Theory, Vol. IT-38*, No. 1 January 1992.

[919] P.P. Vaidyanathan and Vincent C. Liu. Classical sampling theorems in the context of multirate and polyphase digital filter bank structures. *IEEE Trans. Acoust., Speech, Signal Processing, Vol. ASSP-36, No. 9*, pages 1480–1495, 1988.

[920] P.H. Van Cittert. Zum einfluss der splatbreite auf die intensitatswerteilung in spektrallinien II. *Z. Phys., Vol. 69*, pages 298–308, 1931.

[921] E. Van der Ouderaa and J. Renneboog. Some formulas and application of nonuniform sampling of bandwidth limited signal. *IEEE Trans. Instrumentation & Measurement, Vol. IM-37, No. 3*, pages 353–357, 1988.

[922] P.J. Van Otterloo and J.J. Gerbrands. A note on sampling theorem for simply connected closed contours. *Inform. Contr., Vol. 39*, pages 87–91, 1978.

[923] G.A. Vanasse and H. Sakai. Fourier spectroscopy. In *Progress in Optics*, E. Wolf, editor. North-Holland, Amsterdam, 1967.

[924] A. VanderLugt. Optimum sampling of Fresnel transforms. *Appl. Opt., Vol. 29, No. 23*, pages 3352–3361, 1990.

[925] A.B. VanderLugt. Signal detection by complex spatial filtering. *IEEE Trans. Information Theory, Vol. IT-10*, pages 2–8, 1964.

[926] R.G. Vaughan, N.L. Scott, and D.R. White. The theory of bandpass sampling. *IEEE Trans. Signal Processing, Vol. SP-39, No. 9*, pages 1973–1984, 1991.

[927] R. Veldhuis. *Restoration of Lost Samples in Digital Signals*. Prentice Hall, New York, 1990.

[928] G.A. Viano. On the extrapolation of optical image data. *J. Math. Phys., Vol. 17*, pages 1160–1165, 1976.

[929] H. Voelker. Toward a unified theory of modulation. *Proc. IEEE, Vol. 54*, page 340, 1966.

[930] L. Vogt. Sampling sums with kernels of finite oscillation. *Numer. Funct. Anal., Optimiz., Vol. 9*, pages 1251–1270, 1987/88.

[931] M. von Laue. Die freiheitsgrade von strahlenbundeln. *Ann. Phys.*, Vol. *44*, pages 1197–1212, 1976.

[932] L.A. Wainstein and V.D. Zubakov. *Extraction of Signals from Noise.* Prentice-Hall, Englewood Cliffs, NJ, 1962.

[933] J.F. Walkup. Space-variant coherent optical processing. *Opt. Eng.*, Vol. *19*, pages 339–346, 1980.

[934] G.G. Walter. A sampling theorem for wavelet subspaces. *IEEE Trans. Information Theory*, Vol. *38*, pages 881–884, March 1992.

[935] P.T. Walters. Practical applications of inverting spectral turbidity data to provide aerosol size distributions. *Appl. Opt.*, Vol. *19*, pages 2353–2365, 1980.

[936] A. Walther. The question of phase retrieval in optics. *Opt. Acta*, pages 41–49, 1962.

[937] A. Walther. Gabor's theorem and energy transfer through lenses. *J. Opt. Soc. Am.*, Vol. *57*, pages 639–644, 1967.

[938] A. Walther. Radiometry and coherence. *J. Opt. Soc. Am.*, Vol. *58*, pages 1256–1259, 1968.

[939] H.S.C. Wang. On Cauchy's interpolation formula and sampling theorems. *J. Franklin Inst.*, Vol. *289*, pages 233–236, 1970.

[940] G. Watson. *Treatise on the Theory of Bessel Functions*, 2nd ed. Cambridge University Press, Cambridge, 1962.

[941] H.J. Weaver. *Applications of Discrete and Fourier Analysis.* Wiley, New York, 1983.

[942] A. Weinberg and B. Liu. Discrete time analysis of nonuniform sampling first-and-second order digital phase lock loop. *IEEE Trans. Communications*, pages 123–137, 1974.

[943] P. Weiss. Sampling theorems associated with Sturm-Liouville systems. *Bull. Amer. Math. Soc.*, Vol. *63*, page 242, 1957.

[944] P. Weiss. An estimate of the error arising from misapplication of the sampling theorem. *Amer. Math. Soc. Notices 10-351*, (Abstract No. 601-54), 1963.

[945] A.L. Wertheimer and W.L. Wilcock. Light scattering measurements of particle distributions. *Appl. Opt.*, Vol. *15*, pages 1616–1620, 1976.

[946] E.T. Whittaker. On the functions which are represented by the expansions of the interpolation theory. *Proc. Royal Society Edinburgh*, Vol. *35*, pages 181–194, 1915.

[947] J.M. Whittaker. On the cardinal function of interpolation theory. *Proc. Math. Soc. Edinburgh*, Vol. *2*, pages 41–46, 1927-1929.

[948] J.M. Whittaker. The "Fourier" theory of the cardinal function. *Proc. Math. Soc. Edinburgh*, Vol. *1*, pages 169–176, 1929.

[949] J.M. Whittaker. *Interpolatory Function Theory.* Cambridge Tracts in Mathematical Physics, No. 33, Cambridge University Press, Cambridge, 1935.

[950] E.T. Whittaker and G.N. Watson. *A Course of Modern Analysis.* Cambridge University Press, Cambridge, 1927, Chaps. 20 and 21.

[951] N. Wiener. *Extrapolation, Interpolation, and Smoothing of Stationary Time Series.* Wiley and Sons Inc., New York, 1949.

[952] N. Wiener. *Collected Works, Vol. III* (ed. by P. R. Masani). M. I. T. Press, Cambridge, MA, 1981.

[953] E. Wigner. On the quantum correction for thermodynamic equilibrium. *Phys. Rev., Vol. 40*, pages 749–759, 1932.

[954] R.G. Wiley, H. Schwarzlander, and D.D. Weiner. Demodulation procedure for very wideband FM. *IEEE Trans. Commun., Vol. COM-25*, pages 318–327, March 1977.

[955] R.G. Wiley. On an iterative technique for recovery of bandlimited signals. *Proc. IEEE*, pages 522–523, 1978.

[956] R.G. Wiley. Recovery of bandlimited signals from unequally spaced samples. *IEEE Trans. Communications, Vol. COM-26*, pages 135–137, 1978.

[957] N.P. Willis and Y. Bresler. Norm invariance of minimax interpolation *IEEE Trans. Information Theory, Vol. 38, No. 3*, pages 1177–1181, 1992.

[958] T. Wilson and C. Sheppard. *Theory and Practice of Scanning Microscopy.* Academic Press, London, 1978.

[959] D.J. Wingham The reconstruction of a band-limited function and its Fourier transform from a finite number of samples at arbitrary locations by singular value decomposition *IEEE Trans. Signal Processing, Vol. SP-40*, pages 559–570, March 1992.

[960] J.T. Winthrop. Propagation of structural information in optical wave fields. *J. Opt. Soc. Am., Vol. 61*, pages 15–30, 1971.

[961] E. Wolf. Coherence and radiometry. *J. Opt. Soc. Am., Vol. 68*, pages 6–17, 1982.

[962] E. Wolf. New theory of partial coherence in the space-frequency domain. Part I: Spectra and cross spectra of steady-state sources. *J. Opt. Soc. Am., Vol. 72*, pages 343–351, 1982.

[963] E. Wolf. Coherent-mode propagation in spatially bandlimited wave fields. *J. Opt. Soc. Am. A, Vol. 3*, pages 1920–1924, 1986.

[964] H. Wolter. Zum grundtheorem der informationstheorie, insbesondere in der optik. *Physica*, pages 457–475, 1958.

[965] H. Wolter. On basic analogies and principal differences between optical and electronic information. In *Progress in Optics*, E. Wolf, editor. North-Holland, Amsterdam, 1961, pages 155–210.

[966] P.M. Woodward. *Probability and Information Theory with Applications to Radar.* Pergamon Press Ltd., London, 1953.

[967] J.M. Wozencraft and I.M. Jacobs. *Principles of Communication Engineering.* John Wiley & Sons, Inc., New York, 1965.

[968] X. Xia and Z. Zhang. A note on difference methods for the prediction of band-limited signals from past samples. *IEEE Trans. Information Theory, Vol. IT-37, No. 6*, pages 1662–1665, 1991.

[969] A.D. Yaghjian. Simplified approach to probe-corrected spherical near-field scanning. *Electron. Lett.*, Vol. 20, pages 195–196, 1984.

[970] A.D. Yaghjian. Antenna coupling and near-field sampling in plane-polar coordinates. *IEEE Trans. Antennas & Propagation*, Vol. 40, pages 304–312, March 1992.

[971] M. Yamaguchi, N. Ohyama, T. Honda, J. Tsujiuchi, and S. Hiratsuka. A color image sampling method with suppression of Moiré fringes. *Opt. Comm.*, Vol. 69, pages 349–352, 1989.

[972] G.G. Yang and E. Leith. An image deblurring method using diffraction gratings. *Opt. Comm.*, Vol. 36, pages 101–106, 1981.

[973] K. Yao. *On some representations and sampling expansions for bandlimited signals.* Ph.D. thesis, Dept. of Electrical Eng., Princeton University, Princeton, NJ, 1965.

[974] K. Yao. Applications of reproducing kernel Hilbert spaces – Bandlimited signal models. *Inform. Contr.*, Vol. 11, pages 429–444, 1967.

[975] K. Yao and J.B. Thomas. On truncation error bounds for sampling representations of bandlimited signals. *IEEE Trans. Aerospace & Electronic Systems*, Vol. AES-2, page 640, 1966.

[976] K. Yao and J.B. Thomas. On some stability and interpolating properties of nonuniform sampling expansions. *IEEE Trans. Circuit Theory*, Vol. CT-14, pages 404–408, 1967.

[977] K. Yao and J.B. Thomas. On a class of nonuniform sampling representation. Symp. Signal Trans. Processing (Columbia Univ.), 1968.

[978] B. Yegnanarayana, S. Tanveer Fatima, B.T.K.R. Nehru, and B. Venkataramanan. Significance of initial interpolation in bandlimited signal interpolation. *IEEE Trans. Acoust., Speech, Signal Processing*, Vol. ASSP-37, No. 1, pages 151–152, 1989.

[979] Shu-jen Yeh and Henry Stark. Iterative and one-step reconstruction from nonuniform samples by convex projections. *J. Opt. Soc. Am. A*, Vol. 7, No. 3, pages 491–499, 1990.

[980] J.L. Yen. On the nonuniform sampling of bandwidth limited signals. *IRE Trans. Circuit Theory*, Vol. CT-3, pages 251–257, 1956.

[981] J.L. Yen. On the synthesis of line sources and infinite strip sources. *IRE Trans. Antennas & Propagation*, pages 40–46, Jan. 1957.

[982] D.C. Youla. Generalized image restoration by method of alternating orthogonal projections. *IEEE Trans. Circuits & Systems*, Vol. CAS-25, pages 694–702, 1978.

[983] D.C. Youla. Mathematical theory of image restoration by the method of convex projections, *Image Recovery: Theory and Applications*, H. Stark, ed., Academic Press Inc., Orlando, FL, 1987, pages 29–77.

[984] D.C. Youla and V. Velasco. Extensions of a result on the synthesis of signals in the presence of inconsistent constraints. *IEEE Trans. Circuits & Systems*, Vol. CAS-33, pages 465–468, 1986.

[985] D.C. Youla and H. Webb. Image restoration by the method of convex projection: Part 1 - theory. *IEEE Trans. Medical Imaging, Vol. MI-1*, pages 81–94, 1982.

[986] F.T.S. Yu. Image restoration, uncertainty, and information. *Appl. Opt., Vol. 8*, pages 53–58, 1969.

[987] F.T.S. Yu. Optical resolving power and physical realizability. *Opt. Comm., Vol. 1*, pages 319–322, 1970.

[988] F.T.S. Yu. Coherent and digital image enhancement, their basic differences and constraints. *Opt. Comm., Vol. 3*, pages 440–442, 1971.

[989] E. Yudilevich and H. Stark. Interpolation from samples on a linear spiral scan. *IEEE Trans. Medical Imaging, Vol. MI-6*, pages 193–200, 1987.

[990] E. Yudilevich and H. Stark. Spiral sampling: Theory and application to magnetic resonance imaging. *J. Opt. Soc. Am. A, Vol. 5*, pages 542–553, 1988.

[991] L.A. Zadeh. A general theory of linear signal transmission systems. *J. Franklin Inst., Vol. 253*, pages 293–312, 1952.

[992] J. Zak. Finite translations in solid-state physics. *Phys. Rev. Lett., Vol. 19*, pages 1385–1387, 1967.

[993] J. Zak. Dynamics of electrons in solids in external fields. *Phys. Rev., Vol. 168*, pages 686–695, 1968.

[994] J. Zak. The $kq$-representation in the dynamics of electrons in solids. *Solid State Physics, Vol. 27*, pages 1–62, 1972.

[995] M. Zakai. Band-limited functions and the sampling theorem. *Inform. Contr., Vol. 8*, pages 143–158, 1965.

[996] A. Zakhor. Sampling schemes for reconstruction of multidimensional signals from multiple level threshold crossing. In *Proc. ICASSP '88*, 1988.

[997] A. Zakhor, R. Weisskoff, and R. Rzedzian. Optimal sampling and reconstruction of MRI signals resulting from sinusoidal gradients. In *IEEE Trans. Signal Processing, Vol. SP-39, No. 9*, pages 2056–2065, 1988.

[998] H. Zander. Fundamentals and methods of digital audio technology. II. Sampling process. *Fernseh- und Kino-Tech., Vol. 38*, pages 333–338, 1984.

[999] A. Zayed, G. Hinsen, and P. Butzer. On Lagrange interpolation and Kramer-type sampling theorems associated with Sturm-Liouville problems. *SIAM J. Appl. Math, Vol. 50, No. 3*, pages 893–909, 1990.

[1000] A.J. Zayed. Sampling expansion for the continuous Bessel transform. *Appl. Anal., Vol. 27*, pages 47–65, 1988.

[1001] A.H. Zemanian. *Distribution Theory and Transform Analysis*. McGraw-Hill, New York, 1965.

[1002] A.H. Zemanian. *Generalized Integral Transforms*. Interscience, New York, 1968.

[1003] M. Zoltowski. Global stability of the synchronism of a second-degree phased locked loop with nonuniform sampling. *Rozpramy Elecktrotechniczne, Vol. 34, No. 1 (in Polish)*, pages 37–58, 1988.

# Index